AF842123

Long-Time Predictions in Dynamics

NATO ADVANCED STUDY INSTITUTES SERIES

Proceedings of the Advanced Study Institute Programme, which aims at the dissemination of advanced knowledge and the formation of contacts among scientists from different countries

The series is published by an international board of publishers in conjunction with NATO Scientific Affairs Division

A	Life Sciences	Plenum Publishing Corporation
B	Physics	London and New York
C	Mathematical and Physical Sciences	D. Reidel Publishing Company Dordrecht and Boston
D	Behavioral and Social Sciences	Sijthoff International Publishing Company Leiden
E	Applied Sciences	Noordhoff International Publishing Leiden

Series C – Mathematical and Physical Sciences

Volume 26 – Long-Time Predictions in Dynamics

Long-Time Predictions in Dynamics

Proceedings of the NATO Advanced Study Institute held in Cortina d'Ampezzo, Italy, August 3–16, 1975

edited by

V. SZEBEHELY and B.D. TAPLEY

Dept. of Aerospace Engineering and Engineering Mechanics, University of Texas at Austin, Austin, Tex. 78712, U.S.A.

D. Reidel Publishing Company

Dordrecht-Holland / Boston-U.S.A.

Published in cooperation with NATO Scientific Affairs Division

ISBN 90-277-0692-1

Published by D. Reidel Publishing Company
P.O. Box 17, Dordrecht, Holland

Sold and distributed in the U.S.A., Canada, and Mexico
by D. Reidel Publishing Company. Inc.
Lincoln Building, 160 Old Derby Street, Hingham, Mass. 02043, U.S.A.

All Rights Reserved
Copyright © 1976 by D. Reidel Publishing Company, Dordrecht, Holland
No part of the material protected by this copyright notice may be reproduced or utilized
in any form or by any means, electronic or mechanical, including photocopying,
recording or by any informational storage and retrieval system,
without written permission from the copyright owner

Printed in The Netherlands by D. Reidel, Dordrecht

TABLE OF CONTENTS

IV. Dynamics in Astronomy

V. Abstracts of Seminar-Contributions

PREFACE

This volume contains lectures given at the NATO Advanced Study Institute on Long-Time Predictions in Dynamics conducted in Cortina d'Ampezzo, Italy during August 3-16, 1975.

The lectures were presented in groups, according to the original structure of the Institute. Under "Fundamentals" the general concepts were treated by Contopoulos, DeWitt, Reichl, Stiefel, Szebehely, Bartlett, Kirchgraber, Verhults and Sigrist. This was followed by the series of lectures on "Numerical and Statistical Analysis" offered by Aarseth, Baumgarte and Tapley. The third principal subject was "Three and Many-Body Problems" with Garfinkel, Broucke, Hadjidemetriou, Marchal, Nahon, Waldvogel, Losco, and Markellos as the major speakers. The last group of lectures treated "Dynamics in Astronomy" by Colombo, Message, Ovenden, Vicente, and Douglas.

Some of the outstanding lectures were rather didactic in nature or were published elsewhere or could not meet the deadline for publication. The Editors will be delighted to furnish leads to those interested in these lectures. Some of the lectures were presented in form of seminar-contributions. These are published as Summaries at the end of this Volume.

The Institute was dedicated to the conceptual, analytical, numerical and applied aspects of the problem of long-time prediction in dynamics. This fundamental problem emerged in all lectures: linearization, regularization, stabilization, averaging, estimation, periodic orbits, qualitative aspects, secular variations, resonance, invariants, etc. were some of the subjects treated in depth.

Some conclusions are offered here with the utmost humility and with the advance acknowledgement of the fact that we all hear what we want to hear.

The limitations of numerical integration must be extended if meaningful results are to be obtained with long time validity. The essential issues seem to be linearization, stabilization and regularization.

Qualitative results obtained by the use of the integrals of

motion of non-integrable dynamical systems are probably the most powerful tools to predict long-time behavior. The meager results available today must be sharpened and extended.

The role of periodic motions in the study of dynamical systems is of fundamental importance, whether established analytically or numerically. In fact Poincaré emphasized periodic orbits as the only "breach" through which non-integrable dynamical systems may be penetrated. From the point of view of the numerical analyst only periodic (and asymptotic) orbits may be established for arbitrary long time.

One principle followed by all speakers was the emphasis placed on non-integrable dynamical systems or more precisely put, on not integrated systems. The study of artificially created and neatly integrated problems devoid of any possible connection with actual dynamical systems received little attention unless these problems could be used to generate non-integrable systems.

The first series of lectures during the Institute were dedicated to the intricate connections between theory of numbers, topology and linearization. The last series discussed the origin and evolution of the solar system. This spectrum creates the impression of unboundedness and even of possible divergence unless remembered that the idea of sound analysis of dynamical systems keeps the subjects solidly together. Indeed, after many lectures and discussions (some of them unpredictable in time, lasting well after midnight), after patient attempts to understand pertinent results (and some results with unpredictable pertinency), after clarifying taboo subjects (random and ergodic sea, reversible and irreversible motions and systems, inverse and direct problems, stability), after living together for two weeks embedded in dynamics with incidental short pauses of eating and sleeping, we might declare one grudgingly and not unaminously acceptable conclusion: the study of long-time behavior of non-integrable dynamical systems is of considerable importance and difficulty and it is probably one of the most exciting, rewarding and challenging intellectual pursuits available today.

As Director of the Institute, I wish to thank for the assistance of my Co-Directors, Drs. E. Stiefel and G. Colombo, and for the help received from Drs. B. Tapley, J. Waldvogel, P. Nacozy, T. Feagin and A. Antonelli. Without the generous support of the Scientific Affairs Division of NATO, of the National Research Council of Italy and of The University of Texas at Austin neither the Institute nor this Volume would exist.

V. Szebehely
Director of the Institute

LIST OF PRINCIPAL SPEAKERS

Aarseth, S. J.	(Norwegian)	University of Cambridge, Institute of Astronomy, Madingley Road, Cambridge, England
Bartlett, J. E.	(U.S.A.)	Dept. of Physics and Astronomy, University of Alabama, P. O. Box 1921, University, Alabama, 35486, U.S.A.
Baumgarte, J.	(German)	Technische Universitat Braunschweig Mechanik-Zentrum, Pockelstrasse 4, Postfach der UT 3329, Braunschweig, West Germany
Broucke, R. A.	(Belgian)	CPB-Room 205, Jet Propulsion Laboratory, Pasadena, California 91103, U.S.A.
Colombo, G.	(Italian)	Istituto di Meccanica, Universita di Padova, Padova, Italy
Contopoulos, G.	(Greek)	Astronomical Dept., University of Athens, Athens, Greece
DeWitt-Morette, C.	(French)	Dept. of Astronomy, University of Texas, Austin, Texas, 78712, U.S.A.
Douglas, B. C.	(U.S.A.)	NOAA/NOS/NGS, C12, 6001 Executive Blvd., Rockville, Maryland, 20852, U.S.A.
Garfinkel, B.	(U.S.A.)	Yale University Observatory, New Haven, Conn., 06520, U.S.A.
Goudas, C. L.	(Greek)	University of Patras, Dept. of Mechanics, Patras, Greece

Hadjidemetriou, J.	(Greek)	Dept. of Theoretical Mechanics, University of Thessaloniki, Thessaloniki, Greece
Kirchgraber, U.	(Swiss)	Seminar fur angew. Math., ETH, Clausiusstr. 55, 8006 Zurich, Switzerland
Marchal, C. L.	(French)	O.N.E.R.A., 92320 Chatillon, France
Message, J.	(British)	Dept. of Applied Mathematics, University of Liverpool, P. O. Box 147, Liverpool L69 3BX, England
Nahon, F.	(French)	25 Avenue de l'Europe, F-92310 Sevres, France
Ovenden, M. W.	(Canadian)	The University of British Columbia, Vancouver, B.C., Canada V6T IW5
Reichl, L. E.	(U.S.A.)	Physics Dept., Univeristy of Texas, Austin, Texas, 78712, U.S.A.
Sigrist, N.	(Swiss)	Seminar fur angew. Math., ETH, Clausiusstr. 55, 8006 Zurich, Switzerland
Stiefel, E.	(Swiss)	Seminar fur angew. Math., ETH, Clausiusstr. 55, 8006 Zurich, Switzerland
Szebehely, V.	(U.S.A.)	Dept. of Aerospace Engineering and Engineering Mechanics, The University of Texas at Austin, Austin, Texas, 78712, U.S.A.
Tapley, B. D.	(U.S.A.)	Dept. of Aerospace Engineering and Engineering Mechanics, The University of Texas at Austin, Austin, Texas, 78712, U.S.A.

Verhulst, F.	(Dutch)	Mathematisch Instituut der Ryksuniversiteit. Budapestplaan 6, Utrecht, The Netherlands
Vicente, R. O.	(Portuguese)	R. Mestre Aviz., 30, R/C, Lisboa 3, Portugal
Waldvogel, J.	(Swiss)	Seminar fur angew. Math., ETH, Clausiusstr. 55, 8006 Zurich, Switzerland

LIST OF PARTICIPANTS

Adomian, G.	(U.S.A.)	Dept. of Math., University of Georgia, Athens, Georgia, 30602, U.S.A.
Allen, C.	(Canadian)	Instituto de Astronomio, UNAM, Apartado 70-264, Mexico 20, DF
Baxa, P.	(Italian)	University of Trieste Istituto di Geodesia e Geofisica Via de dell'Universita 7, I-34100 Trieste, Italy
Bell, C.	(U.S.A.)	Jet Propulsion Laboratory, 4800 Oak Grove Drive, Pasadena, California, 91103, U.S.A.
Berg, J. S.	(French)	Laboratoire de mecanique theoretique, Faculte des Sciences de Besancon, 25990 Les Verrieres de Joux, France
Berger, X.	(French)	CERGA, 8bd Emile Zola, F-06130, Grasse, France
Blanchard, D.	(U.S.A.)	Goddard Space Flight Center, Greenbelt, Maryland, 20771, U.S.A.
Borderies, N.	(French)	GRGS/CNES, 18 Avenue Edouard Belin, 31400 Toulouse, France
Bowman, B. R.	(U.S.A.)	Dept. of Geodesy, DMATC, 6500 Brooks Lane, Washington, D. C., 20315, U.S.A.
Braun, A.	(Israeli)	Dept. of Theoretical Physics, The Hebrew University, Jerusalem, Israel
Christides, T.	(Greek)	Dept. of Theoretical Mechanics, University of Thessaloniki, Thessaloniki, Greece

Claes, H.	(Belgian)	Facultes Universitaires de Naumr, Dept. de Math., 8 Rempart de la Vierge B-5000, Namur, Belgique
David, H.	(U.S.A.)	Institute of Applied Systems Analysis, Schloss Laxenburg, A-2361 Laxenburg, Austria
Delmas, C.	(French)	CERGA 8, bd. Emile Zola F-06530, Peymeinade, France
Dormand, J. R.	(British)	Dept. of Mathematics, Teesside Polytecnic, Middlesbrough, Cleveland, TSI 3BA, United Kingdom
Dvorak, R.	(Austrian)	Universitatssternwarte Graz., A-8010 Graz, Austria
Feagin, T.	(U.S.A.)	UTSI, Tullahoma, Tennessee, 37388, U.S.A.
Forti, G.	(Italian)	Largo E. Fermi 5, 50125 Firenze, Italy
Galgani, L.	(Italian)	Istituto di Fisica dell'Universita, Via Cecoria 16, 20133 Millano, Italy
Goossens, M.	(Belgian)	Astronomisch, Instituut Naamsestraat 61, B-3000, Leuven, Belgium
Henrard, J.	(Belgian)	Facultes Universitaires de Namur Dept. de Mathematiques 8, Rempart de la Vierge 8, 5000 Namur, Belgium
Kinoshita, H.q	(Japanese)	S.A.O., 60 Garden Street, Cambridge, Massachusetts, 02138, U.S.A.
Langlois, M.	(French)	Passevant, F-25360, Bouclans, France
Lueneburg, E.	(German)	DFVLR, I 551 D8031, Oberpfaffenhofen, Deutschland

Martinez, B. J.	(Spanish)	Observatoire de Paris, DANOF/ GRGS, 92190 Meudon, France
Melbourne, W. G.	(U.S.A.)	Jet Propulsion Laboratory, Mail Stop CPB-208, 4800 Oak Grove Dr., Pasadena, California, 91103, 91103, U.S.A.
Mertzanides, C.	(Greek)	Astronomical Dept., University of Thessaloniki, Thessaloniki, Greece
Michalomitrakis, M.	(Greek)	University of Thessaloniki, Thessaloniki, Greece
Nacozy, P. E.	(U.S.A.)	Dept. of Aerospace Engineering and Engineering Mechanics, The University of Texas, Austin, Texas, 78712, U.S.A.
Osorio, J. J.de S. P.	(Portuguese)	Observatorio Astronomico,"Prof. Manuel de Barros", Universidade do Porto Monte de Virgem, Vila Nova de Gaia, Portugal
Puel, F.	(French)	Observatoire de Besancon, F-25000 Besancon, France
Rand, R. H.	(U.S.A.)	Dept. Theo. Appl. Mechanics, Thurston Hall, Cornell University, Ithaca, New York, 14850, U.S.A.
Roels, J.	(Belgian)	Universite de Louvain, Institut de Mathematique Pure et appliquee, Chemin du Cyclotron 2, B-1348 Louvain - La Neuve, Belgium
Schmidt, D.	(German)	2003 Avenel Road, Adelphi, Maryland, 20783, U.S.A.
Schmidt, K.	(German)	Ruhr-Universitat Bochum, Bereich Extraterrestische Physik Abt.XII, D-463, Bochum Postfach 2148, Buscheystrasse

Schutz, B. E.	(U.S.A.)	Dept. of Aerospace Engineering and Engineering Mechanics, The University of Texas, Austin, Texas, 78712, U.S.A.
Schwelm, G.	(German)	Ruhr-Universitat Bochum, Bereich Extraterristische Physik, D 463, Bochum Universitatsstr. 150, Geb. NA 01/133
Simo, C.	(Spanish)	Pje. Marimon 11, Barcelona 11 Spain
Standish, E. Myles	(U.S.A.)	CPB-104, Jet Prouplsion Laboratory, Pasadena, California, 91103, U.S.A.
Stokes, A.	(U.S.A.)	Code 582, Goddard Space Flight Center, Greenbelt, Maryland, 20771, U.S.A.
Thiry, Y.	(French)	4 Rue du Marche, F-92160 Antony, France
Van der Burgh, A.	(Dutch)	Dept. of Mathematics, University of Technology, Julianalaan 123, Delft, The Netherlands
Velez, C. E.	(U.S.A.)	NASA Goddard Space Flight Center, Code 582, Greenbelt, Maryland, 20771, U.S.A.
Vinti, J. P.	(U.S.A.)	Bldg. W 91-202, M.I.T. Mass. Inst. Tech., Cambridge, Mass., 02139, U.S.A.
Wisskirchen, P.	(German)	Bonn, West Germany
Zare, K.	(Iran)	Dept. of Aerospace Engineering and Engineering Mechanics, The University of Texas, Austin, Texas, 78712, U.S.A.

PART I

FUNDAMENTALS

FROM THE THEORY OF NUMBERS VIA GYROSCOPES AND LIE ALGEBRAS TO LINEAR CELESTIAL MECHANICS

E. Stiefel

Institut fur Angewandte Mathematik, Eidge. Technische Hochschule, Zurich, Switzerland

ABSTRACT. The impact of Euler's identity-concerning sums of four squares - on some fields of modern applied mathematics is described. Beginning with number theory the importance of abstract algebraic structures (groups, Lie algebras) for applied work is emphasized. In particular gyroscopes and near parabolic Keplerian motion are discussed.

It is the author's feeling that algebraic structures (such as groups, algebras, in particular Lie algebras) will play a more and more important role also in celestial mechanics as they do already in quantum mechanics.

1. ALGEBRAS, NUMBER THEORY

Consider two complex planes $z = x_1 + ix_2$, $u = u_1 + iu_2$ and the conformal mapping (also called Levi-Civita transformation)

$$z = u^2, \quad \text{thus } r = |z| = |u|^2 . \tag{1}$$

In real notation

$$x_1 = u_1^2 - u_2^2, \quad x_2 = 2u_1u_2, \quad x_1^2 + x_2^2 = (u_1^2 + u_2^2)^2. \tag{2}$$

V. Szebehely and B.D. Tapley (eds.), Long-Time Predictions in Dynamics, 3–15. *All Rights Reserved. Copyright © 1976 by D. Reidel Publishing Company, Dordrecht-Holland.*

These are the well-known formulae for Pythagorean triples provided u_1, u_2 are chosen as integers, for instance $u_1 = 2$, $u_2 = 1$:

$$x_1 = 3, \quad x_2 = 4, \quad 3^2 + 4^2 = 5^2.$$

The complex numbers form a structure called an algebra by pure mathematicians. This name is given to a vector space supplemented by a rule for multiplication of vectors which obeys the distributive law with respect to the vectorial addition. It is of interest that this multiplication satisfies the <u>rule of norms</u>.

$$|u\ v| = |u|\ |v|. \tag{3}$$

Another important 4-dimensional algebra is defined by the quaternions. Let 1, i, j, k be the base-vectors of the 4-dimensional space and define the multiplication by

$$i^2 = j^2 = k^2 = -1, \quad ij = -ji = k, \quad \text{and cyclic permutation.} \tag{4}$$

A quaternion is thus

$$u = u_o + u_1 i + u_2 j + u_3 k$$

and it's <u>conjugate</u> is defined by

$$\overline{u} = u_o - u_1 i - u_2 j - u_3 k\ . \tag{5}$$

The rule

$$\overline{uv} = \overline{v}\ \overline{u} \tag{6}$$

is obvious for the base-vectors and holds true for general quaternions by virture of the distributive law. The rule of norms can be proved as follows. We have

$$u\ \overline{u} = u_o^2 + u_1^2 + u_2^2 + u_3^2 = |u|^2, \tag{7}$$

hence

$$|uv|^2 = (uv)\ (\overline{uv}) = (uv)\ (\overline{v}\ \overline{u}) = u(v\overline{v})\ \overline{u}$$
$$= u\ |v|^2\ \overline{u} = (u\overline{u})\ |v|^2 = |u|^2\ |v|^2.$$

(The multiplication is associative). In real notation there follows <u>Euler's identity</u>.

$$(u_o^2 + u_1^2 + u_2^2 + u_3^2)\ (v_o^2 + v_1^2 + v_2^2 + v_3^2)$$
$$= (u_o v_o - u_1 v_1 - u_2 v_2 - u_3 v_3)^2 + (u\ v_1 + u_1 v_o + u_2 v_3 - u_3 v_2)^2$$
$$+ (u_o v_2 - u_1 v_3 + u_2 v_o + u_3 v_1)^2 + (u_o v_3 + u_1 v_2 - u_2 v_1 + u_3 v_o)^2. \tag{8}$$

Let us return to number theory by assuming the u and v to be integers. Then the integers

$$a = u_0^2 + u_1^2 + u_2^2 + u_3^2, \quad b = v_0^2 + v_1^2 + v_2^2 + v_3^2$$

are decomposed into 4 squares and the identity makes clear that also the product $(a\ b)$ is decomposable into four squares. Example:

$$a = 25 = 0^2 + 0^2 + 3^2 + 4^2$$

$$b = 79 = 3^2 + 3^2 + 5^2 + 6^2$$

$$ab = 1975 = 39^2 + 2^2 + 21^2 + 3^2 .$$

By making use of this identity Euler proved the famous statement of Lagrange that every integer is decomposable into four squares. The proof is classical and uses a method of descent invented by Fermat. The deeper reason of this phenomenon is obviously the rule of norms of the quaternions.

Proceeding along this line A. Hurwitz established in 1896 a theory of numbers of quaternions. It may be mentioned that Jacobi computed the number of decompositions of an integer into 4 squares by clever application of the theory of elliptic functions (more precisely ϑ-functions.

Are there more algebras with a valid rule of the norm?

If the algebra is required to be associative, the answer is no. (Frobenius)

There is an 8-dimensional non-associative algebra obeying the rule of norms, the so-called Cayley-algebra. According to a result of Hurwitz the realm of norm-product algebras is, with this Caley-algebra, exhausted.

A more general question is raised by the following consideration. In an algebra with rule of norms a product $u\ v$ can only vanish if one of the factors (or both) vanish. From $uv = 0$ there follows indeed

$$|u|\ |v| = |u\ v| = 0, \text{ thus } |u| = 0 \text{ or } |v| = 0,\ u = 0 \text{ or } v = 0.$$

The rule of norms is thus weakened and becomes a rule of the vanishing product or rule of divisors of zero. Consequently the problem is posed to find all the algebras which obey the rule of the vanishing product. The answer was given fairly recently by Kervaire after preliminary results of the author. There are no additional algebras of that kind. It is of interest that no algebraic proof is known of this statement. The proofs make use of deep topological aids.

2. TOPOLOGY

Without going too far into topology we may throw some light on the role which topology plays in such problems which are of course a little remote from celestial mechanics. For this purpose we consider the unit vectors p in our 4-dimensional quaternion space.

Their arrow-heads form a 3-dimensional sphere S^3. Denoting the components of p by p_0, p_1, p_2, p_3 we attach at the point p the vector $(-p_1, p_0, p_3, -p_2)$ which is orthogonal to the position vector p and is thus tangential to the sphere S^3. To sum up, we have constructed a continuous field of vectors $\neq 0$ tangential to S^3, that is a field of directions. The study of vector fields on manifolds is an important branch of topology. A classical result is due to L. E. Brouwer: There is no continuous tangential direction field on the 2-dimensional sphere $x_1^2 + x_2^2 + x_3^2 = 1$. This is of course the deeper reason for the fact that no 3-dimensional algebra exists which satisfies the rule of the vanishing product, since such an algebra would generate a vector field on S^2, similar to the vector fields on S^3 constructed by the quaternions above.

3. GYROSCOPES

We proceed to the discussion of gyroscopes and rigid body motion. Let u be a fixed quaternion of absolute value equal to 1 and let p be any vector in the 3-dimensional space, that is a quaternion of the type

$$p = p_1 i + p_2 j + p_3 k. \tag{9}$$

Let us consider the relation

$$q = \overline{u}\, p\, u. \tag{10}$$

From $\overline{p} = -p$ there follows

$$\overline{q} = \overline{u}\, \overline{p}\, u = -\overline{u}\, p\, u = -q.$$

Hence q is also a quaternion of real part = 0 and is consequently a vector in our 3-space. Thus Eq. (10) establishes a linear mapping of the 3-space onto itself with conservation of the length of vectors as follows from

$$|q|^2 = q\,\overline{q} = \overline{u}\, p\, u\, \overline{u}\, \overline{p}\, u = \overline{u}\, p\, \overline{p}\, u = |p|^2 (u\, \overline{u}) = |p|^2.$$

The mapping $p \rightarrow q$ is thus a rotation (the determinant being +1).

Its matrix is orthogonal and is obtained by putting $p = i, j, k$ consecutively. The result is

$$\begin{bmatrix} (u_0^2 + u_1^2 - u_2^2 - u_3^2) & 2(u_1u_2 + u_ou_3) & 2(-u_ou_2 + u_1u_3) \\ 2(u_1u_2 - u_ou_3) & (u_o^2 - u_1^2+u_2^2-u_3^2) & 2(u\ u_1 + u_2u_3) \\ 2(u_ou_2 + u_1u_3) & 2(-u_ou_1 + u_2u_3) & (u_o^2-u_1^2 - u_2^2+u_3^2) \end{bmatrix} \qquad (11)$$

To each set of 4 parameters u_i satisfying $u_o^2+u_1^2+u_2^2+u_3^2 = 1$ is attached a rotation. Observe that (11) rotates the basic cartesian frame into a frame of general position. Its base vectors are the columns of the matrix (11). This frame can be interpreted as the system of the principal axis of a rigid body and consequently the u are the well known Cayley- or Rodrigues-parameters of a gyroscope. We draw the attention of the reader to the fact, that these parameters are always well defined by the position of the body (up to the sign of the u_i). Thus the Cayley parameterization is well suited for the description of the motion of a rigid body. In contrast to that fact the Eulerian angles are not always defined by the position. One of these angles is defined indeed by the line of the node, that is the intersection of the equatorial plane of the body with a fixed plane of reference. This line is no longer defined in the special position where the two planes coincide. This fact is reflected by singularities which occur in the differential equations of the spinning body. (Denominators containing the sine of the inclination). It is a basic philosophy of modern computational practice to avoid such geometric singularities which are introduced artificially by unfit parameters.

But such unfit parameters are sometimes unavoidable. Let us consider, for instance, the classical elements of Keplerian motion. They display similar geometric singularities as the Eulerian angles. These elements are used in almost all analytical perturbation theories since they represent the closed solution of the unperturbed motion. In order to get rid of the involved singularities U. Kirchgraber established a transformation theory permitting to reformulate results of perturbation theory in terms of coordinates (or regular parameters) in cases where these results were obtained by making use of singular elements.

4. GROUPS

The rotations established in (11) form an algebraic structure called a group.

By a (transformation) group is understood in general a set of transformations such that the composition of two transformations is also a member of the set. Also the identity and the inverse must be contained in the set.

The law of composition of rotations is the rule of multiplication of quaternions since by virtue of the definition we may write:

Assumption: $q = \bar{u}\, p\, u, \qquad r = \bar{v}\, q\, v.$

Statement: $r = \bar{v}\, \bar{u}\, p\, u\, v = (\overline{uv})p\,(uv).$

5. REPRESENTATIONS

Before going to celestial mechanics we touch briefly the theory of representation of groups and quantum mechanics. Let us attach to a quaternion $u = u_o + u_1 i + u_2 j + u_3 k$ the matrix

$$U = \begin{pmatrix} u_o + iu_1 & u_2 + iu_3 \\ -u_2 + iu_3 & u_o - iu_1 \end{pmatrix}, \qquad \sum_{(j)} u_j^2 = 1. \tag{12}$$

An easy verification leads to the fact that the multiplication of two quaternions is equivalent to the multiplication of the corresponding matrices. As a corollary it follows that the composition of two rotations (11) is also given by the multiplication of the matrices (12). The group of rotations in 3-space is thus represented by the group of the unitary matrices (12). Kustaanheimo made use of this situation for his reformulation of celestial mechanics. He is interested in the linear transformation

$$\begin{pmatrix} w_1 \\ w_2 \end{pmatrix} = \begin{pmatrix} u_o + iu_1 & u_2 + iu_3 \\ -u_2 + iu_3 & u_o - iu_1 \end{pmatrix} \begin{pmatrix} z_1 \\ z_2 \end{pmatrix} \tag{13}$$

attached to the matrix (12). The 2-dimensional complex vector z_1, z_2 is called a spinor.

Let us, at last, consider the one-parametric family of spinor rotations

$$\begin{pmatrix} 1 + iu_1 t & (u_2 + iu_3)t \\ (-u_2 + iu_3)t & 1 - iu_1\, t \end{pmatrix} \tag{14}$$

The corresponding infinitesimal transformation is obtained by differentiation with respect to t:

$$J = \begin{pmatrix} iu_1 & u_2 + iu_3 \\ -u_2 + iu_3 & -iu_1 \end{pmatrix} \tag{14a}$$

By inserting the base vectors (1, 0, 0), (0, 1, 0), (0, 0, 1) we obtain the celebrated Pauli-matrices

$$e_1 = \begin{pmatrix} i & 0 \\ 0 & -i \end{pmatrix}, \quad e_2 = \begin{pmatrix} 0 & 1 \\ -1 & 0 \end{pmatrix}, \quad e_3 = \begin{pmatrix} 0 & i \\ i & 0 \end{pmatrix} \tag{15}$$

They represent the spin of the electron in quantum mechanics whereas the matrix (11) represents the rotation of a rigid body in classical mechanics.

6. THE CONCEPT OF A LIE-ALGEBRA

The general infinitesimal transformation J in (14a) is a linear combination of the Pauli-matrices:

$$J = u_1 e_1 + u_2 e_2 + u_3 e_3 \tag{16}$$

Evidently a 3-dimensional space is generated with base-vectors e_1, e_2, e_3. It is not an algebra with respect ot matrix-multiplication since for instance

$$e_1^2 = \begin{pmatrix} -1 & 0 \\ 0 & -1 \end{pmatrix}$$

which is not a vector of the space under consideration. In order to remove that difficulty, we modify the rule of multiplication by introducing the commutator-product of two matrices A, B:

$$[A, B] = AB - BA. \tag{17}$$

Then $[e_1, e_1] = 0$ which is a vector of our space. By straightforward arithmetic it is seen that

$$[e_1, e_2] = 2e_3, \quad [e_2, e_3] = 2e_1 \quad [e_3, e_1] = 2e_2. \tag{18}$$

In general we understand by a Lie-algebra a vector space with a definition of a commutator $[u, v]$ which is distributive with respect to the vectorial addition and which satisfies the laws

$[v, u] = -[u, v]$, Jacobi's identity:

$$\left[u, [v,w]\right] + \left[v, [w,u]\right] + \left[w, [u,v]\right] = 0. \tag{19}$$

It is easy to verify these laws in case of the definition (17). Is it possible to construct the finite transformations (12) from the infinitesimal transformations (14a)? The answer is

$$U = e^{Jt}. \tag{20}$$

We do verify this only in two examples.

1. For $J = e_1$ we get

$$U = e^{e_1 t} = \begin{pmatrix} e^{it} & 0 \\ 0 & e^{-it} \end{pmatrix}$$

This is of the type (12) with

$$u_o = \cos t,\ u_1 = \sin t,\ u_2 = u_3 = 0.$$

2. For $J = e_2$ one finds

$$U = e^{e_2 t} = \begin{pmatrix} \cos t & \sin t \\ -\sin t & \cos t \end{pmatrix}$$

since the differential equation of the exponential

$$\frac{dU}{dt} = e_2 U$$

is satisified.

In order to rebuild the multiplicative structure of the group of the finite transformations (the law of composition) we have to learn to compute

$$e^{J_1} e^{J_2} = e^{J_3}.$$

In general $J_3 \neq J_1 + J_2$ but the so-called Campbell-Hausdorff formula

$$J_3 = J_1 + J_2 + \tfrac{1}{2}[J_1, J_2] + \frac{1}{12}\Big[[J_1, J_2],\ J_2\Big] + \frac{1}{12}\Big[[J_2, J_1],\ J_1\Big] + \ldots$$

is valid. It should be emphasized, that all terms are commutators.

The generation of a finite transformation U by an infinitesimal transformation J by means of

$$U = e^{Jt} = 1 + Jt + \frac{1}{2!} J^2t^2 + \ldots$$

is called a Lie-series. This method is practical since it is easy to compute the inverse

$$U^{-1} = e^{-Jt}.$$

Dynamical perturbation theories make very much use of transformation and Hori was the first to propose the application of Lie series for analytical perturbation theories of v. Zeipel's type.

A very important Lie-structure in dynamics is produced by the Poisson-brackets. Let $f(p_i, q_i)$ and $g(p_i, q_i)$ be two functions of coordinates q_i and momenta p_i and define

$$[f, g] = \sum_{(i)} \left(\frac{\partial f}{\partial q_i} \frac{\partial g}{\partial p_i} - \frac{\partial f}{\partial p_i} \frac{\partial g}{\partial q_i} \right).$$

The laws (19) are evidently satisfied. As an example let us take the vector space of the angular momentum u of a particle.

$$u_1 = q_2p_3 - q_3p_2, \quad u_2, u_3 \text{ by cyclic permutation} \tag{21}$$

or $u = q \times p$ for short.

It follows

$$[u_1, u_2] = u_3 \text{ (and cyclic permutation).} \tag{22}$$

Thus a 3-dimensional Lie-algebra is born. The corresponding group of finite transformations is the group of rotations.

We close that discussion by an example in celestial dynamics. Let a particle move according to Kepler's laws. In addition to the angular momentum u it is of interest to study the Laplace-vector:

$$v = (p \times u) - \frac{q}{r} \tag{23}$$

(r = distance). Both vectors u, v are elements (or integrals of motion); they are constant during a Keplerian motion. u determines indeed the position of the orbital plane. v is a vector directed to the second focus of the Keplerian orbit and it determines the shape of the orbit.

Recently J. Baumgarte observed that u and v together form a 6-dimensional Lie-algebra if subjected to the Poisson-bracket as law of multiplication.

More precisely one must introduce the vector

$$w = \frac{1}{\sqrt{-2H}} v,$$

where H is the Hamiltonian. We have then

$$[u_1, u_2] = u_3, \quad [w_1, u_2] = w_3, \quad [w_1, w_2] = u_3$$

and the rules obtained by cyclic permutation. The impact of this Lie structure on the integration of astrodynamical problems is not yet fully understood.

As a conclusion we may observe that the integrals of motion of dynamical systems have the tendency to form Lie-algebras and that this Lie structure is fit to classify these integrals.

7. CELESTIAL MECHANICS

Let u_o, u_1, u_2, u_3 be again a point in the 4-space R^4 and let us map it onto the point in the 3-space which is given by the first column of the matrix (11)

$$x_1 = u_o^2+u_1^2-u_2^2-u_3^2, \quad x_2 = 2(u_1u_2 - u_ou_3), \quad x_3 = 2(u_ou_2 + u_1u_3). \tag{24}$$

As was seen above it follows $x_1^2+x_2^2+x_3^2 = 1$ provided $u_o^2+u_1^2+u_2^2+u_3^2 = 1$. The unit sphere S^3 is mapped therefore onto the unit sphere S^2. The topologist <u>H. Hopf</u> has studied and classified all continuous mappings of S^3 onto S^2. In the terminology of mechanics we may say that the position of a spinning gyroscope is replaced by the direction of its axis.

In celestial mechanics the transformation (24) is known as the KS-transformation. It describes the position and motion of a particle (x_1, x_2, x_3) by the position and motion of four parameters u_o, u_1, u_2, u_3.

The use of redundant generalized coordinates is understandable since from our discussion of algebras we have seen that the dimension three is likely to generate singularities. Since KS-transformation was described [1] in extenso at the Cortina Institute 1972 we do not offer the general theory but we mention the fact that Keplerian motion in the x-space is transformed into harmonic oscillation

$$u_i'' + \frac{h}{2} u_i = 0. \qquad i = 0, 1, 2, 3. \tag{25}$$

h is the negative total energy and an accent denotes differentiation with respect to an independent variable which is linked to the time t by the differential relation $t' = r$ = distance in x-space.

For this Cortina session we draw the attention of the audience to an application of the KS-approach which is, at least remote, connected with the topic "long time predictions in dynamics." Without loss of generality we restrict ourselves to the 2-dimensional case $u_0 = u_3 = 0$. Hence Eq. (24) is reduced to

$$x_1 = u_1^2 - u_2^2, \qquad x_2 = 2u_1u_2. \tag{26}$$

This is exactly the set (2) of formulae of Pythagorean triples and thus we add the equation for the distance

$$r = \sqrt{x_1^2 + x_2^2} = u_1^2 + u_2^2. \tag{27}$$

Consider now a particle moving in the u_1, u_2-plane on a straight line ℓ with constant velocity β. Without loss of generality we adopt for ℓ the vertical at abscissa $u_1 = \alpha$. The motion is thus

$$u_1 = \alpha, \qquad u_2 = \beta s, \tag{28}$$

where s is a parameter on ℓ. The transformation (26) produces the motion

$$x_1 = \alpha^2 - \beta^2 s^2, \qquad x_2 = 2\,\alpha\,\beta\,s, \qquad r = \alpha^2 + \beta^2 s^2 \tag{29}$$

of the image of the particle in the x_1, x_2 plane. From

$$r = 2\alpha^2 - x_1 \tag{30}$$

it follows that the general point on this path has equal distance from the origin and from the vertical $x_1 = 2\alpha^2$. Hence the path is a parabola focused at the origin and this situation is in agreement with Kepler's first law. But also the second law is satisfied provided the appropriate relation between time t and parameter s is established. The increment of t should be proportional to the increment of the area swept by the vector of position:

$$t' = \frac{dt}{ds} = \begin{vmatrix} x_1 & x_2 \\ x_1' & x_2' \end{vmatrix} = \begin{vmatrix} \alpha^2-\beta^2 s^2 & 2\alpha\beta s \\ -2\beta^2 s & 2\alpha\beta \end{vmatrix} = 2\alpha\beta(\alpha^2 + \beta^2 s^2)$$

$$t' = 2\alpha\beta r, \qquad t = 2\alpha\beta\, s\,(\alpha^2 + \frac{1}{3}\beta^2 s^2). \tag{31}$$

This leads to the following conclusion.

A uniform rectilinear motion in the u-space is KS-mapped onto a parabolic Keplerian motion in the x-space.

Corollary: During a pure parabolic motion the rectangular coordinates x_1, x_2, x_3 are quadratic polynomials in s and the time is a cubic polynomial.

In the terminology of celestial mechanics the parameters u_i are elements of the parabolic motion since they vary linearly as functions of s. They obey the differential equations

$$u_i'' = 0, \qquad i = 0, 1, 2. 3. \tag{32}$$

In order to apply these results to the numerical integration of orbits we mention the trivial but important fact that any reasonable numerical routine does integrate the differential equations (32) without truncation error. This leads to the expectation that also in a perturbed and near parabolic case the integration of the u-equations is better than the integration of the x-equations. The former are

$$u_i'' + \frac{h}{2} u_i = \text{perturbation}, \qquad t' = r \tag{33}$$

as is suggested by Eq. (25).

This conjecture was corroborated by numerical experiments. We compared the following two methods:

1. The x-method. Dependent variables x_1, x_2, x_3, t, independent variable s, where $dt = r \cdot ds$.

2. The u-method. Integration of the equations (33).

Always a Runge-Kutta 7/8 integrator was used. If oblateness J_2 and lunar perturbation were adopted the u-method was always more accurate than the x-method. As a rule of thumb a gain of 3 decimals may be expected. (The initial eccentricity of the orbit was chosen in the interval $0.95 \le e \le 1.02$.) If drag occurs both methods are equally accurate for perigees lower than 300 km. For higher perigees the u-method is better.

A final remark concerning the integration of the time t is in order. In the u-method the dependent variables (unknown functions) are u_0, u_1, u_2, u_3, t.

Whereas the u_i are elements, the time t is not an element. Recently V. R. Bond from Johnson Space Flight Center invented a time element τ[2]. During a pure Keplerian (not necessarily parabolic) motion the element τ is constant. In perturbed cases τ is a function of s subjected to a differential equation. The time t is computed from τ by an explicit closed formula. Bond's element was also tested by numerical experiments with the u-method. In many cases it was slightly better than the integration of $t' = r$; but this advantage is outweighted a little by an increase of the computing time.

REFERENCES

1. B. D. Tapley, V. Szebehely (editors): Recent Advances in Dynamical Astronomy. Proceedings of the NATO Advanced Study Institute in Dynamical Astronomy held in Cortina d'Ampezzo, 1972, 3-20.

2. V. R. Bond: The Uniform, Regular Differential Equations of the KS Transformed Perturbed Two-Body Problem. Celestial Mechanics, 10 (1974) 303-318.

LECTURES ON LINEARIZING TRANSFORMATIONS OF DYNAMICAL SYSTEMS

V. Szebehely

Department of Aerospace Engineering and Engineering Mechanics, University of Texas at Austin, Austin, Texas

CONTENTS

ABSTRACT. Linearization of ordinary differential equations is investigated by transformations of the dependent and independent variables. The direct problem investigates the transformations of non-linear equations of interest in dynamics. The inverse problem establishes the totality of non-linear differential equations which are reducible to linear equations by transformations. The transformations of the independent variable relate the differentials of the old and new variables by functions of the dependent variables.

V. Szebehely and B.D. Tapley (eds.), Long-Time Predictions in Dynamics, 17–42. *All Rights Reserved. Copyright © 1976 by D. Reidel Publishing Company, Dordrecht-Holland.*

INTRODUCTION

These four lectures corresponding to the four parts of this paper may be considered independently of each other though the lectures are cross-referenced. Abstract, Introduction, Definition of the Problem, Solution, and Cases of Special Interest are the chapter headings for the lectures.

Linearizations or quasi-linearizations are never ceasing ambitions of the theoretician. Of the two kinds of linearizations, exact and approximate, (legal and illegal), interest is directed in these lectures to the first kind. By exact linearization, we mean transformations resulting in linear equations, completely and exactly replacing the original non-linear systems. The approximate linearization is the result of a first order expansion and, therefore, it approximates the system instead of replacing it.

These lectures deal with linearization as a first step in a somewhat more sophisticated program in long-term predictions in dynamics. Differential equations with solutions of finite power series are generalizations of linear differential equations in several important aspects. The solution is once more exact and regular, but furthermore, the requirements might be relaxed when the approximations are considered. Numerical integration techniques using Taylor series or power series approximations become exact when the solution is a finite power series. Consequently, if the solution may be represented in strongly convergent (infinite) power series, the numerical integration becomes highly efficient. Therefore, the long-term program is to establish transformations which result in equations, the solutions of which, may be approximated by strongly convergent power series in terms of the new variables. It is expected that the general ideas and approaches presented in these lectures are also applicable, with some modifications, to this, more general problem of long-term prediction in dynamics.

All four lectures utilize the same type of transformation of the independent variable. Poincaré's and Sundman's uniformization offer the basic idea when the differentials of the new and old independent variables are connected with an arbitrary function of the dependent variables. Indeed, this concept may be generalized and many degrees of freedom systems can be treated by the introduction of several new independent variables, all of which, in turn, may be related to the original independent variable or to each other. Such transformations are presented elsewhere (Szebehely and Bettis, 1971, Szebehely 1975) and in these lectures only the single independent variable transformation is used.

Harmonic oscillators form a special case of linear systems.

It is realized that perturbed linear systems behave differently when the corresponding unperturbed linear systems have real or imaginary frequencies, therefore, when describing linearizations, this aspect is carefully pointed out.

It is hoped that these lectures are applicable to a variety of dynamical systems, in fact to a variety of problems which may be described by ordinary differential equations. The fact that the applications given are slanted toward celestial mechanics should be considered more a limitation of the author than of the method.

1. One-degree of freedom dynamical systems. The direct problem.

The reduction of a single degree of freedom dynamical system to a linear differential equation is accomplished in the most general case. This result is applied to the motion of two bodies with perburbations. It is shown that if the time and coordinate transformations are of the form x^n , where x is the distance, then there are only six possible combinations of such transformations giving linear differential equations for the unperturbed problem of two bodies. Two of the six linear differential equations represent harmonic oscillators with real frequencies for negative total energy. Two of the six transformations accomplish the linearization by the use of time transformation only, without associated coordinate transformations. Position and velocity-dependent perturbations are discussed.

1.1 Introduction

Because of the well known stability properties for numerical integration of perturbed harmonic oscillators as opposed to the Liapunov unstable behavior of the problem of two bodies, it is of considerable interest in dynamics and in celestial mechanics to investigate transformations leading to such equations. For a recent thorough discussion see Stiefel and Scheifele (1971).

The available tools are time and coordinate transformations and reduction of the equations of motion by integrals. In the one-dimensional conservative case there is only one integral that is available for reduction. In the three-dimensional case the possible use of the energy and angular momentum integrals offers a considerably greater variety.

1.2 The General Problem and its Solution

The differential equation of interest is of the form

$$\ddot{x} = h(x) \quad . \tag{1}$$

As an example of interest, the one-dimensional case of the unperturbed problem of two bodies may be mentioned:

$$\ddot{x} = -\frac{\mu}{x^2} \quad , \tag{2}$$

where x is the distance between the bodies, dots denote derivatives with respect to the time t, and the constant μ is the product of the gravitational constant and the total mass.

Position dependent perturbations, $\varepsilon h_1(x)$, may be incorporated into $h(x)$ and this is assumed for the subsequent general treatment. Velocity and position dependent perturbations, $\varepsilon h_2(x,\dot{x})$ will be treated separately.

Introducing

$$H(x) = 2\int h(x)dx, \tag{3}$$

the integral of Equation (1) becomes

$$\dot{x}^2 = H(x) + 2c_e \quad , \tag{4}$$

where c_e is a constant.

For the unperturbed problem of two bodies $H=2\mu x^{-1}$ and the equation

$$\dot{x}^2 = 2\left(\frac{\mu}{x} + c_e\right) \tag{5}$$

is known as the integral of energy.

The type of time transformation of principal interest is a generalization of the change of the independent variable, orginally proposed by Sundman (1912):

$$s = \int_{t_0}^{t} \frac{dt}{g(y)} \quad , \tag{6}$$

where s is the new independent variable (pseudo-time), and y is the new coordinate defined by

$$x = f(y) \ . \tag{7}$$

In general, then, the problem is the reduction of Equation (1) to a form

$$y'' = Ay + B \ , \tag{8}$$

where primes denote derivatives with respect to s, and A and B are constants. If $A<0$, Equation (8) represents a harmonic oscillator. In the process of reduction the transformations, i.e., the selections of the functions $g(y)$ and $f(y)$ as well as the available integral, i.e., Equation (4) play the principal roles.

The general approach is to transform Equation (2) as follows. The old and new "velocities" are related by

$$\frac{dx}{dt} = \frac{df}{dy}\frac{dy}{ds}\frac{ds}{dt} \,, \quad \text{or} \quad \dot{x} = \frac{f'}{g} y' \tag{9}$$

where $f' = df/dy$ and $y' = dy/ds$.

The integral, Equation (4), transforms into

$$y'^2 = (\frac{g}{f'})^2(H + 2c_e). \tag{10}$$

Since in the case of a one-dimensional motion the equation of motion is related to the energy-integral, provided $\partial h_2/\partial\dot{x} = 0$ the transformed form of Equation (1) may be obtained from Equation (10) by differentiation. (It is trivial, but of fundamental importance that this is not the case for the two or three-dimensional problems). Taking derivatives with respect to s, we obtain from Equation (10):

$$y'' = \frac{d}{dy}\left[\frac{1}{2}(\frac{g}{f'})^2(H+2c_e)\right] \,, \tag{11}$$

which, when compared to Equation (8) gives

$$(\frac{g}{f'})^2(H+2c_e) = Ay^2 + 2By + C = Q(y), \tag{12}$$

where C is a constant and $H = H\,[x(f(y))]$. Note that the exclusion of y' in Eqn. (8) avoids the appearance $\int y'\,dy$ in Eqn. (12).

For a given "potential" H, Equation (12) is the condition which the functions g and f must satisfy in order to linearize the equation of motion. In fact, selecting the coordinate transformation $f(y)$, we have for the time transformation,

$$g = f'\left[\frac{Q}{H+2c_e}\right]^{1/2} \,. \tag{13}$$

When transformation of the coordinate is not used, $x=y$, $f'=1$ and

$$g = \left[\frac{Q(x)}{H+2c_e}\right]^{1/2} , \tag{14}$$

which for the unperturbed gravitational case $(H = 2\mu x^{-1})$ becomes

$$g = \left[\frac{xQ(x)}{2(\mu+c_e x)}\right]^{1/2} . \tag{14 bis}$$

Further specializing, one may select $g(y) = y = x$, which is the original form of Sundman's transformation. The well-known result,

$$x'' = 2c_e x + \mu , \tag{15}$$

is obtained when the coefficients of equal powers of x are equated in Equation (14 bis). The next section offers a systematic discussion of the general result given by Equation (12).

At this point it is instructive to point out the importance of the use of the integral of energy. To this subject reference is made in the sequel.

Consider $\ddot{x} = h(x) + \varepsilon h_2(x,\dot{x})$, in which case the perturbed motion does not possess an integral of energy. Proceeding as before, we compute $\dot{x} = f'y'/g$ as well as

$$\ddot{x} = \frac{f'}{g^2}\left[y'' - y'^2\left(\frac{g'}{g} - \frac{f''}{f'}\right)\right] .$$

Consequently,

$$y'' = y'^2 \frac{d}{dy} \ln g/f' + \frac{g^2}{f'}(h + \varepsilon h_2) \tag{11 bis}$$

and for linearization of the unperturbed equation, we have $f' = kg$ with

$$\frac{g^2 h}{f'} = \frac{gh}{k} = Ay + B , \tag{12 bis}$$

corresponding to Equation (12). Therefore, for a given function h, the time transformation is determined by Equation (12 bis) and the coordinate transformation is given by $f' = kg$. Note that this choice, $(f' = kg)$ eliminates the term quadratic in y' by making its factor zero. Other possibilities of linearization introduce the new independent variable in the relation between f and g.

To emphasize this restricted choice, assume as above, that there is no coordinate transformation, i.e., $x=y$ and $f'=1$. Then $g=1/k$ and $h=(Ax+B)k^2$, i.e., only a linear force field may be treated . If we wish to use Sundman's transformation, $g(y)=y$, $yh=k(Ay+B)$ and consequently only fields proportional to $x^{-1/2}$ may be treated. Finally, and more realistically, if $h=-\mu x^{-2}= -\mu f^{-2}$, Equation (12 bis) gives: $f=2\mu/(k^2\dot{Q})$.

The corresponding time-transformation is governed by

$$g = -\frac{4\mu}{k^3}\frac{Ay+B}{Q^2} .$$

Note that there is no free choice for f or g in this case.

1.2 Special Cases of Interest

The result given by Equation (12) is now applied to the gravitational problem, $H=2\mu/f(y)$, using a generalization of Sundman's transformation, $g(y)=y^n$ and similar transformation of the coordinate, $x=f(y)=y^m$. Note that in this way we have

$$ds = \frac{dt}{x^{n/m}} . \tag{16}$$

If these relations are substituted in Equation (12), we have

$$(\mu+c_e y^m)\, y^{2n-3m+2} = \frac{m^2}{2}(Ay^2+2By+C),$$

from which all possible combinations of n and m may be obtained. (Note that the value, $m=0$, according to Equation (12), is not allowed since this corresponds to $f'\equiv 0$).

The selection of possible (m,n) combinations may proceed as follows. Let $2n-2m+2=\alpha$ and $2n-3m+2=\beta$ or $m-\alpha-\beta$ and $n = \frac{3}{2}\alpha-\beta-1$. Equation (17) now becomes

$$c_e y^\alpha + \mu y^\beta = \frac{m^2}{2}(Ay^2+2By+C), \tag{18}$$

from which we see that the possible combinations are $\alpha=0,1,2$ and $\beta=0,1,2$ and $\alpha\neq\beta$, or in other words, there are 6 possibilities. For instance, if $\alpha=1$ and $\beta=2$, we have $m=-1$, $n=-3/2$ and by equating coefficients of equal powers of y we obtain $A=2\mu$, $B=c_e$, and $C=0$. Similarly, the values corresponding to the other five (α,β) combinations may be obtained and are shown in Table I.

The results may be presented in a less concise but perhaps simpler form, aiding possible applications. We begin with the simplest case, $m=n=1$. The transforamtions in this case are $ds=dt/y$ and $x=y$ or simply $ds=dt/x$ and there is no coordinate transformation. The coefficients occurring in the equation of the harmonic oscillator may be found in Table I, i.e., $A=2c_e$, $B=\mu$, $C=0$ and consequently we have $x''=2c_e x+\mu$ as shown before, see Equation (15). Now consider the case $m=n=2$. The transformations are $ds=dt/y^2$ and $x=y^2$ or $ds=dt/x$. The time transformation is the same as before $(m=n=1)$ but it is associated with a quadratic coordinate transformation. The equation of motion is $y''=c_e y/2$. Note that the ratio n/m governs the time transformation as shown by Equation (16) and, therefore, the above two cases ($m=n=1$ and $m=n=2$) furnish the same time transformation. This is also the case when $m=-2$, $n=-3$, and $m=-1$, $n=-3/2$ since in both cases $n/m=3/2$ and so $ds=dt/x^{3/2}$. The corresponding coordinate transformations are of course different; $x=y^{-2}$ and $x=y^{-1}$, respectively.

There is no coordinate transformation when $m=1$. Two such cases are shown in Table I, $m=n=1$ and $m=1$, $n=1/2$. The first gives $ds=dt/x$ as mentioned above, the second offers $ds=dt/\sqrt{x}$.

Note that $A<0$ when $m=n=2$ or $m=n=1$ and $c_e<0$. These are the only cases giving harmonic oscillators with real frequencies.

Returning to the "non-conservative" case, i.e., when the integral of energy is not utilized, we have $f'=kg$, $m-1=n$ and $m=k$ if, as before the substitutions $f=y^m$ and $g=y^n$ are used. Furthermore, from Equation (12 bis) we obtain

$$y^{-(m+1)} = -\frac{m}{\mu}(Ay + B),$$

which allows only two possibilities:

(i) $m = -2$, $n = -3$, $A = \mu/2$, $B = 0$ and

(ii) $m = -1$, $n = -2$, $A = 0$, $B = \mu$.

These cases correspond to the first and second columns in Table I and accomplish linearization but do not give harmonic oscillators.

1.3 Summary

The results may be summarized according to the value of n/m as follows:

For $n/m=1/2$ we have $ds = dt/\sqrt{x}$ and no other transformation is needed. The equation of motion is $x'' = c_e$.

For $n/m = 1$, we have $ds = dt/x$ and either the transformation $x = y^2$, $(y'' = \frac{c_e y}{2} + \frac{\mu}{2})$ or no coordinate transformation $(x'' = 2c_e x + \mu)$. In both cases harmonic oscillators with real frequencies appear for $c_e < 0$. The coordinate transformation is known as Levi-Civita's transformation.

For $n/m = 3/2$, we have $ds = dt/x^{3/2}$ and either the transformation $x = y^{-1}$ or $x = y^{-2}$ is needed, giving $y'' = 2\mu y + c_e$ or $y'' = \mu y/2$.

For $n/m = 2$, we have $ds = dt/x^2$ and the needed coordinate transformation is $x = y^{-1}$, with $y'' = \mu$.

Note that coordinate transformation is not needed when using $dt/\sqrt{x}$ or dt/x. All other time transformations require an appropriate (say $x = 1/y$) coordinate transformation.

Table I. The Six Linearizing Transformations

α	0	0	1	1	2	2
β	1	2	0	2	0	1
m	-1	-2	1	-1	2	1
n	-2	-3	1/2	-3/2	2	1
n/m	2	3/2	1/2	3/2	1	1
A	0	$\mu/2$	0	2μ	$c_e/2$	$2c_e$
B	μ	0	c_e	c_e	0	μ
C	$2c_e$	$c_e/2$	2μ	0	$\mu/2$	0

2. One-degree of freedom dynamical systems. The inverse problem.

The inverse problem is to establish the totality of (non-linear) differential equations representing dynamical systems which are reducible to linear equations by transformations of the dependent and independent variables. The problem is solved in the most general case and three functions of the dependent variable are introduced by means of the two transformations. These three functions represent the variable coefficients of the quadratic and linear damping terms and the forcing function in the non-linear dynamical system. The results are applied to the special cases treated in Part I and it is shown that problems with force functions including gravitational effects and second order harmonics are reducible to linear differential equations.

2.1 Introduction

The inverse problem is a natural and straightforward extension of the direct problem for dynamical systems a with single degree of freedom. We allow damping and do not use any integrals of the motion in order to obtain the most general solution.

2.2 The General Problem and its Solution

We ask what the totality of differential equations is, that may be transformed into

$$y'' + ay' + by = c, \tag{1}$$

by the transformations $x = f(y)$ and $dt = g(y)\,ds$, corresponding to Equations (6) and (7) in Part 1. Note that Equation (1) with the constants a, b, c is a slightly generalized form of Equation (8) of Part I; the additional term might represent damping promoting numerical stability.

The solution to this inverse problem is

$$\ddot{x} + \alpha(x)\dot{x}^2 + \beta(x)\dot{x} + \gamma(x) = 0, \tag{2}$$

where

$$\alpha(x) = \frac{d}{dx} \ln GF' \quad , \tag{3-a}$$

$$\beta(x) = aG^{-1} \, , \tag{3-b}$$

$$\gamma(x) = \frac{bF - c}{F'G^2} \tag{3-c}$$

The proof is similar to that given in Part I and therefore it is omitted.

In Equations (3-a,b,c) $F(x)$ and $G(x)$ correspond to $f(y)$ and $g(y)$, i.e., the inverse of the coordinate transformation is $y = F(x)$ and $G(x) = g[F(x)]$. Note that these two functions define the transformation and also the functions α, β and γ, but not vice-versa. The relation between the three functions α, β and γ is obtained as follows.

By integrating Equations (3-a) we obtain

$$GF' = k_1 \exp(\textstyle\int \alpha(x)\,dx), \tag{4}$$

where k_1 is a constant. From Equation (3-b) we have $G = a\beta^{-1}$ and from Equation (3-c)

$$\frac{F'}{bF - c} = \frac{1}{\gamma G^2} = \frac{\beta^2}{\gamma a^2} ,$$

or after integration, $bF - c = k_2 \exp\left(\frac{b}{a^2}\int \frac{\beta^2}{\gamma}\right)$, where k_2 is a constant.

Computing from this F' and substituting the result into Equation (4) we have the desired relation between the three functions α, β and γ:

$$\frac{k_2\ \beta}{k_1 a\gamma} = \exp\left[\int\left(\alpha - \frac{b\beta^2}{a^2\gamma}\right) dx\right] . \qquad (5)$$

2.3 Special Cases of Interest

An example illustrates these results. It has been shown in Part 1 that the transformations $g(y) = y^{-2}$ and $x = f(y) = y^{-1}$ give a linear equation when applied to the gravitational problem of two bodies. In fact, in this case $m = -1$, $n = -2$, and $y'' = \mu$ as shown in Table I of Part 1. This case in the inverse notation corresponds to $F(x) = x^{-1}$ and $G(x) = x^2$. Substitution in Equations (3-a,b and c) gives $\alpha(x) \equiv 0$, $\beta(x) = ax^{-2}$ and

$$\gamma(x) = \frac{c}{x^2} - \frac{b}{x^3} .$$

The differential equation in the x system

$$\ddot{x} = a\frac{\dot{x}}{x^2} + \frac{c}{x^2} - \frac{b}{x^3} = 0 \qquad (6)$$

Therefore, may be transformed to Equation (1).

If the hyperbolic damping $a\dot{x}x^{-2}$ (corresponding to large damping at close approaches) and the higher order harmonic (corresponding to the term b/x^3) are eliminated, Equation (1) reduces to

$$y'' = c = \mu .$$

In general, using the transformations specified in Part 1 with undetermined exponents, i.e., $x = y^m$ and $g = y^n$, we have $F(x) = x^{1/m}$ and $G(x) = x^{n/m}$. Consequently, $\alpha = (n + 1 - m)\,(xm)^{-1}$ according to Equation (3-a), $\beta = ax^{-n/m}$ according to Equation (3-b) and

$$\gamma = bmx^{\frac{m-2n}{m}} - cmx^{\frac{m-1-2n}{m}}$$

according to Equation (3-c).

The resulting differential equation, corresponding to Equation (2) is

$$\ddot{x} + \left(\frac{n+1-m}{m}\right)\frac{\dot{x}^2}{x} + a\frac{\dot{x}}{x^{n/m}} + \frac{bm}{x^{2n/m-1}} - \frac{cm}{x^{\frac{1+2n}{m}-1}} = 0 \,. \tag{7}$$

From this it follows that if one of the terms is to represent gravitational attraction, then either

$$\text{(i)}\ \frac{2n}{m} - 1 = 2, \quad \text{or (ii)}\ \frac{1+2n}{m} - 1 = 2, \quad \text{that is}$$

$$\text{either (i)}\ n = \tfrac{3}{2}m, \text{ or } \ \text{(ii)}\ n = \frac{3m-1}{2} \,.$$

Rewriting Equation (7), we have for case (i):

$$\ddot{x} + \frac{m+2}{2m}\,\frac{\dot{x}^2}{x} + a\frac{\dot{x}}{x^{3/2}} + \frac{bm}{x^2} - \frac{cm}{x^{2+1/m}} = 0. \tag{7-i}$$

and for case (ii):

$$\ddot{x} + \frac{m+1}{2m}\,\frac{\dot{x}^2}{x} + a\frac{\dot{x}}{x^{\frac{3}{2}-\frac{1}{2m}}} + \frac{bm}{x^{2-1/m}} - \frac{cm}{x^2} = 0. \tag{7-ii}$$

Note that no integral is assumed to exist, consequently the term $\dot{x}^2/x$ is not eliminated unless in case (i) $m=-2$ (and $n=-3$, $m/n = 2/3$) or in case (ii) $m=-1$ (and $n=-2$, $m/n=1/2$). For these cases Equation (7-i) is

$$\ddot{x} + a\frac{\dot{x}}{x^{3/2}} - \frac{2b}{x^2} + \frac{2c}{x^{3/2}} = 0, \quad \text{for } \ m = -2, \tag{8}$$

$$\ddot{x} - \frac{\dot{x}^2}{2x} + a\frac{\dot{x}}{x^{3/2}} - \frac{b}{x^2} + \frac{c}{x} = 0, \quad \text{for } m = -1, \tag{9}$$

and Equation (7-ii) becomes

$$\ddot{x} + \frac{\dot{x}^2}{4x} + a\frac{\dot{x}}{x^{7/4}} - \frac{2b}{x^{5/2}} + \frac{2c}{x^2} = 0, \text{ for } \ m = -2, \tag{10}$$

$$\ddot{x} + a\frac{\dot{x}}{x^2} - \frac{b}{x^3} + \frac{c}{x^2} = 0, \text{ for } m = -1. \tag{11}$$

Naturally, attention is directed to Equation (11), which shows that when $m=-1$, $n=-2$, and $a=0$, not only the term c/x^2 but also b/x^3 might be linearized by the transformations $y = x^{-1}$ and $g = y^{-2}$.

Another approach is to stipulate $\alpha(x) = k_1$ = constant. If this is combined with a gravitational field, $\gamma(x) = k^2x^{-2}$ then Equation (3-c) becomes

$$\frac{k^2}{x^2} = \frac{bF-c}{k_2^2\, e^{2k_1x}} F' ,$$

since according to Equation (3-a) we have $GF' = k_2e^{k_1x}$ with k_2 = constant. Selecting now, only for the sake of simplicity $b = 0$, we have

$$F =- \frac{(k\,k_2)^2}{c} \int [x^{-1}\exp k_1x]^2\, dx$$

and

$$G = - \frac{cx^2}{k^2k_2} \exp(-k_1x).$$

The differential equation of motion is

$$\ddot{x} + k_1\dot{x}^2 + k_3\frac{\dot{x}}{x^2}\exp k_1x + \frac{k^2}{x^2} = 0,$$

where $k_3 = -ak_2k^2c^{-1}$.

Further specialization may be $\alpha(x) = k_1$ and $\beta(x) = k_2$, but it must be noted that as seen from Equation (3-b) these conditions do not allow transformation of the independent variable, i.e., $G(x) = ak_2^{-1}$ = constant. Nevertheless, we note that the coordinate transformation becomes for this case simply $y = k_3e^{k_1x}$ and the equation of motion is

$$\ddot{x} + k_1\dot{x}^2 + k_2\dot{x} + k_2e^{-k_1x} + k_4 = 0 ,$$

where all k_i's are constants. This equation transforms into

Equation (1) with primes replaceable by dots.

We note, finally, that if there is no coordinate transformation i.e., $F(x) = k_1 x + k_2$, the function $G(x)$ will completely determine the equation of motion:

$$\ddot{x} + \frac{G'}{G}\dot{x}^2 + \frac{a\dot{x}}{G} + \frac{bx+k_3}{G^2} = 0 ,$$

where $k_3 = (bk_2-c)/k_1$ = constant. This equation may be transformed by $dt = G(x)\,ds$ into Equation (1).

3. N-degrees of freedom dynamical systems. The direct problem

Transformations of non-linear differential equations describing dynamical systems of n-degrees of freedom are introduced in order to obtain linear representations. The dependent as well as the independent variables are transformed and it is shown that in general such linearizations cannot be accomplished without the use of integrals of the motion.

3.1 Introduction

The simplicity of the transformation theory of one-degree of freedom dynamical systems is lost and the conditions for linearization become somewhat subtle. The negative result announced follows from the non-existence of the inverse of the transformation matrix and while the result is expected, its not at all trivial proof is not without interest.

3.2 The General Problem and its Solution

The dynamical system is represented by the system of differential equations

$$\ddot{x}_i = h_i(x_j,\dot{x}_j,t), \tag{1}$$

where $i = 1(1)n$, and dots denote derivatives with respect to the time t.

The transformation of the independent variable is a straightforward generalization of Sundman's (1912) transformation,

$$ds = \frac{dt}{g(y_i)} . \tag{2}$$

The coordinate transformation is a generalization of the one used for the one-degree of freedom case,

$$x_i = f_i(y_j) \ . \tag{3}$$

Note that the functions h_i and f_i are known as vector-vector functions, while the function g is a scalar-vector function. The corresponding transformation functions pertinent for the regularization of the planar restricted problem of three bodies (Szebehely, 1967) are: a real function g of a complex variable and a complex function f of a complex variable. For instance, Levi-Civita's (1903) transformation is defined by $g = |y|^2$ and $f = y^2$, where y is a complex variable.

The direct problem is defined as finding the functions g and f_i so that the equations of motion using the new system of variables (y_i, s) will be linear, i.e.,

$$y_i'' + a_{ij}y_j' + b_{ij}y_j + c_i = 0, \tag{4}$$

where primes denote derivatives with respect to s and the coefficients, a_{ij}, b_{ij} and c_i are independent of the dependent variable y_i and its derivatives, but in general may depend on s. The summation convention is used here and later on, unless otherwise specified, i.e., we sum with respect to repeated subscripts.

Note that the inverse problem treated in the sequel starts with Equation (4) and establishes the totality of the functions h_i. The reciprocity of the two problems and the strong correspondence between their solutions cannot escape the reader.

The transformation of Equation (1) presents no difficulty but requires some care. First $\dot{x}_i$ and $\ddot{x}_i$ are expressed as functions of y_i, y_i' and y_i'' as follows:

$$\dot{x}_i = \frac{\partial f_i}{\partial y_j} \frac{dy_j}{ds} g^{-1},$$

or

$$\dot{x}_i = f_{i,j}y'_j g^{-1} \tag{5}$$

$$\text{and} \quad \ddot{x}_i = g^{-2}[f_{i,j}y_j'' + (f_{i,jk} - f_{i,j}\, g_{,k}\, g^{-1})\, y_j'y_k'] \ , \tag{6}$$

where

$$f_{i,jk} = \frac{\partial^2 f_i}{\partial y_j \partial y_k} \quad \text{and} \quad g_{,i} = \frac{\partial g}{\partial y_i} \ .$$

Equating now Equations (6) and (1) in the form

$$\ddot{x}_i = h_i\ (f_i,\ f_{i,j}\ y_j'\ g^{-1},\ \int g\ ds), \tag{7}$$

gives the equation of motion in term of the new variables:

$$g^2 h_i = f_{i,j}\ y_j'' + g(f_{i,j}\ g^{-1}),_k y_j'\ y_k' \quad . \tag{8}$$

In order to compare this result with Equation (4) we solve Equation (8) for y". For this purpose it is assumed that the inverse of the transformation $x_i = f_i(y_j)$ exists and it is given by

$$y_i = F_i(x_j) \tag{9}$$

and consequently

$$\frac{\partial f_j}{\partial y_k}\frac{\partial F_i}{\partial x_j} = \frac{\partial f_i}{\partial y_j}\frac{\partial F_j}{\partial x_k} = \delta_{ik} \tag{10}$$

where δ_{ik} is Kronecker's delta.

Multiplication of Equation (8) by $F_{\ell,i} = \partial F_\ell/\partial x_i$ gives

$$y_\ell'' + gF_{\ell,i}(f_{i,j}\ g^{-1}),_k\ y_j'\ y_k' - g^2\ F_{\ell,i}h_i = 0. \tag{11}$$

Note that in Equation (11) the term $F_{\ell,i}$ is to be interpreted as

$$F_{\ell,i}(x_j) = \partial F_\ell(x_j)/\partial x_i = H_{\ell\ i}\ [f_j(y_k)] = G_{\ell i}\ (y_k). \tag{12}$$

Equation (11) is the result of the derivation. It is the basis for the study aimed at the selection of the functions g and f_i.

At this point the complete generality of the function h_i must be restricted in order to proceed. If h_i as defined by Equation (1) does not depend on the velocities $(\dot{x}_j)$, then only the second term of Equation (11) contains y_i' and in order to obtain a linear system in y_i', we must have

$$F_{\ell,i}\ (f_{i,j}\ g^{-1}),_k = 0\ . \tag{13}$$

On the other hand if the velocity $\dot{x}_i$ occurs in h_i and consequently h_i, according to Equation (7) contains y_i' , then a combination of the second and third terms in Equation (11) may result in linearity in y_i'. That this is indeed the case will be shown in connection with the inverse problem in Part 4.

Furthermore, in Equation (11) the quadratic term in y_i' may be eliminated by the use of integrals of the dynamical system. The existence of such isolating integrals depends, once again, on the function of h_i. The use of the integrals in gravitational systems is treated elsewhere (Szebehely, 1976).

In the following the consequences of Equation (13) are investigated. During the attempt to find the conditions regarding the functions g and f_i which result in the elimination of the quadratic terms, the singularity of the transformations matrix is proved.

3.3 Conditions for the Elimination of Non-linear Velocity Terms

Expanding Equation (13) we obtain

$$F_{\ell,i}\, f_{i,jk} = \delta_{\ell j}(\ell n\ g)_{,k} \tag{14}$$

Since $f_{i,jk} = f_{i,kj}$, we have

$$\delta_{\ell j} g_{,k} = \delta_{\ell k} g_{,j} \tag{15}$$

and if $\ell = j \neq k$ one might show that

$$\delta_{\ell\ell} g_{,k} = \delta_{jk} g_{,j}$$

or

$$(n-1) g_{,k} = 0.$$

Therefore, either g = constant and there is no time transformation or $n=1$ and linearization is possible only in the one-dimensional case.

Consider now the solution of Equation (13) in a more restrictive sense and let

$$(f_{i,j}\, g^{-1}) = a_{ij} = \text{constant} \tag{14 bis}$$

Similarly to Equation (15) one has

$$a_{ij}\, g_{,k} = a_{ik}\, g_{,j} \tag{15 bis}$$

and consequently Equation (14 bis) may be solved for f_i, after Equation (15 bis) was solved for g. We will show, however, that the matrix a_{ij} is singular, (and so is $f_{i,j}$) and consequently $F_{i,j}$ does not exist. In fact if the inverse of a_{ij} exists and it is b_{ij}, then $a_{ij}b_{jk}=\delta_{ik}$ and multiplying Equation (15 bis) by $b_{\ell i}$ we have again Equation (15). Equation (15 bis) shows that the ratio of $g,_k$ and $g,_j$ must be independent of i. Since Equation (15 bis) is a scalar equation,

$$\frac{a_{ij}}{a_{ik}} = \frac{a_{1j}}{a_{1k}} = \frac{a_{2j}}{a_{2k}} = \ldots \frac{a_{nj}}{a_{nk}} = \frac{g,_j}{g,_k} = S_{jk} = \text{constant},$$

where, of course, there is no summation. In other words, the ratio of the elements in the j-th and k-th columns of any row is a constant, independently of which row (what i) is being considered. Therefore,

$$a_{ij} = a_i b_j$$

and

$$b_j g,_k = b_k g,_j \tag{16}$$

where a_i and b_i are constant.

Prior to writing the solution of Equation (16), attention is directed to some simple properties of $S_{jk} = b_j/b_k$:

(i) $S_{kj} = S_{jk}^{-1}$, and therefore $S_{kk} = 1$ (no summation).

(ii) $S_{kj} = S_{k\ell}S_{\ell j}$ (no summation).

(iii) $S_{i_1 i_n} = S_{i_1 i_2} S_{i_2 i_3} \ldots S_{i_{n-1} i_n}$ (no summation).

This is a straightforward generalization of (ii). In the above formulae describing the properties of S, the term S_{kj} is an element of S, not S itself.

The solution of the linear partial differential equation (16) for the function $g(y_i)$ is

$$g = g(z), \tag{17}$$

where the argument is $z = b_i y_i$ and the function g is arbitrary. Substitution into Equation (16) results in the identity $b_j\, g'\, b_k = b_k\, g'\, b_j$, where $g' = dg/dz$.

In order to obtain the functions f_i, we consider the solution of Equation (14 bis)

$$f_{i,j} = g\, a_i b_j ,$$

and obtain

$$f_i = a_i \int g(z)\, dz + c_i . \tag{18}$$

3.4 Conclusions

Combined transformations of the time and of the coordinates of an n-degree of freedom dynamical system can not lead to harmonic oscillators without the introduction of the integral(s) of the system into the equations of motion.

If no coordinate transformation is used

$$f_{i,j} = F_{i,j} = \delta_{ij}$$

and Equation (11) becomes

$$y''_\ell - (\ell n\, g)_{,k}\, y'_\ell\, y'_k - g^2 h_\ell = 0.$$

This shows that in order to linearize the quadratic terms in the velocity we must have g=constant. However, in this case, unless the force field is linear the equation of motion is non-linear, and the original equation of motion is reconstructed.

If no time transformation is used, Equations (11) or (14) give $f_{i,j} = a_{ij}$ = constant, consequently only linear coordinate-transformations are allowed and the original differential equation is reconstructed once again.

4. N-degrees of freedom dynamical systems. The inverse problem.

The totality of non-linear systems reducible to linear oscillators is established and examples are given. The utilization of the integrals of the motion is relegated to the sequel where the three-dimensional gravitational problem is treated.

4.1 Introduction

The concept of inverting the basic transformation problem is described in considerable detail in Part 2. The negative result of general validity obtained in Part 3 is circumvented when the inverse problem is studied since the terms quadratic in the velocities are desirable in the inverse problem, while terms of the very same origin prevent the solution of the direct problem.

4.2 The General Problem and its Solution

We begin with an n-degree of freedom coupled linear system where the independent variables are $\bar{y}(y_1, y_2, \ldots y_n)$ or simply y_i and the dependent variable is s. This system may be described by the equation of motion:

$$y_i'' + a_{ij}\, y_j' + b_{ij} y_j + c_i = 0\,, \tag{1}$$

where primes denote derivatives with respect to s, a_{ij}, b_{ij} and c_i are matrices, independent of y_i and its derivatives, and $i=1(1)n$ and $j=1(1)n$. Here and everywhere the summation convention is used, i.e., we sum with respect to repeated subscripts.

Subjecting Equation (1) to the transformations

$$y_i = F_i(x_j) \text{ and } ds = \frac{dt}{g(y_i)}\,, \tag{2}$$

the totality of equations reducible to Equation (1) is obtained:

$$\ddot{x}_\ell + \alpha_{\ell jk}(x_i)\,\dot{x}_j\,\dot{x}_k + \beta_{\ell k}(x_i)\dot{x}_k + \gamma_\ell(x_i) = 0, \tag{3}$$

where x_i is the new dependent variable, dots denote derivatives with respect to the new independent variable t, and the matrices denoted by Greek letters are defined as follows:

$$\left.\begin{aligned} \alpha_{\ell jk} &= G^{-1} f_{\ell,i}\,(GF_{i,j})_{,k} = \delta_{\ell j}(\ell n\; G)_{,k} + F_{i,jk}\, F_{\ell,i} \\ \beta_{\ell k} &= a_{ij}\, F_{j,k}\, f_{\ell,i}\, G^{-1}\,, \\ \text{and } \gamma_\ell &= f_{\ell,i}(b_{ij}F_j + c_i)\, G^{-2}. \end{aligned}\right\} \tag{4}$$

Here $g(y_i) = G[F_i(x_j)] = G(x_i)$ is the function governing the time transformation, after the coordinate transformation has been introduced. Similarly $\partial f_i/\partial y_j = \phi_{ij}[F_k(x_\ell)] = \phi_{ij}(x_\ell)$. The abbreviated notation for partial derivatives is used according to which $G_{,k} = \partial G/\partial x_k$ and $F_{i,j} = \partial F_i/\partial x_j$. The inverse of the transformation $y_i = F_i(x_j)$ is $x_i = f_i(y_j)$ and consequently

$$\frac{\partial f_i}{\partial y_j} \frac{\partial F_j}{\partial x_k} = \delta_{ik} , \tag{5}$$

where δ_{ik} is Kronecker's delta corresponding to the unit matrix.

In the following Equations (3) and (4) are first verified and then some examples are offered. With straightforward compuation we have

$$y_i' = \frac{\partial y_i}{\partial x_j} \frac{dx_j}{dt} \frac{dt}{ds}$$

or

$$y_i' = F_{i,j} \dot{x}_j G. \tag{6}$$

Furthermore,

$$y_i'' = G^2 \left\{ F_{i,j}\ddot{x}_j + \left[F_{i,jk} + F_{i,j} (\ln G)_{,k} \right] \dot{x}_j\dot{x}_k \right\} . \tag{7}$$

Another form, convenient for evaluations is

$$y_i'' = G^2 F_{i,j}\ddot{x}_j + G(GF_{i,j})_{,k}\dot{x}_j\dot{x}_k. \tag{7 bis}$$

Substituting Equations (6) and (7) into (1) and multiplying by $G^{-2}f_{\ell,i}$ results in Equation (3) and in the definitions of the functions α, β and γ as given by Equation (4).

The solution of the inverse problem is Equation (3); i.e., equations of this form may be transfered to linear equations. Notes:

(1) The functions α, β, and γ forming the coefficients of Equation (3) are not independent. In fact, the two transformation functions, G and F_i determine all three coefficients, but not vice-versa.

(2) The linearity of Equation (1) is not destroyed if the coefficients depend on s. After the transformation β and γ must depend on t, but the $s(t) = s$ relation is not known until the problem is solved, i.e., until $y_i(s)$ is found, see Equation (2). Consequently, it is assumed, for the time being that the coefficients a, b, c are independent of s and consequently the coefficients α, β, γ will not depend on t.

(3) If the coefficient determining the damping, a_{ij} is the unit-matrix, $a_{ij} = \delta_{ij}$, then from the second of Equations (4) we have $\beta_{\ell k} = \delta_{\ell k} G^{-1}$ and once selecting the function G, the

elements of β are determined.

(4) The reductions of Equations (3) and (4) in the one-degree of freedom case is simple by supressing all subscripts:

$$\ddot{x} + \alpha(x)\dot{x}^2 + \beta(x)\dot{x} + \gamma(x) = 0\ , \qquad (3\text{ bis})$$

$$\left.\begin{aligned} \alpha &= (G\,F')^{-1}(GF')'\ , \\ \beta &= a\,G^{-1}, \\ \gamma &= \frac{bF + c}{F'G^2}\ , \end{aligned}\right\} \qquad (4\text{ bis})$$

since now $f = f_i$, $F = F_i$, $a = a_{ij}$, $b = b_{ij}$, $c = c_i$, $\alpha = \alpha_{ijk}$, $\beta = \beta_{ij}$, $\gamma = \gamma_i$, $\frac{df}{dy} = \left(\frac{dF}{dx}\right)^{-1}$. These equations are in complete agreement with Equations (3-a,b,c) found in the one-degree of freedom case discussed in Part 2.

4.3 Special Cases of Interest

(1) If there is no coordinate transformation, $y_i = x_i$ and $F_{i,j} = f_{i,j} = \delta_{ij}$. The transformed (non-linear) equations of motion become

$$\ddot{x}_\ell + \dot{x}_\ell\dot{x}_k(\ell n\ G),_k + G^{-1}a_{\ell k}\ \dot{x}_k + G^{-2}(b_{\ell k}x_k + c_\ell) = 0, \qquad (8)$$

since

$$F_{i,jk} = 0,\quad \alpha_{\ell jk} = \delta_{\ell j}(\ell n\ G),_k,$$

$$\beta_{\ell k} = a_{ij}\delta_{jk}\delta_{\ell i}\ G^{-1} = a_{\ell k}\ G^{-1}$$

and

$$\gamma_\ell = \delta_{\ell i}(b_{ij}x_j + c_i)G^{-2} = (b_{\ell j}x_j + c_\ell)\ G^{-2}.$$

(2) If in addition the time transformation is a generalization of Sundman's (1912) transformation:

$$g = G = Ar^n = A(x_k\ x_k)^{n/2} = A(y_k\ y_k)^{n/2},$$

then $G,_k = An\ r^{n-2}x_k$ and the non-linear equation becomes

$$\ddot{x}_\ell + n\,\dot{x}_\ell\,\frac{x_k\dot{x}_k}{r^2} + \frac{a_{\ell k}}{A}\,\frac{\dot{x}_k}{r^n} + \frac{b_{\ell k}x_k + c_\ell}{A^2 r^{2n}} = 0. \tag{9}$$

Sundman's value, $n=1$, allows for a gravitational field of force if $b_{\ell k}=0$.

(3) Using the value of $n=3/2$ and omitting the damping terms by writing $a_{\ell k}=0$, we have from Equation (9):

$$\ddot{x}_\ell + \frac{3}{2}\,\dot{x}_\ell\,\frac{x_k\dot{x}_k}{r^2} + b_{\ell j}\,\frac{x_j}{r^3} + \frac{c_\ell}{r^3} = 0\ , \tag{10}$$

where $x_k\dot{x}_k = \bar{r}\,\dot{\bar{r}} = r\,\dot{r}$ and $x_j r^{-3}$ corresponds to $\bar{r}/r^3$, considering that $\bar{r} = \bar{r}(x_1 \ldots x_n)$ and $|\bar{r}| = (x_k x_k)^{1/2} = r$. The last two terms of Equation (10) may represent gravitational attraction and a higher order harmonics.

(4) A generalization of the above discussed special cases is accomplished when both the time transformation and the coordinate transformation are operational. Introduce

$$G = r^n \quad \text{and} \quad y_i = F_i(x_j) = x_i r^m, \tag{11}$$

where n and m are arbitrary constants and $r = (x_i\ x_i)^{1/2}$ is the distance in the x_i space. The corresponding (transformed) distance in the y_i space is $\rho = (y_i y_i)^{1/2} = (x_i x_i)^{1/2}\ r^m = r^{m+1}$. The inverse coordinate transformation follows from the second of Equations (11):

$$x_i = f_i(y_j) = y_i \rho^\mu\ \ , \tag{12}$$

where $\mu = -m(m+1)^{-1}$ and therefore $\rho^\mu r^m = 1$.

If there is no time transformation, $n = 0$ and $ds = dt$. If there is no coordinate transformation, $m = 0$ and $y_i = x_i$. Therefore, ther parameters of the transformation are m and n. These parameters now allow the computation of the coefficients α, β, γ in Equation (3). The basic formulas to be used in these computations are as follows:

$$F_{i,j} = \delta_{ij}\ r^m + x_i x_j m r,^{m-2}$$

$$f_{i,j} = \delta_{ij}\ r^{-m} + x_i x_j \mu r,^{-(m+2)}$$

and $$G,_k = n r^{n-2} x_k.$$

Evaluation of the coefficients given by Equations (4) and their substitution into Equation (3) is straightforward, though, not at all trivial. The coefficients become:

$$\alpha_{\ell jk} = \frac{1}{r^2}\left[(m+n)\delta_{\ell j}x_k + m\delta_{\ell k}x_j - \frac{1}{r^4}\frac{m(2+m)}{1+m}x_jx_kx_\ell\right],$$

$$\beta_{\ell k} = \frac{1}{r^n}\left[a_{\ell k} + \frac{m}{r^2}x_i\left(a_{\ell i}x_k - \frac{1}{1+m}a_{ik}x_\ell\right) - \frac{1}{r^4}\frac{m^2}{1+m}a_{ij}x_ix_jx_kx_\ell\right],$$

and $$\gamma_\ell = \frac{1}{r^{2n}}\left[b_{\ell j}x_j + \frac{c_\ell}{r^m} - \frac{1}{r^2}\frac{m}{1+m}x_ix_\ell\left(b_{ij}x_j + \frac{c_i}{r^m}\right)\right].$$

The equation of motion becomes

$$\ddot{x}_\ell + \frac{2m+n}{r}\dot{r}\dot{x}_\ell + \frac{m}{1+m}\frac{x_\ell}{r^2}\left[\bar{v}^2 - (m+2)\dot{r}^2\right] +$$
$$+ \frac{1}{r^n}\left(\dot{x}_j + \frac{m\dot{r}}{r}x_j\right)\left(a_{\ell j} - \frac{m}{1+m}\frac{1}{r^2}a_{ij}x_ix_\ell\right) +$$
$$+ \frac{1}{r^{2n}}\left(b_{ij}x_j + \frac{c_i}{r^m}\right)\left(\delta_{\ell i} - \frac{m}{1+m}\frac{1}{r^2}x_ix_\ell\right) = 0, \qquad (13)$$

where the square of the velocity vector is $\bar{v}^2 = \dot{x}_k\dot{x}_k = (\dot{\bar{r}})^2$ and for the other symbols, see Case 3.

Equation (13) offers generous opportunities to study special cases. No attempt will be made here to divest the reader from such pleasures, nevertheless, a few, carefully selected cases will be treated.

(A) The non-linear terms in the velocities are $\dot{r}\dot{x}_\ell$, $\bar{v}^2$ and $\dot{r}^2$. In the case of a single degree of freedom dynamical system these three terms are equal and therefore they may be eliminated if their coefficient is zero, i.e., if $(n+m)(1+m) = 0$. The $m=-1$ choice is not allowed since the coordinate transformation given by Equation (11) may not be inverted in this case. Therefore, we have $n=-m$ as found in Part 2. If the original n-degree of freedom system is investigated, the non-linear velocity terms may be eliminated only if $2m+n = 0$ and $m = 0$, i.e., if $m=n=0$ which is trivial since neither the time, nor the coordinates are transformed in this case. This finding is in complete agreement with the result of Part 1.

(B) Note that studying the one-degree of freedom "Inverse Problem" the natural notation is different from the multi-degree of freedom case. A comparison of the results is not possible without the proper relation between the parameters of the transformations. In fact in the single degree of freedom case we have

$x = f(y) = y^{m_1}$ and $g = y^{n_1}$ while for the present case $x_i = y_i\rho^{\mu}$ and $g = r^n$. Consequently, when the multi-dimensional problem is reduced to a single-degree of freedom case, we have $x = y^{1+\mu}$ and and $g = x^n = y^{n(1+\mu)}$. Therefore, the relations between the corresponding parameters are: $m_1 = 1 + \mu$ and $n_1 = n(1 + \mu)$ or $m_1 = (1 + m)^{-1}$ and $n_1 = n(1 + m)^{-1}$. In view of this, the previously found results, $n + m = 0$ corresponds to $n_1 + 1 - m_1 = 0$ as shown by Equation (7) of Part 2.

Consider now the last term of Equation (13), corresponding to γ_ℓ. This term may be identified with the force field since it is the only term in Equation (13) not containing either accelerations or velocities, and it represents b_{ij} and c_i which correspond to the spring constant and to a constant force in the original linear equation (Equation (1)). Since x and r are dimensionally equivalent and all the other symbols in γ_ℓ are constants, we have only two essentially different ways to introduce gravitational forces:

(i) $2n - 1 = 2$ or (ii) $m + 2n = 2$.

The first choice corresponds to transforming the linear spring forces of the (y,s) space into the gravitational field and the second transforms the constant force field acting on the oscillator into the gravitational field in the (x,t) space. Case (i) uses $n=3/2$ and case (ii) corresponds to $n=(2-m)/2$. Making the selection of $n=3/2$ one force is gravitational, while the other is proportional to $r^{-(3+m)}$ and therefore, if there is no coordinate transformation $(m = 0)$, then the force law is described by the combination of terms proportional to r^{-2} and r^{-3}. Selecting the parameters of the transformation according to case (ii), we have again the gravitational field superimposed on a field proportional to r^{-2n+1} and, therefore, if $n = 2$, the two fields are similar to those found in case (i). In fact

for case (i) we have $\dfrac{b_{ij}x_j}{r^3} + \dfrac{c_i}{r^{3+m}}$ or $\dfrac{b_{ij}x_j}{r^3} + \dfrac{c_i}{r^3}$

and for case (ii): $\dfrac{b_{ij}x_j}{r^{2n}} + \dfrac{c_i}{r^2}$ or $\dfrac{b_{ij}x_j}{r^4} + \dfrac{c_i}{r^2}$.

Note that the 3/2 power law for the time transformation plays a rather distinguished role, once again.

(D) Significant simplification of Equation (13) occurs when $a_{ij} = 0$, i.e., no utilization of damping in Equation (1) is desired. Furthermore, if the first non-linear velocity term is to be eliminated, we have $2m + n = 0$. This might be more helpful than it appears at first since the other terms quadratic in the velocities may be modified by the use of some integrals. Let us furthermore assume that $b_{ij} = 0$ and that the force field (now governed only by c_i) is gravitational. Then $2n + m = 2$, which together with the previous equation gives $n = 4/3$, $m = -2/3$ for the values of the parameters governing the transformations. The equation of motion for this case becomes considerably simpler than Equation (13):

$$\ddot{\bar{r}} = 2\frac{\bar{r}}{r^2}(\bar{v}^2 - \frac{4}{3}\dot{r}^2) - \frac{\bar{c}}{r^2} - \frac{2(\bar{c}\bar{r})\bar{r}}{r} .$$

REFERENCES

The following references are applicable to the four parts.

Levi-Civita, T. (1903), Ann. Math. [2], 9, 1.

Stiefe, E. and Scheifele, G., (1971), Linear and Regular Celestial Mechanics, Springer, Publ., N.Y.

Sundman, K., (1912), Acta Math. 36, 106.

Szebehely, V. (1967-a), Theory of Orbits, Academic Press, N.Y.

Szebehely, V. (1967-b), Astom. J., 72, 370.

Szebehely, V. (1976), "Linearization of Dynamical Systems Using Integrals of the Motion", Cel. Mech, in Print.

STRONGLY PERTURBED DYNAMICAL SYSTEMS

G. Contopoulos

Department of Astronomy, University of Athens, Greece.

1. INTRODUCTION

In calculating the behavior of a dynamical system over long times we encounter three types of quantities:

(1) Quantities that remain constant, like the classical integrals of motion. Even if numerical errors in the calculations produce variations in these quantities, these are usually small.

(2) Quantities varying linearly in time, like the phase variables. In such cases the inaccuracies also increase linearly in time.

(3) Quantities varying exponentially in time, like the distance of two nearby orbits in an unstable case. The inaccuracies increase also exponentially. In many cases the increase is so fast that any prediction over long times is practically impossible.

As an example, let us consider the orbit of a periodic comet. In general its major axis changes only a little at each approach near the sun. On the other hand the time of its return to perihelion becomes more and more uncertain. The uncertainty increases linearly in time. However, if the comet approaches one of the

V. Szebehely and B.D. Tapley (eds.), Long-Time Predictions in Dynamics, 43–56. *All Rights Reserved. Copyright* ©1976 *by D. Reidel Publishing Company, Dordrecht-Holland.*

major planets the prediction of its behavior becomes sometimes extremely difficult. The comet may be captured, or its orbit may become hyperbolic, and the transition from one case to another is rather abrupt. In such cases the uncertainty is large.

There are many cases in orbit theory where we encounter large instabilities. Suppose that an orbit starts at a distance ε from an unstable periodic orbit. After time t the deviation becomes

$$\Delta x = \varepsilon 10^{at} , \tag{1}$$

where a is proportional to the characteristic exponent. If $\varepsilon = 10^{-8}$ (the error of single-precision calculations), the time needed to reach a large uncertainty ($\Delta x=1$) is equal to

$$t = 8/a . \tag{2}$$

If an orbit is highly unstable, then the predicted orbit is very soon quite wrong. No essential improvement is done by using double precision arithmetic, or by improving the integration scheme.

In such cases it is often useful to go to the other extreme and assume complete chaos. E.g., no one tries to predict the exact orbit of a molecule of gas in a room. However, the assumption of molecular chaos allows the accurate prediction of some average quantities like the pressure and temperature of the gas.

Thus, we come to the problem of the transition of a dynamical system to a statistical mechanical one. We will only sketch the main points here, referring to other papers for more details on the subject (Contopoulos 1975, 1976, and references there).

2. TRANSITION FROM QUASI-PERIODIC TO ERGODIC MOTION

Let us consider, first, systems of two degrees of freedom, with Hamiltonians of the form

$$H = \omega_1 I_1 + \omega_2 I_2 + \varepsilon \text{ (higher order terms, trigonometric in } \delta_1, \delta_2), \tag{3}$$

where $I_1, I_2, \delta_1, \delta_2$ are action-angle variables and ω_1, ω_2 are the unperturbed frequencies.

If the higher order terms are missing, or if they are functions of I_1, I_2 only, the system is separable, and the actions I_1, I_2 are integrals of motion.

If we have only one combination of angle variables, say $(m\delta_1 - n\delta_2)$, the system is again separable, but in a different way; in such a case the combination of the actions $(I_2 + \frac{n}{m} I_1)$ is an integral of motion independent of H. The same is true if all trigonometric terms have the same value of the ratio n/m.

If, however, there are two or more different trigonometric terms we have an "interaction of resonances", whose main manifestations are:

(1) The appearance of infinite types of resonant periodic orbits. These periodic orbits have frequencies $\bar{\omega}_1$ and $\bar{\omega}_2$, near ω_1 and ω_2, such that $\bar{\omega}_1/\bar{\omega}_2 = n/m$ (ratio of two integers). The range of values of n/m, near ω_1/ω_2, for which we observe resonant periodic orbits is of order h, where h is the numerical value of the Hamiltonian (3), i.e.,

$$\frac{\omega_1}{\omega_2} - \frac{n}{m} = O(h). \qquad (4)$$

The stable periodic orbits are represented, on a surface of section, by stable invariant points, surrounded by closed islands. There are also second order periodic orbits, with frequencies $\bar{\bar{\omega}}_1$, $\bar{\bar{\omega}}_2$ integer multiples of $\bar{\omega}_1$, $\bar{\omega}_2$, surrounded by the corresponding island third order periodic orbits, etc.

(2) The appearance of homoclinic points. The unstable periodic points are the points of intersection of two corresponding asymptotic curves. In an integrable case the asymptotic curves are closed. However, if we have an interaction of resonances (non-integrable case) the asymptotic curves intersect each other at an infinite number of "homoclinic" points.

(3) The appearance of heteroclinic points. These are points of intersection of asymptotic curves corresponding to two different unstable periodic orbits. The orbits starting at such points approach asymptotically two, or more, periodic orbits of different multiplicities, therefore, they are the most conspicuous manifestations of the "interaction of resonances". The existence of homoclinic points implies also the existence of infinite heteroclinic points (Moser 1973).

(4) The successive points of intersection of an orbit by a surface of section (consequents), starting near an unstable orbit, look scattered in a random way in an "ergodic" region (called by Birkhoff (1927) a "zone of instability"). Such ergodic regions appear in every finite domain of phase space, although their total measure may be small.

If the energy is large the ergodic region becomes large and is called sometimes an "ergodic sea". The transition from a case where the ergodic region is insignificant to a case where it is very important is rather abrupt, therefore, one can define a critical, or transition energy E_c. The transition energy can be predicted theoretically (Contopoulos 1967, Walker and Ford 1969).

In the ergodic regions we often see "islands of stability" surrounding stable invariant points.

It is not certain whether for large enough energies these islands disappear, or whether small islands always appear, in which case the system never becomes really ergodic (Contopoulos 1975).

(5) If the energy is enough some particles may escape to infinity. However, if there are closed invariant curves, no escape appears for orbits starting inside them. One may define empirically a tough boundary, such that orbits outside it escape while orbits inside it never escape. This boundary is usually near the asymptotic curves of an unstable invariant orbit.

Similar considerations apply to cases of more than two degrees of freedom. In such cases, however, there is one further problem concerning stability. In a system of N-degrees of freedom the integral surfaces surrounding the stable periodic orbits are N-dimensional while the phase-space for a constant value of the energy, h, has 2N-1 dimentions. Therefore, the existence of integral surfaces does not imply stability. The fact that the zones of instability appear in every finite domain of phase space available (for a given h), may allow the diffusion of particles over all the available phase space. This is called Arnold diffusion (Arnold and Avez 1967). However, no Arnold diffusion has been found in various recent numerical experiments (Contopoulos 1976), therefore, its role is uncertain.

There is empirical evidence that systems of many degrees of freedom are not ergodic if their energy is small enough (assuming the energy of equilibrium equal to zero). It seems that in most cases one can define a transition energy, E_c, between ordered motion, when most initial conditions give quasi-periodic motion, and ergodic motion. The most important problem is the behavior of E_c for large N. Galgani and Scotti(1972) and Galgani(1975) have found the remarkable result that the ratio E_c/N is constant for N between 8 and at least 1000. It seems, therefore, reasonable to assume that this ratio remains constant for $N \to \infty$. This conclusion has far reaching consequences. E.g., one can derive several quantum-mechanical results by classical methods (Galgani 1974).

3. PERIODIC ORBITS IN HIGHLY PERTURBED SYSTEMS

The periodic orbits are the main landmarks in the totality of orbits of a dynamical system. This is true, in particular, in the case of highly perturbed systems, where it is quite difficult to have theoretical information about the orbits.

In the following we distinguish two types of periodic orbits (Contopoulos 1970a):

(1) Regular periodic orbits, that can be produced from the orbits of the unperturbed problem by varying continuously their shape as a function of the perturbation. As we will see below, this type of orbit includes also the periodic orbits close to homoclinic points.

(2) Irregular periodic orbits, that are independent of the above, appearing only above a certain minimum perturbation

A. REGULAR PERIODIC ORBITS

We have made several numerical experiments in a Hamiltonian of the form

$$H = \frac{1}{2}(\dot{x}^2 + \dot{y}^2 + \omega_1^2 x^2 + \omega_2^2 y^2) - \varepsilon x y^2 = h. \qquad (5)$$

In particular, we found several families of periodic orbits, by keeping ω_1, ω_2, h constant and varying the perturbation ε

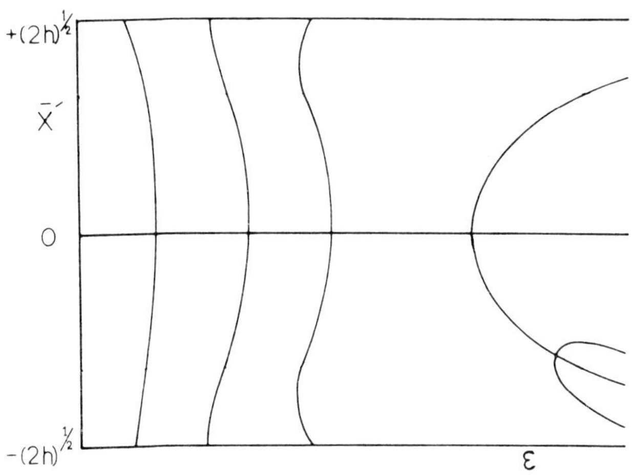

Figure 1. The forms of characteristics curves if $\frac{\omega_1}{\omega_2} < \frac{2}{3^{1/2}}$ (schematically).

(Contopoulos 1970a). The characteristics of these families give the value of $\bar{x} = \omega_1 x$ as a function of ε, whenever $\dot{x} = 0$ (Figures 1 and 2).

For relatively small ε the properties of the characteristics can be derived by an approximate theory as follows.

We bring the Hamiltonian to a normal form, i.e., we change the variables in such a way that the new Hamiltonian is a function of $\Phi'_{10} = \frac{1}{2}(\dot{x}'^2 + \omega_1^2 x'^2)$ and $\Phi'_{20} = \frac{1}{2}(\dot{y}'^2 + \omega_2^2 y^2)$ only (Contopoulos and Hadjidemetriou 1968). This can be done by means of von Zeipel's method. The new Hamiltonian is

$$H' = \Phi'_{10} + \Phi'_{20} + \frac{\varepsilon^2 \Phi'_{20}}{\omega_1^2 \omega_2^4 (\omega_1^2 - 4\omega_2^2)} \left[\frac{1}{4}(8\omega_2^2 - 3\omega_1^2)\Phi'_{20} + 2\omega_2^2 \Phi'_{10} \right] + \ldots = h. \qquad (6)$$

In general an orbit is quasi-periodic with frequencies

$$\bar{\omega}_i = \omega_i \frac{\partial H'}{\partial \Phi'_{i0}} . \qquad (7)$$

If the ratio $\bar{\omega}_1/\bar{\omega}_2$ is rational, n/m, we have a periodic orbit. Thus

$$\frac{n}{m} = \frac{\omega_1 \frac{\partial H'}{\partial \Phi'_{10}}}{\omega_2 \frac{\partial H'}{\partial \Phi'_{10}}} = \frac{\omega_1}{\omega_2} \left\{ 1 + \frac{\varepsilon^2}{2\omega_2^4(\omega_1^2 - 4\omega_2^2)} \left[\left(3 - 4\frac{\omega_2^2}{\omega_1^2}\right) h - 3\Phi'_{10} \right] + \ldots \right\} . \qquad (8)$$

This formula is only approximate, because the normal form of the Hamiltonian is not valid in a region around a periodic orbit. However, this region is of $O(\varepsilon^{1/2(n+m-4)})$, therefore it is very small if ε is small and $n+m>4$*.

---

*An accurate method to find the periodic orbits in all resonant cases has been developed by Contopoulos (1968). However, the above simple method gives results which are qualitatively correct unless n/m = 1, or 2, or 1/2.

If we set $\dot{x}' = 0$ we find $\Phi'_{10} = \frac{1}{2}\bar{x}'^2$. If we introduce this value in Equation (8) we find the equation of the characteristic curves. It is obvious that such curves exist only if the difference between n/m and ω_1/ω_2 is small, of $O(\varepsilon^2)$.

The forms of the characteristic curves depend on the value of $(3 - 4\frac{\omega_2^2}{\omega_1^2})$. If this is negative (case $\frac{\omega_1}{\omega_2} < \frac{2}{3^{1/2}}$) the maximum of ε^2 occurs for $\bar{x}' = 0$. Thus the characteristic curves, for small ε^2, are concave towards the line $\varepsilon = 0$ (Figure 1). In this case we have solutions for small ε^2 only if $n/m > \omega_1/\omega_2$. For large ε^2 the characteristics become convex towards the line $\varepsilon = 0$. This is due to higher order terms in the normalized Hamiltonian.

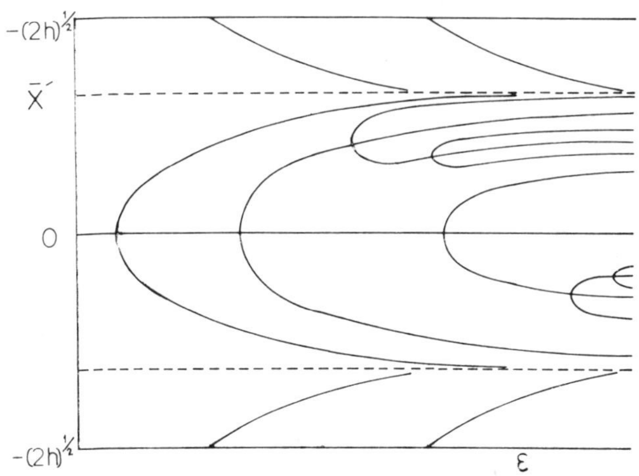

Figure 2. The forms of characteristic curves if $\frac{\omega_1}{\omega_2} > \frac{2}{3^{1/2}}$ (schematically).

If $(3 - 4\frac{\omega_2^2}{\omega_1^2})$ is positive (case $\frac{\omega_1}{\omega_2} > \frac{2}{3^{1/2}}$) the characteristics are represented in Figure 2. All of them approach asymptotically the lines

$$\bar{x}'^2 = 2\left(1 - \frac{4\omega_2^2}{3\omega_1^2}\right)h. \tag{9}$$

The axis $\overline{x}'=0$ is characteristic of the "central" family (or 1/1 family) where

$$\overline{x} = \overline{x}_c = \frac{4h\varepsilon}{\omega_1(4\omega_2^2-\omega_1^2)} + \ldots \qquad (10)$$

This family is stable up to a certain ε; then a new 1/1 family branches off and the original family becomes unstable.

The limiting lines $\overline{x}' = \pm(2h)^{1/2}$ represent the family of periodic orbits y=0.

Several families of periodic orbits with different n/m branch off the central periodic orbit and the orbit y=0, when these are stable. Some of the new families are stable and they generate higher order periodic orbits, closing after 2m, 3m, etc. intersections of the x-axis. These are the periodic orbits of the second "genre" of Poincaré.

In general the new characteristics have their minimum ε at the parent orbit. Only in the case $\frac{n}{m} = 2/3$ was it found that ε_{min} occurs outside the characteristic of the central periodic orbits.

We notice that two characteristics of different multiplicities (different ratios n/m) cannot intersect, except if one ratio n/m is equal to 1. Thus, the various characteristics can terminate only at the axis $\overline{x}'= 0$, the boundaries $\overline{x}'= \pm(2h)^{1/2}$(i.e. y=0), or at infinity.

This "termination principle" can be compared with Strömgren's termination principle for the restricted three-body problem where we have termination at the primaries (infinite potential), at asymptotic periodic orbits around L_4, L_5 (infinite period), or at infinity (infinite dimensions and period). Here we do not have singularities in the potential, nor any equilibrium points except the origin. This is why a family can either join another family or go to infinity.

It seems probable that the characteristic curves which are found to extend to large ε are not closed but extend up to $\varepsilon = \infty$. An argument in favor of this conjecture, based on the existence of irregular periodic orbits, is given below. Furthermore, we notice that the "central" periodic orbit approaches the orbit y = 0 as ε becomes large (Fig. 3). Thus, it seems probable that as $\varepsilon \to \infty$ the "central" periodic orbit tends to the axis y=0. Then probably all periodic orbits tend to the same orbit (y=0) for $\varepsilon \to \infty$. In all our numerical experiments with the Hamiltonian(5) we never found any periodic orbit tending to become escaping for large ε.

Birkhoff (1927) proved the existence of an infinite number of periodic orbits near every homoclinic point. Danby (1973) found numerically such periodic orbits and followed them for smaller energies (i.e., smaller ε in our case) in order to find how they originate. The result of this investigation is very interesting. It seems that all these orbits originate from stable regular families of the type considered above. Namely, a stable periodic orbit, represented by a certain number of invariant points around the central invariant point, becomes unstable, as ε increases, and at the same time it generates another stable family, and so on. Thus, a cascade of unstable families is formed, which may even have infinite members if all the families become unstable beyond a certain ε. The corresponding invariant points approach closer and closer some homoclinic points. The families cannot terminate at such homoclinic points, because these are not invariant points corresponding to periodic orbits. This is a further indication that these families do not terminate for a finite ε, but extend up to $\varepsilon = \infty$.

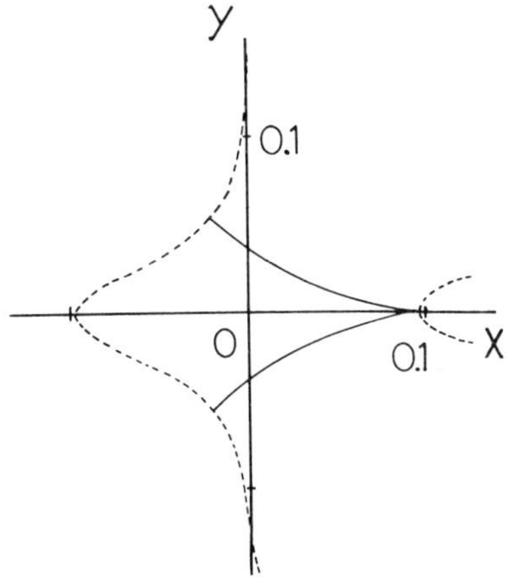

Fig. 3. The central periodic orbit for $\varepsilon = 100$, $h = 0.00765$, $\omega_1^2 = 1.6$, $\omega_2^2 = 0.9$. The dashed lines represent the curve of zero velocity.

The change of stability of periodic orbits was studied orginally by Poincaré (1899). If we vary the perturbation ε continuously, we find that a stable periodic orbit may become unstable at a critical value ε_0, or vice versa. At such a point either a stable characteristic joins an unstable one, and ε is maximum or minimum, or more than one characteristics cross each other. The rule is that if the numbers of stable and unstable branches joining at ε_o, are s_-, u_- and s_+, u_+ for $\varepsilon < \varepsilon_o$ and $\varepsilon > \varepsilon_o$ respectively, then

$$s_- - u_- = s_+ - u_+ . \tag{11}$$

More detailed studies of the change of stability were made by Hénon (1965), Deprit and Henrard (1968), and Contopoulos (1970 b). An important application is that is we have a stable family for $\varepsilon < \varepsilon_o$*, which becomes unstable for $\varepsilon > \varepsilon_o$, then a new stable family is formed (represented by two branches), that has its minimum at ε_o. Thus, the stability of the original family is not lost but inherited by the new family and its "descendants".

It is possible, however, that the ranges of ε over which the new families are stable, are smaller and smaller, so that the total range of ε over which the totality of the "descendants" of the original families is stable is finite. Thus, we may have a critical value of ε, beyond which all families generated from the periodic orbits of the unperturbed problem become unstable. This conclusion seems probable if we look at the numerical ranges of stability of the various calculated families of regular periodic orbits. However, we cannot conclude from this that there are no stable periodic orbits at all beyond this critical ε, because there are other families for larger ε independent of the regular families considered above, which may be stable. These are the irregular families.

B. IRREGULAR PERIODIC ORBITS

The irregular periodic orbits belong to families whose characteristics do not intersect the characteristics of the regular periodic orbits. Therefore, such orbits cannot be continuously transformed to the orbits of the unperturbed system.

These families form certain "sets of families". Namely, there is a pair of irregular families starting at a minimum $\varepsilon = \varepsilon_{min}$, one of them stable and the other unstable. The stable family becomes unstable for some larger ε, generating a new stable family, and so on. All the families generated in this way form a set. It is possible, as in the case of the regular families, that a set contains an infinite number of families, all becoming unstable beyond a critical ε.

We have found many sets of families of irregular orbits in the model problem (5). For every set there is a range of values of ε,

---

*We assume that for $\varepsilon < \varepsilon_o$ there is no other family joining this family at ε_o. In this case $u_-=0$, $s_-=1$, $u_+=1$, therefore $s_+=2$.

near the minimum perturbation ε_{min}, where the new set produces islands of stability in the ergodic sea, or in the escape region.

The number of new sets is larger for larger ε, and it may be that this trend continues up to arbitrarily large ε. If this is true, then there are islands of stability for every ε and the system never reaches complete ergodicity, or, if the curves of zero velocity are open, there is always a set of orbits (periodic and non-periodic) of positive measure, which do not escape.

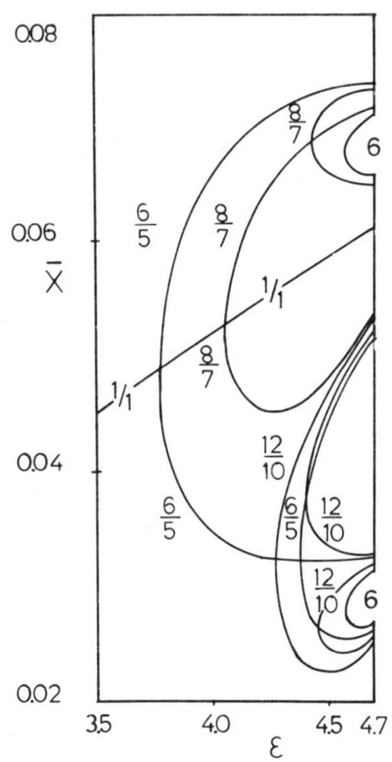

Fig. 4. Characteristics of regular and irregular families of periodic orbits (case $\omega_1^2 = 1.6$, $\omega_2^2 = 0.9$, $h = 0.00765$).

Many irregular periodic orbits play a major role in producing ergodicity. Such orbits join together regions of phase space that seem to "belong" to quite different types of regular periodic orbits. Namely, in the diagram $(\varepsilon, \bar{x})$ (Fig. 4) certain regions are populated by families corresponding to the same ratio of frequencies $\omega_1/\omega_2 =$

n/m = 2n/2m = 3n/3m.... This is the case, in Figure 4, with the regions with $\omega_1/\omega_2=6/5$ (region-5, containing periodic orbits intersecting the x-axis 5,10,... times) and $\omega_1/\omega_2 = 8/7$ (region -7). Thus, it was somewhat unexpected when we found a family of periodic orbits intersecting the x-axis 6 times, whose characteristic had two branches, one in the region -5 and another one in the region -7.

However, all these orbits are in the "ergodic sea", therefore, there are no closed invariant curves separating the region -5 from the region -7. The orbit of multiplicity 6 that connects the two regions is simply one further manifestation of the interaction of resonances.

Most irregular orbits do not have a unique rotation number* associated with them. In other words, we cannot say unambiguously that while the periodic orbit makes m oscillations up and down (m is the number of invariant points) it makes n oscillations to the left and right.

In Figure 5 we give an example of an irregular periodic orbit, represented by 7 invariant points (7 points of intersection with the axis y=0). The successive points of intersection are marked. However, it is not clear what are the successive angles between the invariant points. If we measure the various angles from the "central" invariant point C we must consider the points along the solid line which surrounds C. Then the successive angles between the points 01234567 are "near" $\frac{2}{7}, \frac{1}{7}, \frac{3}{7}, \frac{2}{7}, \frac{3}{7}, \frac{1}{7}, \frac{2}{7}$, and their average value is 2/7. If on the other hand we measure the angles from the origin O, we have to consider the points along the dashed line, which surrounds O. Then the successive angles are "near" $\frac{5}{7}, \frac{1}{7}, \frac{4}{7}, \frac{1}{7}, \frac{4}{7}, \frac{1}{7}, \frac{5}{7}$, and their average value is 3/7. Thus, the rotation number is not unambiguously defined.

This is related to the fact that such an orbit joins two quite dissimilar regions of phase space, namely a region -5 and a region -3. The irregular orbits are probably related to the periodic orbits near heteroclinic orbits. The existence of infinitely many periodic orbits near each heteroclinic point can be proved by a modification of the theorem of Birkhoff (1927) about the existence of such orbits

*The rotation number is the average angle between two successive consequents, measured from a centrally located point, with the circle as unit. If measured clockwise, it is equal to $\left(\frac{\omega_1}{\omega_2} = \frac{n}{m}\right)$ (modulo 1).

near homoclinic points. The multiplicities of such orbits are larger than a minimum value and go up to infinity.

Let us consider a heteroclinic point produced by the intersection of the asymptotic curves of two unstable periodic orbits, say, one of multiplicity 5 and one of multiplicity 7. The periodic orbits starting near this heteroclinic point approach both unstable periodic orbits. Therefore, such orbits should have invariant points both in the region -5 and in the region -7, at least if their multiplicity is large enough. Because of that these periodic orbits cannot be regular.

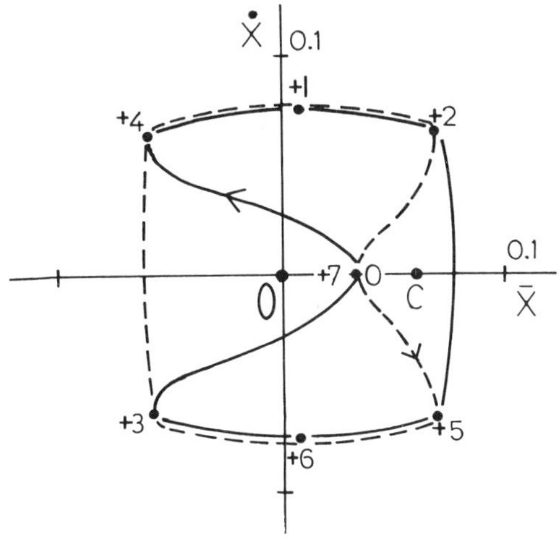

Fig. 5. A set of 7 invariant points representing an irregular period orbit. C represents the central invariant point.

We conjecture that if we follow such orbits for smaller ε we should reach a stable irregular orbit.

If there are irregular periodic orbits continuing for arbitrarily large ε, as indicated by the numerical experiments, then both regular and irregular families cannot terminate for finite ε. E.g., if the irregular family of multiplicity 6, discussed above, exists for all ε, the characteristics of the orbits of multiplicities 5 and 7 cannot terminate at the central periodic orbit or at the boundary. Otherwise they would cross first a characteristic of multiplicity 6, which is impossible. It seems, therefore, that all families, regular and irregular, extend to $\varepsilon = \infty$, tending to the same orbit $y = 0$.

It would be of interest to compare the above case with other similar cases. In particular, one should explore that happens to the various families, regular and irregular, as the ratio ω_1/ω_2 changes over a wide range of values.

Further, one should compare the above results with those of more complicated systems like the restricted three-body problem. Thus one would be able to understand in more detail the general characteristics of non-integrable dynamical systems.

REFERENCES

1. Arnold, V. I., and A. Avez, 1967, Problèmes Ergodiques de la Mécanique Classique, Gauthier-Villars, Paris.
2. Birkhoff, G. D., 1927, Acta Math. 50, 359.
3. Contopoulos, G., 1967, Bull. Astron. Ser. 3, 2, 223.
4. Contopoulos, G., 1968, Astrophys. J. 153, 83.
5. Contopoulos, G., 1970 a, Astron. J. 75, 96.
6. Contopoulos, G., 1970 b, Astron. J. 75, 108.
7. Contopoulos, G., 1975, IAU Symp. 69.
8. Contopoulos, G., 1976, in N. R. Lebovitz (ed.). Theoretical Principles in Astrophysics and Relativity, Chicago.
9. Danby, J.M.A., 1973, Celes. Mech. 8, 273.
10. Deprit, A., and J. Henrard, 1968, Advances Astron. Astrophys., 6, 1.
11. Galgani, L., 1974, Classical Mechanics and Quantum Mechanics (preprint).
12. Galgani, L., 1975 (private communication).
13. Galgani, L., and A. Scotti, 1972, Rev. Nuovo Cimento 2, 189.
14. Hénon, M., 1965, Ann. d'Astrophys, 28, 992.
15. Moser, J., 1973, Stable and Random Motions in Dynamical Systems, Princeton Univ. Press.
16. Poincaré, H.,1899, Les Méthodes Nouvelles de la Mécanique Céleste, III, Gauthier-Villars, Paris.
17. Walker, G. H., and J. Ford, 1969, Phys. Rev. 188, 416.

CATASTROPHES IN LAGRANGIAN SYSTEMS -- A

Cécile DeWitt-Morette

Department of Astronomy and Center for Relativity
University of Texas at Austin, Austin, Texas 78712

INTRODUCTION. The dynamical systems discussed in these lectures are systems with a finite number of degrees of freedom. We shall consider both classical and quantum mechanical systems whose dynamical equations can be obtained from a Lagrangian $L(f(t), \dot{f}(t), t)$. The paths f map a time interval $\mathbb{T} = [t_a, t_b]$ into the configuration space $\mathbb{M}$ of the system. For nearly all physical systems $\mathbb{M}$ is a Riemannian manifold often multiply connected. A classical path $q : \mathbb{T} \to \mathbb{M}$ is a solution of the Euler Lagrange equation satisfying some boundary conditions which are chosen here to be $q(t_a) = a$, $q(t_b) = b$, unless otherwise specified. For example $L = \frac{1}{2} \dot{f}^2(t) - V(f(t))$

$$\begin{matrix} dq(t)/dt = v(t) \\ dv(t)/dt = -V'(q(t)) \end{matrix} \quad \longleftrightarrow \quad dX(t)/dt = F(X(t))$$

<u>The role played by equilibrium points in the study of classical stytems is analoguous to the role played by classical paths in quantum systems</u>. Indeed:
An equilibrium point $\bar{X}$ of the system is a particular "motion" of the system which satisfies the equation $F(\bar{X})=0$; an equilibrium point is a critical point of the potential function $V : \mathbb{M} \to \mathbb{R}^n$ $V'(\bar{X})=0$

The nature of an equilibrium point is best investigated by studying the long time behavior of the nearby motions $\{X(t)\}$. This, in turn, is determined to a large extent by the nature of the

V. Szebehely and B.D. Tapley (eds.), Long-Time Predictions in Dynamics, 57–65. *All Rights Reserved. Copyright © 1976 by D. Reidel Publishing Company, Dordrecht-Holland.*

eigenvalues of the derivative $F'(\bar{X})$ of F at the equilibrium point $\bar{X}$, i.e., by the Hessian $V''(\bar{X})$ of V at $\bar{X}$.

A similar pattern appears in the Feynman formalism of quantum mechanics: Let $S(f) = \int_{\mathbb{T}} L(f(t), \dot{f}(t), t)\, dt$ be the action of the system. Let $C_{ab}(\mathbb{T})$ be the space of continuous paths $f : \mathbb{T} \to \mathbb{M}$ such that $f(t_a) = a$, $f(t_b) = b$, $S : C_{ab}(\mathbb{T}) \to \mathbb{R}$. A classical path q is a critical point of the action

$$S'(q) = 0$$

The nature of a critical point of the action S is determined to a large extent by its Hessian $S''(q)$.

This pattern differs from the previous one in the following respect: Whereas V is a function on the finite dimentional space $\mathbb{M}$, S is a function on the infinite dimensional space $C_{ab}(\mathbb{T})$. The critical points $\bar{q}$ of V are solutions of the algebraic equation $V'(\bar{q})=0$, the critical points q of S are solutions of the Euler Lagrange differential equation $S'(q)=0$. The time evolution of a system whose "equilibrium point" satisfies a differential equation is governed by a partial differential equation, namely the Schrödinger equation

$$-ih\, \partial\psi(x,t)/\partial t = H\psi(x,t)$$

$\psi(x,t)$ is the probability amplitude of finding the particle at (x,t). Here again the nature of the eigenvalues of H determine the longtime behavior of the system: the sinks are now called "bound states" and the sources are now called "scattering states". The concept of "nearby motions" (understand quantum states) of a given critical point (understand classical limit) is much more difficult to analyze than in the case of classical dynamical systems. This concept can be said to be the core of the path integral formalism of quantum mechanics. There, the probability amplitude $K(b,t_b,a,t_a)$ for a transition from a state (a,t_a) to a state (b,t_b) is given by a "sum" over all paths $f \in C_{ab}(\mathbb{T})$ of $\exp \frac{i}{\hbar} S(f))$

$$K(b,t_b; a,t_a) = \text{"}\Sigma\text{"} \exp \frac{i}{\hbar} S(f)$$

$$\psi_t(x) = \int_{\mathbb{M}} K(x,t; x',t_a)\, \psi_{t_a}(x')\, dx' \quad \text{where} \quad \psi_t(x) = \psi(x,t)$$

If we scale time by $\hbar$-1, the classical limit q of a quantum system can be obtained either by letting t tends to ∞ or letting $h \to 0$. We shall not discuss quantum systems any further since our aim is not to investigate quantum systems but to obtain more information on classical systems by considering them as the limits of quantum systems.

We shall discuss the catastrophes of systems where equilibrium points are given either by $V'(\bar{q})=0$ or by $S'(q)=0$. We shall call them respectively a system V or a system S. Such systems are called gradient systems. We shall give an example for each case, define catastrophes and finally give a criterion for catastrophes of Lagrangian systems (systems S). The example chosen for the catastrophes of an S system is given in a separate paper and can be read before paragraph 4.

2. ZEEMAN CATASTROPHE MACHINE.[1] Example of V systems Consider a disk rotating around its center x, two rubber gands zy of length ℓ_1 and zc of length ℓ_2 are attached to its rim at z. y is a fixed point, c is a movable point which controls the system. The motion of the disk is given by $\theta(t)$ where $\theta(t)$ is for instance the angle yxz, its equilibrium is an angle $\bar{\theta}$, critical point of a potential function $V(.,c)$ which depends on c

$$V'_\theta(\bar{\theta},c) = 0$$

V depends on θ^4, V' depends on θ^3.

Catastrophe theory is the theory of "jumps" which may occur when the control point $\underline{c}$ moves smoothly on the control space, here R^2. Let us plot $\bar{\theta}$ as a function of c. The equation $V'_\theta(\bar{\theta},c)=0$ has 1, 2, or 3 solutions. The surface $\bar{\theta}(c)$ (figure 1) has two folds, the area in grey corresponds to maxima of V, the area in white to minima of V. As c moves across the control space, $\bar{\theta}(c)$ moves on the surface, when it arrives at a fold, it "jumps" from one sheet of stable equilibrium positions to the other sheet of stable equilibria. A continuous change of the control factors produces a discontinuous change in the state of the system.

3. DEFINITION OF A CATASTROPHE[2] Let $\mathbb{R}^k$ be the control space, k finite; let Σ be the space of the equilibrium points of the system, solutions of $V'(\bar{q},c)=0$ or $S'(q,c)=0$ as the case may be.

Thom's Theorem: Σ is a k-dimensional manifold. It is remarkable that this is true both for V systems where Σ is a subset of a finite dimensional space $\mathbb{M} \times \mathbb{R}^k$ and for S systems where Σ is a subset of an infinite dimensional space $\mathcal{C}_{ab}(\mathbb{T}) \times \mathbb{R}^k$.

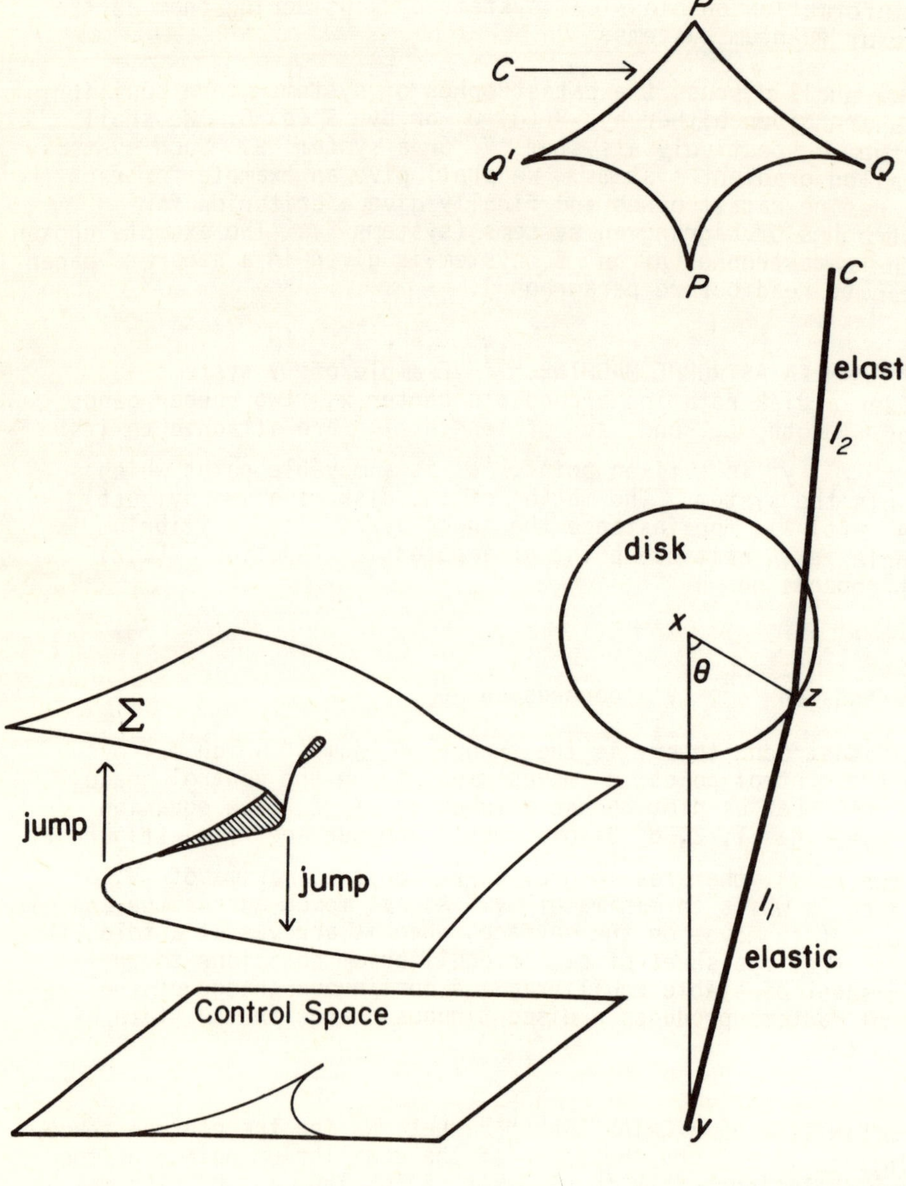

Figure 1. The theory of catastrophe analyzes the discontinuous behavior of a gradient system caused by continuous changes in the factors which control the system

It follows that the nature of catastrophes depends only on the dimensionality of the control space

Thus if k =	1	2	3	4	5
there can be	1	2	5	7	11

different kinds of elementary catastrophes. When $k \geq 6$, there are an infinite number of different kinds of catastrophe. Hence regardless of how complicated a system is, if its control space is of low dimensionality, one can construct a simple model having the same kind of catastrophes. For this reason catastrophe theory has been applied to a great variety of complicated systems: biological systems, defense analysis, statistical mechanics, etc.

Thom's Definition: Catastrophes are the singular points of the projection may $\chi : \Sigma \to \mathbb{R}^k$; i.e., the points $\sigma \in \Sigma$ such that $\chi'(\sigma)$ is not one-one. In the example §2, , σ is a singular point of χ if $T_\sigma \Sigma$ is "vertical".

4. A CATASTROPHE CRITERION FOR LAGRANGIAN SYSTEMS. The behavior of a Lagrangian system is determined by

- the boundary conditions.
- and the critical values q of the action S defined on the space of paths f which satisfy the chosen boundary conditions.

Theorem: Let $\overline{\beta}(u,v)$ be the 2n parameter family of solutions of the Euler-Lagrange equation of a system S. Set $\overline{\beta}(u,v)(t) = \beta(u,v,t)$. Let I be a set of n initial conditions defining v as a function of u. Let F be a set of n final conditions defining u as a function of t. Let the control space of the system be the space of $(m+1)$ tuples

$$(\beta(u,v(u),t),t) \in \mathbb{M} \times \mathbb{T}$$

Catastrophes occur if and only if the final conditions F are such that the $(n+1)$ triple (u,t) satisfies the equation

$$\det \frac{D(\beta(u,v(u),t)}{Du} = 0 \tag{1}$$

The proof proceeds in three steps, each one of them proves a theorem interesting in its own right.

a) The Euler-Lagrange equation $Lq(t)=0$, and Jacobi equation alias the variational equation alias the equation of geodetic deviation alias the small disturbance equation $J_q x(t)=0$, correspond respectively to the first and the second variation of action

$$S'(q)x = \int_{\mathbb{T}} (Lq(t))\, x(t)\, dt$$

$$S''(q)xx = \int_{\mathbb{T}} (J_q x(t))x(t)\, dt$$

$$L\beta(u,v,t) = 0 \qquad \text{for every } u \text{ and } t$$

$$0 = \frac{d}{du^i}(L\beta(u,v,t)) = J_{\overline{\beta}(u,v)} \frac{\partial\beta}{\partial u^i}(u,v,t)$$

hence

Proposition a: The general solution of the Jacobi equation $J_{\overline{\beta}(u,v)}x(t) = 0$ is

$$x(t) = \lambda_i \frac{\partial\beta}{\partial u^i}(u,v,t) + \mu_i \frac{\partial\beta}{\partial v^i}(u,v,t) \qquad \text{(n equation)}$$

where $\{\lambda_i, \mu_j, ; i, j, = 1,\ldots n\}$ are 2n constants of integrations.

Let the initial data I be a mixed set of initial positions and velocities, for instance, right ascension, declination, radial velocity. In general let them be

$$(I)\begin{cases} \beta^\alpha(u,v,t_a) = a^\alpha & \text{for some } k \text{ components of } a \\ \partial\beta^\beta(u,v,t_a)/\partial t_a = v^\beta & \text{for some } n\text{-}k \text{ components of } v. \end{cases}$$

We shall consider that the initial data determine implicitly v as a function of u.

Then the n Jacobi fields obtained by variation of the n parameter family $\overline{\beta}(u,v(u))$ satisfying these initial conditions can be labeled

$$x_i(u,t) \overset{\text{def}}{=} \frac{\partial\beta}{\partial u^i}(u,v(u)t) + \frac{\partial\beta}{\partial v^j}(u,v(u)t)\frac{\partial v^j}{\partial u^i}$$

They satisfy the following initial conditions

$$I'\begin{cases} x_i^\alpha(u,t_a) = 0 & \text{for the same } k \text{ components as the fixed components of } a \\ dx_i^\beta(u,t_a)/dt_a = 0 & \text{for the same } (n\text{-}k) \text{ components as the fixed components of } v_a \end{cases}$$

The particular cases $k=n$ and $k=0$ are included in this discussion and are easier to vizualize than the general case.

Proposition b: The n parameter family $\overline{\beta}$ $(u, v(u))$ has a caustic relative to the chosen initial conditions I if and only if

$$\det\left(D(\beta(u, v(u),t))/Du\right) = 0 \qquad (1)$$

Proof:
$\beta(u, v(u),t)$ is, by definition, conjugate to $\beta(u, v(u),t_a)$ along $\bar{\beta}(u,v(u))$ if and only if there is a non zero Jacobi field x with $2n$ vanishing boundary conditions - or, equivalently, if and only if there are less than $2n$ linearly independent Jacobi fields along $\bar{\beta}(u, v(u))$. If equation 1 is satisfied, there is a linear combination $\Sigma\lambda_i x_i(u,t)$ of the n Jacobi fields $\{x_i(u,t)\}$ satisfying the initial conditions I' and such that

$$\Sigma\lambda_i\, x_i\,(u,t) = 0 \qquad \text{for a non-zero set } \{\lambda_i\}$$

Equation 1 gives t as a function of u ; for a given u, i.e., for a given solution $\bar{\beta}\,(u,v(u))$ of the Euler Lagrange equation, equation 1 gives the set of points $\{\beta(u, v(u), t)\}$ which is the caustics relative to the initial conditions I.

Proposition c: The caustics of a family of classical solutions $\bar{\beta}(u,v(u))$ are the projection of the catastrophe set of an integral manifold Σ of the Hamilton Jacobi equation compatible with the chosen initial conditions, on the control space $\mathbb{M} \times \mathbb{T}$

Proof:
Let $(u,t) \mapsto (u,t,v)$ be the mapping defined by the initial data I; let $(u,t,v) \mapsto (u, t, v, \beta\,(u, v))$ be the mapping defined by $S'(\bar{\beta}(u,v)) = 0$. Let $\bar{\beta}(u,v)(t)=\beta(u,v,t)$ and let $\bar{S}(\beta(u, v, t), t)$ be equal to the action S along the path $\bar{\beta}(u,v)$ from t_a to t. Note that the equation

$$\bar{S}(\beta(u,v,t),t) = S(\bar{\beta}(u,v))$$

does not define a function $\bar{S}$ in the proper sense of "function" (although $\bar{S}$ is commonly called the action function) since $\bar{S}$ may be a multivalued mapping of its arguments. It defines however an n+1 dimensional manifold Σ, namely the integral manifold of the Hamilton Jacobi equation satisfying the n given conditions I. The mapping χ defined by the sequence of mappings

$$(u,t) \overset{1}{\mapsto} (u, t, v) \overset{2}{\mapsto} (u, t, u, \beta(u, v,)) \overset{3}{\mapsto} (u, t, v, S(\beta(u, v))$$
$$\overset{4}{\mapsto} (\beta\,(u, v, t), t, \bar{S}\,(\beta(u, v, t);t) \overset{\chi}{\mapsto} \beta(u, v, t), t)$$

is singular if and only if equation 1 is satisfied.

The construction of Σ as an integral manifold of Hamilton Jacobi equation gives an intuitive understanding of proposition

c. Indeed Σ is generated by the bicharacteristics from a given initial n dimensional Cauchy manifold. The bicharacteristics are given by the solutions of the Hamilton's equation which contain 2n constants of integration, n of them are determined by the initial Cauchy data. The bicnaracteristics are thus an n parameter family of solutions of the Euler-Lagrange equation. If this family has an envelope, it generates Σ up to but not beyond its envelope.

5. NOTES ON CATASTROPHE THEORY. Since catastrophe theory is a new subject which is growing in many different directions [3], a few comments may help set in a broader context the catastrophe criterion for Lagrangian systems derived in this paper.

(1) In physics, catastrophe theory brings new insights in:

a) Static. Significant examples (crystal deformation, bridge buckling), as well as references to other applications, can be found in a recent article of J. M. T. Thomson [4].

b) Dynamic. The results derived in our paper have been used [5] to compute path integrals between conjugate points where the Van Vleck determinant, becomes infinite. L. S. Schulman [6] has analyzed catastrophes in path integrals in terms of the eigen functions and eigen values of the small disturbance operator J_q.

c) Thermodynamic. Phase transitions are catastrophes [7]. A statistical system is generally different from a Lagrangian system: the thermodynamical variables which control the system are not solutions of differential equations, but the average result of behavior of very many particles whose motion is governed by differential equations.

(2) In mathematics, the works relevant to our paper include:

a) Classification of catastrophes. Beautiful pictures of catastrophes can be found in the book of A. E. R. Woodcock and T. Poston [8].

b) Analysis of the stability of dynamical systems. An excellent review, with numerous references, entitled "Oscillatory integrals, Lagrange immersions and unfoldings of singularities" [9] has been written by J. J. Duistermaat. Particularly relevant is the paper of J. Guckenheimer on "Catastrophes and Partial Differential Equations" [10]. There he investigates various types of stability which can meaningfully be introduced and is lead naturally to define a category of catastrophes which he calls c-catastrophes. The catastrophes considered in our paper are c-catastrophes. The theorem derived in paragraph 4 is a very practical criterion for c-catastrophes.

Acknowledgement. A very pleasant stay at the Mathematics Institute of the University of Warwick, and stimulating conversations with E. C. Zeeman, K. D. Elworthy and J. Eells are at the origin of this work. J. Lees and S. E. Newhouse have brought to my attention many of the references presented in section 5.

REFERENCES

1. Zeeman, E. C., "A Catastrophe Machine", in Towards a Theoretical Biology 4 edit. by Waddington, C. H. Edinburgh, Edinburgh University Press, 1972.

2. Thom, R., Stabilite Structurelle et Morphogenese, New York, Benjamin 1972, Reviewed by J. Guckenheimer, Bull. Am. Math Soc., 79, 878-890, (1973).

3. Zeeman, E. C., "Applications of Catastrophe Theory", Tokyo International Conference on Manifolds, 1973.

4. Thompson, J. M. T., Nature 254, 392-295, 1975.

5. DeWitt-Morett, C., "Path Integrals in Riemannian Manifolds", to appear in the Proceedings of the International Symposium of Mathematical Problems in Theoretical Physics. Kyoto, 1975, to be published in the Springer-Verlag Lecture Notes in Physics, and "The Semi Classical Expansion" Preprint University of Texas at Austin. Note that proposition 2 in "Path Integrals in Riemannian Manifolds" is wrong. It is corrected in "The Semi-Classical Expansion".

6. Schulman, L. S., "Caustics and Multivaluedness. Two Results of Adding Path Amplitudes", to appear in the Proceedings of the International Conference on Functional Integration and its Applications, London 1974.

7. Schulman, L. S., and Revzen, M., Collective Phenomena 1, 43-49, 1972.

8. Woodcock, A. E. R., and Poston, T., A Geometrical Study of the Elementary Catastrophes, Springer Verlag, Lecture Notes in Mathematics 373, 1974.

9. Duisdermatt, J. J., Comm. Pure Appl. Math. 27. 207-281, 1974.

10. Guckenheimer, J., Annales Inst. Fourier 23, 31-59, 1973.

CATASTROPHES IN LAGRANGIAN SYSTEMS: AN EXAMPLE -- B

Cécile DeWitt-Morette

Department of Astronomy and Center for Relativity
University of Texas at Austin, Austin, Texas 78712

Pete Tschumi

Department of Physics, University of Texas at
Austin, Austin, Texas° 78712

ABSTRACT. The classic example of a soap bubble held by wire loops is used to illustrate the propositions of the previous paper. It is shown that when the soap bubble breaks it is a catastrophe in the technical sense of the word.

The classic problem of the surface area of a soap film held by 2 wire loops has been used since Jacobi to illustrate the fundamental theorems of the calculus of variations. We use it again in this paper to illustrate the catastrophe criterion established in the previous paper.

Problem: Dip two wire circles of radius r and R in a soap solution, and remove them. Assume that the two circles, originally concentric, are gradually pulled apart so that their planes remain perpendicular to the axis joining their centers. The soap film forms a surface of revolution of minimum area and eventually breaks into two circular disks bounded by the wire circles. Show that the breaking of the soap bubble is a catastrophe.

Answer: The area of the surface of revolution generated by curve $f: [0,\ell] \to \mathbb{R}$, is

V. Szebehely and B.D. Tapley (eds.), Long-Time Predictions in Dynamics, 67–70. *All Rights Reserved. Copyright © 1976 by D. Reidel Publishing Company, Dordrecht-Holland.*

$$S(f) = 2\pi \int_0^{\ell} f(t)(1+\dot{f}^2(t))^{1/2}\, dt$$

Set $f(t) = q(t) + x(t)$

$$S(f) = S(q) + S'(q)x + \frac{1}{2}S''(q)\, xx + \sum(q,x)$$

The Euler-Lagrange equation is $-q(t)\ddot{q}(t) + \dot{q}^2(t)+1 = 0$. Its solution q, such that $q(0) = r$ and $q(\ell) = R$, generates the surface of minimal area. The Jacobi equation is

$$J_q x(t) = (1+\dot{q}^2(t))^{-3/2}(-q(t)\ddot{x}(t) + 2\dot{q}(t)\dot{x}(t) - \ddot{q}(t)\, x(t))$$

The general solution of the Euler-Lagrange equation is the 2 parameter family of catenaries.

$$\beta(u,v,t) = u \cosh((t-v)/u) \quad \text{with } u,v, > 0$$

If instead of characterizing a solution by the values of u and v we wish to characterize it by the radii r and R of the loops and their distance apart ℓ, we remark that for a given set (r,R,ℓ) we may have 2 solutions, 1 solution or 0 solution of the Euler-Lagrange equation. Hence the set of equations

$$q(t) = u \cosh((t-v)/u)$$
$$r = u \cosh(v/u)$$
$$R = u \cosh((\ell-v)/u)$$

does not necessarily define a unique curve q.

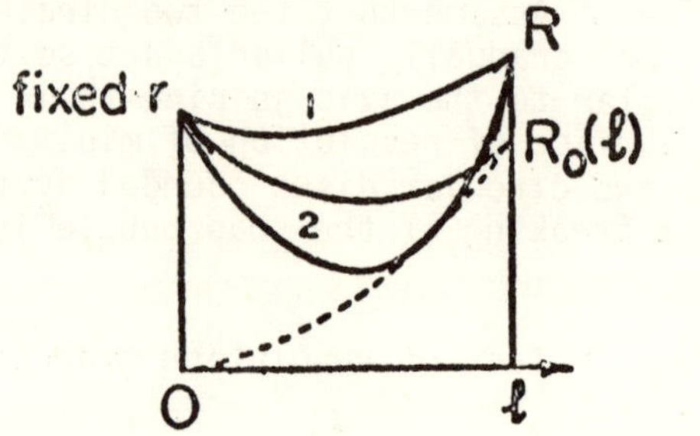

$R_0(\ell)$ is the envelope of $q(t)$

For given values of ℓ and r, there are 2 solutions of the Euler-Lagrange equation if $R > R_o(\ell)$; one of them minimizes S, the other one is an extremum which is neither a local minimum nor a local maximum [1]. When $R = R_o(\ell)$, the two solutions coalesce into one solution. When $R < R_o(\ell)$, the Euler-Lagrange equation has no solution. Alternatively, for given values of R and r, there are 2 solutions for $\ell < R_o^{-1}(R)$, a physical one which minimizes S and an "unphysical" one - unphysical but very important physically as we shall see. There is one solution when $\ell = R_o^{-1}(R)$, and no solution when $\ell > R_o^{-1}(R)$. For a given R, $R_o^{-1}(R)$ is the breaking point.

We shall choose the control space of the system to be the space of pairs (R,ℓ), i.e., the positive quadrant of $\mathbb{R}^2$. The radius r is kept fixed. Since the curve q that generates the surface of minimal area between the loops r and R is a point of an infinite dimensional space $C_{rR}([o,\ell)$, we cannot draw the space $\{(R,\ell,q)\}$. But we can draw the surface Σ defined by the triples $\{(R,\ell,\bar{S}(R,\ell)\}$ where

$$\bar{S}(R,\ell) = S(q)$$

Note that this equation does not define a "function" $\bar{S}$ on the positive quadrant of R^2 in the strict sense of the word, since the mapping $(R,\ell) \to \bar{S}(R,\ell)$ may be multivalued. $\bar{S}(R,\ell)$ is a

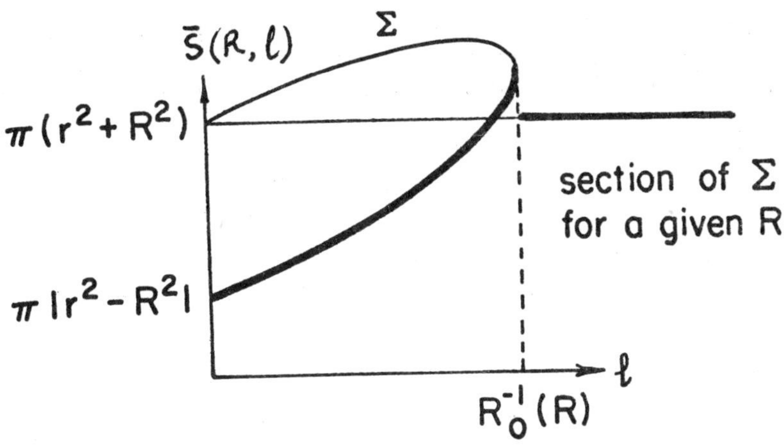

solution of the Hamilton-Jacobi equation. That it is equal to $S(q)$ is one of the most beautiful equations of dynamics.

The castastrophé set $\{\sigma;\sigma\varepsilon\Sigma\}$ is the set of points where the projection map $X:\Sigma\to\mathbb{R}^2$ is singular. We can find it by studying the catastrophe set of the integral manifolds of the Hamilton-Jacobi equation [2] or more simply by studying the sequence of mappings:

$$(u,\ell) \overset{1}{\mapsto} (u,\ell,v) \overset{2}{\mapsto} (u,\ell,v,q) \overset{3}{\mapsto} (u,\ell,v,S(q)) \overset{4}{\mapsto} (R,\ell,\overline{S}(R,\ell)) \overset{\chi}{\mapsto} (R,\ell)$$

where the mapping 1 is defined by $r = u \cosh(v/u) \quad u,v>0$,
the mapping 2 by $q(t) = u \cosh((t-v)/u)$,
the mapping 3 by $S(q) = \int L(q(t), \dot{q}(t))dt$,

the many-one mapping 4 by

$$R = q(\ell) \quad , \quad \overline{S}(R,\ell) = S(q)$$

$\chi'(\sigma) : T_\sigma\Sigma \to T_{\chi(\sigma)}\mathbb{R}^2$ is singular when it is of rank less than 2 i.e., when

$$\det\left(\frac{D(R,\ell)}{D(u,\ell)}\right) = 0 \Longleftrightarrow \frac{\partial R}{\partial u} + \frac{\partial R}{\partial v}\frac{dv}{du} = 0$$

Since r is fixed, v is a function of u defined by $r = u \cosh(v/u)$, and $q(t)$ is a one parameter family of curves parametrized by u. The set of equations

$$\begin{cases} R = u \cosh((\ell-v)/u) \\ \dfrac{\partial R}{\partial u} + \dfrac{\partial R}{\partial v}\dfrac{dv}{du} = 0 \end{cases}$$

defines both the envelope $R = R_0(\ell)$ of this one parameter family of curves and the catastrophe set of the soap bubble. This is an example of the theorem which is proved in the previous paper.
Theorem: The caustics of a family of classical solutions are the projection of the catastrophe set of the Hamilton-Jacobi equation on the control space $\mathbb{M} \times \mathbb{T}$.

A soap bubble breaks when $R = R_0(\ell)$, i.e., when the control point (R,ℓ) crosses the projection of the catastrophe set.

REFERENCES

1. These classic results can be found in several books. They are readily available in the excellent book, "Calculus of Variations" by G. A. Bliss. Mathematical Association of America. Carus Mathematical Monographs 1925.
2. Guckenheimer, J., Annales Inst. Fourier 23 31-59, (1973).

STATISTICAL BEHAVIOR IN CONSERVATIVE CLASSICAL SYSTEMS

L. E. Reichl

Center for Statistical Mechanics and Thermodynamics
University of Texas at Austin

1. INTRODUCTION

In the following series of lectures, we will discuss how instability can arise in conservative classical systems, and how a small uncertainty in initial conditions can be amplified to complete uncertainty if a system exhibits mixing behavior. The methods we shall discuss were originally developed to treat problems in statistical mechanics and therefore, we will indicate briefly their history.

Celestial Mechanics and Statistical Mechanics are both concerned with the behavior of conservative classical systems after a long time. In statistical mechanics

(1) One normally deals with systems with many degrees of freedom and attempts to derive the laws of thermodynamics (for equilibrium systems) or hydrodynamics (for systems which evolve with time) in terms of the underlying microscopic equations of motion for various degrees of freedom. The larger the number of degrees of freedom the smaller will be the fluctuations away from thermodynamic or hydrodynamic behavior.

(2) In statistical mechanics one usually knows very little about the initial conditions although one may know some average properties of the system. For example, for a dilute gas of hard spheres in a box, one can measure the volume, the local average

V. Szebehely and B.D. Tapley (eds.), Long-Time Predictions in Dynamics, 71–98. *All Rights Reserved. Copyright © 1976 by D. Reidel Publishing Company, Dordrecht-Holland.*

kinetic energy (temperature), the local average velocity, the hydrostatic pressure, or the local average particle density.

(3) Often one must know the time evolution of certain quantities for all times. (For example, to obtain expressions for the transport coefficients one must obtain expressions for various autocorrelation functions valid for all times).

One of the central problems for statistical mechanics since the 19th century has been that of irreversibility. Many systems in nature are found to exhibit irreversible behavior. For example

(1) If a gas of particles in a box is initially distributed inhomogeneously in space and initially has a non Maxwell-Boltzmann distribution in velocity, it will evolve until it becomes homogeneously distributed in space and attains a Maxwell-Boltzmann velocity distribution. Then, on the average it will remain that way for all time (there will be fluctuations about the average for systems with a finite number of degrees of freedom).

(2) If we take a finite crystal lattice and add energy to a few of the particles on the lattice, the energy will spread to the other particles in the lattice until the energy becomes partitioned euqally among all of the degrees of freedom. Then, on the average the energy will remain equi-partitioned.

The paradox that confronts one is to try to understand this irreversible behavior when the underlying microscopic laws are reversible.

The first person to make significant progress in understanding the origin of irreversibility was Boltzmann[1]. Boltzmann was able to derive irreversible decay to equilibrium of a dilute gas of hard spheres of mass m given the fact that collisions between particles are elastic and governed by the reversible laws of dynamics. However, Boltzmann had to make the extra assumption that collisions take place at random. He introduced the probability density $f(\vec{p}_1,\vec{r},t)$ where $f(\vec{p}_1,\vec{r},t)d\vec{p}_1 d\vec{r}$ is the probability of finding a particle in volume element $\vec{p}_1 \to \vec{p}_1 + d\vec{p}_1$ and $\vec{r} \to \vec{r} + d\vec{r}$, and obtained the following non-linear equation of motion for $f(\vec{p}_1\vec{r},t)$:

$$\frac{\partial f_1}{\partial t} + \frac{\vec{p}_1}{m} \cdot \frac{\partial f_1}{\partial \vec{r}} = \int d\vec{p}_2 \int d(\cos\theta)\, d\phi\, \frac{g}{m} I(g\theta)\, [f_1 f_2 - f_1' f_2'] \quad (1.1)$$

where $f_1 = f(\vec{p}_1,\vec{r},t)$, $f_1' = f(\vec{p}_1,\vec{r},t)$, etc. Because the gas is dilute, only binary collisions $(\vec{p}_1,\vec{p}_2) \rightleftarrows (\vec{p}_1',\vec{p}_2')$ are considered. $I(g,\theta)$ is the differential scattering cross section and $g = |\vec{v}_1 - \vec{v}_2| = |\vec{v}_1' - \vec{v}_2'|$ is the relative velocity. The assumption of randomness enters in writing the probability densities in product form, $f_1 f_2$ and $f_1' f_2'$. This means that their positions and momenta are completely uncorrelated. The density of particles at a given point $\vec{p}_1$, $\vec{r}$ is simply $N f(\vec{p}_1, r, t)$, where N is the total number of particles.

Boltzmann defined an entropy for the system and using Eq. (I.1) he was able to show that the entropy increased and that the probability density decayed to the Maxwell-Boltzmann distribution

$$f(\vec{p}_1, \vec{r}, t) \rightarrow A\, e^{\frac{p^2}{2mkT}}$$

where A is a normalization constant, k is Boltzmann's constant, and T is the absolute temperature.

Boltzmann's derivation of the irreversible behavior from the reversible laws of dynamics was not rigorous, because he assumed randomness in the collisions between particles. It was the assumption of randomness that gave him his result. Although Boltzmann's derivation was not rigorous, he was the founder of a subject which does attempt to obtain irreversible behavior from the reversible laws of dynamics with no extra assumptions - namely ergodic theory. During the past twenty years much progress has been made in this direction, some of which has relevance for Celestial Mechanics.

Ergodic theory focuses on the behavior of the phase space of a given system as it evolves according to the reversible laws of dynamics. If the flow of points in phase space is such that almost all points flow through every finite region (on a given energy surface) of phase space, then the flow is ergodic. However, ergodic flow may be either smooth (like laminar flow) or it may be 'turbulent'. If the flow is 'turbulent' then it is called mixing flow. A system with mixing flow can exhibit random behavior in a coarse grained sense, and reduced distribution functions can be defined which do decay to equilibrium irreversibly. If a system has mixing flow, then small uncertainties in the initial state of a system can be amplified to complete uncertainty.

In the following sections we shall consider primarily systems with two degrees of freedom because even in these simple systems one begins to see random behavior occuring. We shall always attempt to illustrate ideas with simple examples, rather than become bogged down in complicated mathematics. Our list of references is representative of work done but by no means complete. It serves only as a starting point for those who wish to explore further some of the ideas discussed here.

2. CONSERVATIVE CLASSICAL SYSTEMS

Let us consider a closed classical system with N degrees of freedom. Such a system is completely specified in terms of a set of 2N independent real variables (p^N, q^N) (p^N and q^N denote the sets $p^N = \{p_1, p_2, \ldots p_N\}$ and $q^N = \{q_1, q_2, \ldots q_N\}$, respectively). If the state of the system, $X^N(p^N q^N)$ is known at one time then it is completely determined for any other time. If we can define a Hamiltonian, $H(p^N, q^N, t)$, then the time evolution of the system is given by the set of equations

$$\dot{p}_\ell \equiv \frac{dp_\ell}{dt} = -\frac{\partial H}{\partial q_\ell} \quad \text{and} \quad \dot{q}_\ell \equiv \frac{dq_\ell}{dt} = \frac{\partial H}{\partial p_\ell} \quad (\ell = 1, 2, \ldots N) \tag{2.1}$$

If the Hamiltonian does not depend explicitly on time, then it is a constant of the motion

$$H(p^N, q^N) = E \quad , \tag{2.2}$$

where E is the total energy of the system, and the system is conservative.

Let us now associate to the system a 2N dimensional phase space, Γ . The state of the system, $X(p^N, q^N, t)$, then corresponds to a point in the phase space. As the system evolves in time and its state changes, the system point $X(p^N, q^N, t)$, traces out a trajectory in Γ-space (cf. Fig. I). If there are m constants of motion for the system, the trajectory will be confined to a $2N-m$ dimensional surface in Γ-space corresponding to the intersection of the $2N-1$ dimensional surfaces, S_m, defined by the m constants of motion.

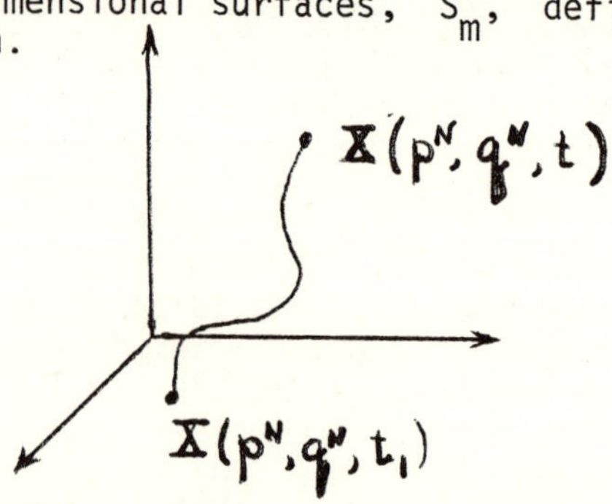

Fig. 1. The state of a system traces out a trajectory in Γ-space as the system evolves in time.

Let us now restrict ourselves to a 2-dimensional phase space. If there is an uncertainty in the initial state of the system, then all we can say is that there is a certain <u>probability</u> that the system point will lie in a given region of phase space. It is therefore convenient to define a probability density, $\rho(p,q)$, where $\rho(p,q)dpdq$ is the probability of finding the point in a volume element of phase space $p \to p+dp$, $q \to q+dq$. Since the system point must lie somewhere in phase space, we have the normalization condition

$$\int_{\Gamma} \rho(pq)\, dpdq = 1 \tag{2.3}$$

where the integration is taken over the entire Γ-space. If we want the probability of finding the system point in a small finite region R of Γ-space then we simply integrate over that region

$$\text{Probability of finding system point in R} = \int_{R} \rho(pq)dpdq \tag{2.4}$$

If at some time there is a small uncertainty in the state of the system, then the probability density will be sharply peaked in some region of phase space and very small in all other regions.

We now wish to know how the probability density evolves in time for a system governed by Hamilton's equations of motion (cf. Eqs. 2.1). This is a problem which is commonly treated in statistical mechanics. To find the equation of motion of the probability density, it is necessary to introduce the concept of an ensemble of η (η is an integer) classical systems, each governed by the same Hamiltonian, but in different dynamical states $X_i(p,q)$ $(i=1,2,\ldots\eta)$. These systems will be represented by η points scattered through out Γ-space. If we let $\eta \to \infty$, these points will form a fluid in Γ-space. If in a given region of Γ-space, the fluid is very dense (many points per unit volume) are in dynamical states coresponding to that region. Alternatively, we can normalize the fluid density to one, and then let the normalized fluid density, $\rho(p, q, t)$, represent the probability density in Γ-space.

If we assume that the total number of systems in our ensemble does not change in time (system points in phase space can't be created or destroyed), then the probability density must satisfy the following conservation equation:

$$\frac{\partial \rho}{\partial t} = - \vec{\nabla}_X \cdot (\rho\, \dot{\vec{X}}) \tag{2.5}$$

where

$$\dot{\vec{X}} = (\dot{p}, \dot{q}) \text{ and } \vec{\nabla}_X = \left(\frac{\partial}{\partial p}, \frac{\partial}{\partial q}\right).$$

We next note that for a system governed by Hamilton's equations, the velocity $\dot{\vec{X}}$ has zero divergence

$$\vec{\nabla}_X \cdot \dot{\vec{X}} = \left(\frac{\partial \dot{p}}{\partial p} + \frac{\partial \dot{q}}{\partial q} \right) = 0 \tag{2.6}$$

(cf. Eq. 2.1). Therefore, Eq. (2.5) can be written

$$\frac{\partial \rho}{\partial t} = - \dot{\vec{X}} \cdot \vec{\nabla}_X \rho = - i L \rho \tag{2.7}$$

where the Liouville operator, L, is defined

$$L = i \left[\frac{\partial H}{\partial q} \frac{\partial}{\partial p} - \frac{\partial H}{\partial p} \frac{\partial}{dq} \right] \tag{2.8}$$

Eq. (2.7) is called Liouville's equation. It is an equation describing the change in the probability density at a given point in Γ-space. If we now look at the total derivative of the probability density, we can use Eq. (2.7) to obtain

$$\frac{d\rho}{dt} = \frac{\partial \rho}{\partial t} + \dot{\vec{X}} \cdot \vec{\nabla}_X \rho = 0 \tag{2.9}$$

Eq. (2.9) is very important because it says that the probability density does not change in the neighborhood of a moving system point.

We may solve the Liouville equation and write the probability density at a given point (p_o, q_o) at time t in terms of its value at the same point at time $t = 0$.

$$\rho(p_o, q_o, t) = U_t \rho(p_o, q_o, 0) = \rho(p_{-t}, q_{-t}, 0) \tag{2.10}$$

where

$$p_{-t} = U_{+t}\, p_o, \quad q_{-t} = U_{-t}\, q_o \tag{2.11}$$

and

$$U_t = e^{-iLt} \tag{2.12}$$

The probability density assigns a weight to the points in phase space. The measure $d\mu_o$ of a set of points in the regions $p_o \to p_o + dp_o$ and $q_o \to q_o + dq_o$ at time $t=0$ is defined

$$d\mu_o = \rho(p_o, q_o, 0)\, dp_o\, dq_o \tag{2.13}$$

The measure of the same set of points at time t is given by

$$d\mu_t = \rho(p_t, q_t, t)\, dp_t\, dq_t \tag{2.14}$$

From Eqs. (2.9) and 2.10) we see that $\rho(p_o, q_o, 0) = \rho(p_t, q_t, t)$

(the probability density is an integral of the motion). The area element $dp_o dq_o$ at time $t=0$ may be written in terms of that at time t, $dp_t dq_t$ by means of the Jacobian $J(t,t_o)$ of the transformation from p_o, q_o to p_t, q_t:

$$dp_t dq_t = J(t,t_o)\, dp_o dq_o \tag{2.15}$$

It can be shown (cf. Refs. 2-4) that for a Hamiltonian system, the Jacobian is one

$$J(t,t_o) = 1 \tag{2.16}$$

Therefore, the measure of a set of points in phase space is preserved in time

$$d\mu_t = d\mu_o \tag{2.17}$$

From the above discussion we conclude that the probability density flows in phase space like an incompressible fluid. A given set of points in phase space does not change its measure although the shape of the region it occupies may change drastically, as we shall see.

3. ERGODICITY

When the energy E is a constant of the motion, trajectories in Γ-space are restricted to the energy surface S_E (provided there are no other constants of the motion). A system is ergodic if the point $X^N(p^N,q^N,t)$ representing the dynamical state of the system, passes through every finite region R_E on S_E, for almost all dynamical states. (The point $X^N(p^N,q^N,t)$ cannot pass through every point because a line cannot fill a surface with dimension 2). If the system is ergodic then the fraction of time it spends in the region R_E will be equal to the fraction of S_E occupied by R_E. A system can only be ergodic on the energy surface S_E, if there are no other isolating constants of the motion besides the energy: i.e., if the system is metrically transitive.

The essential part of the ergodic theorem was proved by Birkhoff. Let us consider an integrable dynamical function $f(X^N)$. We may define a phase average of the function $f(X^N)$ over the energy surface

$$\langle f \rangle_{S_E} \equiv \int_{S_E} f(X^N)\, d^N\Omega_E \tag{3.1}$$

and a time average of $f(X^N)$

$$ {}_T \equiv \lim_{T\to\infty} \frac{1}{T} \int_{t_0}^{t_0+T} f(X^N(t))\, dt \tag{3.2}$$

for all trajectories $X(t)$ for which the time average exists. In Eq. (3.1), the area element $d^N\Omega_E$ is defined

$$ d^N\Omega_E = \frac{dS_E}{|\vec{\nabla}_X H|_{H=E}} \tag{3.3}$$

where dS_E is a Cartesian area element on the surface and $|\vec{\nabla}_X H|_{H=E}$ is the length of the gradient of the Hamiltonian on the energy surface. The area of the points in $d^N\Omega_E$ remains invariant as the system evolves in time.[3] Birkhoff[3] showed that the time average in Eq. (3.2) exists for all integrable dynamical functions of physical interest.

In terms of the above averages, the ergodic theorem may be stated as follows:

A system if ergodic if for all physically interesting dynamical functions $f(X^N)$

(i) The time average f_T exists for almost X^N (all but a set of measure zero)

(ii) when it exists it is equal to the phase average $f_T = f_{S_E}$.

Ergodicity is important for equilibrium statistical mechanics because it can be shown that whenever a system is ergodic, the only ensemble with a probability density which does not vary in time is the microcanonical ensemble[8,9]:

$$ \rho(X^N) = C, \tag{3.4}$$

where C is a constant that can be determined from the normalization condition (cf. 2.3)). Eq. (3.4) is the cornerstone of equilibrium ensemble theory and says that all dynamical states on the energy surface are equally probable.

We can give a simple example[6,9] of time evolution which is ergodic if we consider as our phase space the two dimensional unit square $0<p<1$ and $0<q<1$. We shall assume that the equations of motion are given by

$$\dot{p} = \alpha$$
$$\dot{q} = 1 \qquad (3.5)$$

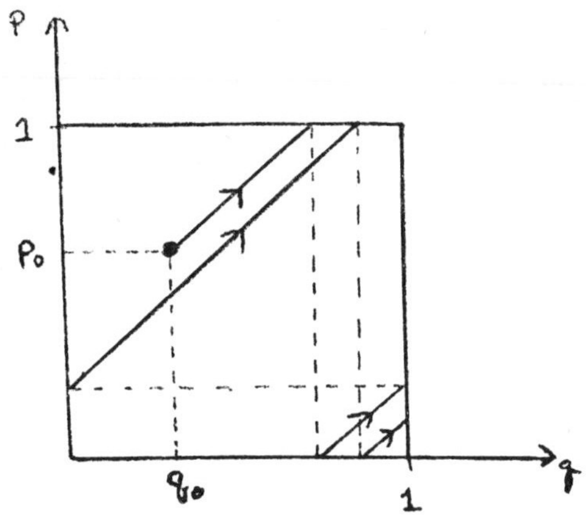

Fig. 2: For α irrational, the trajectory is dense on the unit square.

and we shall impose periodic boundary conditions on the system. The equations of motion are easily solved to give

$$\begin{aligned} p(t) &= p_0 + \alpha t \\ q(t) &= q_0 + t \end{aligned} \quad (\text{mod } 1) \qquad (3.6)$$

The phase trajectory is given by

$$p = p_0 + \alpha(q - q_0) \qquad (3.7)$$

If α is a rational number, $\alpha = \frac{m}{n}$, then the trajectory will be periodic and will repeat itself after a period $t=n$. If α is irrational then the trajectory will be dense on the unit square (but won't fill it). When α is irrational, the system is ergodic (cf. Fig. 2). Let us show this explicitly. Since the phase space is periodic, any integrable function $f(p,q)$ can be expanded in a Fourier series

$$f(p,q) = \sum_{\ell,m=-\infty}^{\infty} A_{\ell m}\, e^{2\pi i(\ell q+mp)} \qquad (3.8)$$

We wish to show that the time average and phase average of $f(p,q)$ are equal for α irrational. The time average is given by

$$f_T = \lim_{T\to\infty} \frac{1}{T} \int_{t_o}^{t_o+T} \sum_{\ell,m} A_{\ell,m} \, e^{2\pi i[\ell(q_o+t)+m(p_o+\alpha t)]} dt$$

$$= A_{o,o} + \lim_{T\to\infty} \frac{1}{T} \sum_{\ell,m\neq 0} A_{\ell,m} \, e^{2\pi i[\ell(q_o+t_o)+m(p_o+\alpha t_o)]}$$

$$\times \frac{(e^{2\pi i(\ell+ m)T}-1)}{2\pi i\ (\ell+\alpha m)}$$

For irrational α, the denominator can never equal zero. Therefore

$$f_T = A_{o,o} \tag{3.9}$$

Similarly, we can show that

$$f_S = \int_0^1 \int_0^1 dpdq\ f(p,q) = Ao,o. \tag{3.10}$$

Hence the system is ergodic. (Note that $dxdy = dx_o dy_o$ so area is preserved).

While ergodicity is an important property of a system in equilibrium, it is not sufficient to cause a system to approach equilibrium. To see this, let us consider a probability density which is not constant over the unit square. For the sake of definiteness assume that at time $t=0$, the probability density is given by

$$\rho(p_o,q_o,0) = \sin\ (\pi p_o)\ \sin\ (\pi q_o)$$

then at time t it will be given by

$$\rho(p_o,q_o,t) = \sin[\pi(p_o-\alpha t)]\ \sin\ [\pi(q_o-t)]$$

The probability is not changed in shape, it is only displaced (cf. Fig. 3)

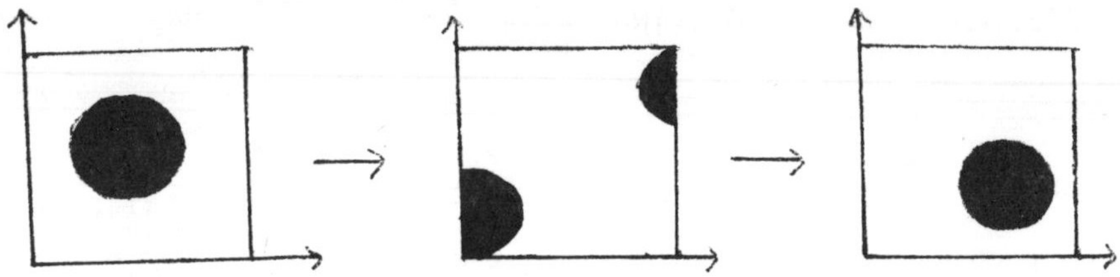

Fig. 3: Ergodicity is not sufficient to change the shape of the probability density

After an infinite length of time it will have wandered uniformly over the entire energy surface.

4. MIXING

A property of conservative classical systems which begins to exhibit some features of irreversibility is that of mixing. We shall give an example of a mixing system for which one can obtain reduced nonequilibrium probability densities which irreversibly approach the microcanonical ensemble. That is, such systems can exhibit irreversibility in a coarse grained sense.

A sufficient condition for a system to be mixing is that for all square integrable functions $f(x^N)$ and $g(X^N)$ on S_E, one must have[6,8,9-12]

$$\lim_{t\to\pm\infty} \int_{S_E} f(X^N)\, g(U_{-t}X^N)\, d\Omega_E$$

$$= \frac{\int_{S_E} f(X^N) d\Omega_E \int_{S_E} g(X^N) d\Omega_E}{\int_{S_E} d\Omega_E} \tag{4.1}$$

Eq. (4.1) is sufficient to ensure that the average value of a dynamical function $f(X^N)$ will approach its equilibrium value in the limit $t\to\pm\infty$. To see this, let $g(X^N) = \rho(X^N)$, where $\rho(X^N)$ is a non-equilibrium probability density. Then

$$f(t) = \int_{S_E} f(X^N)\rho(U_t X^N)d\Omega_E \xrightarrow[t\to\pm\infty]{} \frac{\int_{S_E} f(X^N)d\Omega_E}{\int_{S_E} d\Omega_E} \tag{4.2}$$

Thus, $\langle f(t)\rangle$ approaches its value in the microcanonical ensemble.

It is important to emphasize that mixing gives a coarse grained and not a fine grained approach to equilibrium. The average of the probability density becomes uniform, but the probability density itself does not. This is a result of the fact that $\frac{d\rho}{dt} = 0$ (the probability density doesn't change in a neighborhood of a moving phase point, although it cah change at a given point in space.

We can visualize this if we consider a beaker containing oil and water. We add the oil and water carefully so that they are initially separated, and we assume that they cannot diffuse into one another. We then stir them together (cf. Fig. 4). The local density and the total volume of oil remain constant, but the oil will get stetched into filaments throughout the water. Therefore, on the average the density of oil will become uniform throughout the beaker.

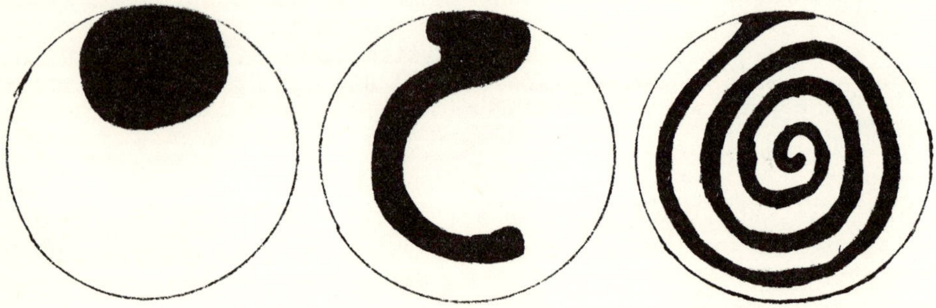

Fig. 4: Stirring oil and water together leads to uniformity on the average.

<u>If we are careful enough</u>, we can also unstir the oil back into its original shape. Therefore, while we get an approach to uniformity, the whole process can be reversed. However, as we shall see, this type of behavior leads to the appearance of random behavior in deterministic systems and coarse grained irreversibility.

The meaning of Eq. (4.1) may therefore be summarized as follows.

Let A and B be two finite arbitrary regions on the surface. S_E. Let us assume that initially all phase points lie in A. If the system is mixing, and we let it evolve in time, then the fraction of points which lie in A or B in the limit $t \to \pm\infty$ will equal the fraction of area S_E occupied by A or B respectively.

A very important example of a classical system which is mixing is the Bernoulli scheme. Before discussing the Bernoulli scheme however, we must generalize the idea of phase space and Riemann integration to the case in which we deal with arbitrary sets. This done with the concept of measure[6, 13, 14]. We will only give a very brief definition of the ideas we will need below.

A non-empty class of subsets a of the set Γ forms a σ-algebra if it contains the set Γ and if it is closed with respect to the operations of (i) forming complements relative to Γ, and (ii) countable unions of subsets.

A measure is a function μ defined on the σ-algebra a. The values of μ are real with range $[0,\infty]$ and are subject to the conditions

a) $\mu(\phi) = 0$, where ϕ is the empty set

b) If the set $\{A_n\}$ is a sequence of pairwise disjoint members of a, then $\mu(\bigcup_{n=1}^{\infty} A_n) = \sum_{n=1}^{\infty} \mu(A_n)$.

c) If $\{B_n\}$ is a sequence of members of a, then one can define a product measure $\mu(\Omega_{n=1}^{\infty} B_n) = \prod_{n=1}^{\infty} \mu(B_n)$.

If a is a σ-algebra in Γ, then Γ is a measure space (Γ, a, μ). The measure preserving group of automorphisms U_t define a measureble flow on (Γ, a, μ). The quantity $\mu(U_t A \cap B)$ is a measurable function of t and $\mu(U_t A) = \mu(A)$.

In terms of measurable functions, the concept of mixing is defined as follows: Given any two measurable subsets A and B of Γ, the group of automorphisms, is mixing if

$$\lim_{t \to \pm\infty} \mu[U_t A \cap B] = \mu(A)\,\mu(B) \tag{4.3}$$

Let us now consider the Bernoulli scheme[(9,15)]. A Bernoulli scheme consists of an 'alphabet' $A = (0,1,\ldots N)$ and a set of doubly infinite sequences of elements

$$S = (\ldots, S_{-2}, S_{-1}, S_0, S_1, S_2, \ldots), \text{ where } (S_k \varepsilon A) \text{ and } (k=0,\pm1,\pm2,\ldots) \tag{4.4}$$

which form the elements of the space Γ . We note that each position, S_k , in the sequence may contain any one of the numbers of the alphabet. Therefore, some sequences will be periodic and others will have members of the alphabet appearing at random in various positions, S_k .

We next define a set of elements $B_k^j = \{S|S_k=j \text{ for } j\varepsilon A\}$. The measure of the set B_k^j is defined to be

$$\mu(B_k^j) = P_j \tag{4.5}$$

and we assume that the measure of the space Γ , which is the disjoint union of the sets B_k^j is normalized to one so

$$\mu(\Gamma) = \sum_{j=0}^{N} \mu(B_k^j) = \sum_{j=0}^{N} p_j = 1. \tag{4.6}$$

If $B_{k_1}^{j_1}, \ldots, B_{k_\ell}^{j_\ell}$ and ℓ distinct sets, $k_1 \neq k_2 \div k_\ell$, the measure of their intersection is the product of their measures

$$\mu\{S|S_{k_1} = j_1, \ S_{k_2} = j_2, \ldots, S_{k_\ell} = j_\ell\}$$

$$= \mu\{B_{k_1}^{j_1} \cap B_{k_2}^{j_2} \cap \ldots \cap B_{k_\ell}^{j_\ell}\} = P_{j_1} P_{j_2} \ldots P_{j_\ell} \tag{4.7}$$

The group of automorphisms, U, which define a flow on Γ is the shift

$$UB_k^j = B_{k+1}^j \tag{4.8}$$

The shift is measure-preserving

$$\mu(UB_k^j) = \mu(B_{k+1}^j) = p_j = \mu(B_k^j). \tag{4.9}$$

The above scheme is called a Bernoulli scheme, and is denoted $B(p_0, p_1, \ldots P_N)$.

We can see more clearly that the Bernoulli scheme exhibits the type of behavior discussed in our oil and water example if we map it into a more familiar phase space. We shall restrict ourselves to one of the simplest Bernoulli schemes, that of the Baker's transformation $B(\frac{1}{2}, \frac{1}{2})$.

For the Baker's transformation[7,8,14] the 'alphabet' consists on only two numbers $A = \{0,1\}$ and the measure is defined so that $p_0=p_1=\frac{1}{2}$. We can map each element $S\varepsilon\Gamma$ into the unit square if we assign to S two numbers

$$p = \sum_{k=0}^{-\infty} S_k 2^{k-1} \quad \text{and} \quad q = \sum_{k=1}^{\infty} S_k 2^{-k} \quad . \tag{4.10}$$

We note that if all $S_k=0$ then $p=q=0$ and if all $S_k=1$ then $p=q=1$ since

$$\sum_{k=1}^{\infty} \frac{1}{2}^k = \frac{\frac{1}{2}}{1-1/2} = 1$$

All other cases fall in between. One may verify that the shift acting on the sequence, S, is equivalent to the following transformation on the unit square

$$U^{-1}(p,q) \to \begin{cases} (2p,\frac{1}{2}q) & 0 \le p < \frac{1}{2} \\ (2p-1,\frac{1}{2}q+\frac{1}{2}) & \frac{1}{2} \le p \le 1 \end{cases} \tag{4.11}$$

The inverse transformation is

$$U(p,q) \to \begin{cases} \frac{1}{2}p + \frac{1}{2} , 2q-1) & \frac{1}{2} \le q < \frac{1}{2} . \\ \frac{1}{2}k, 2q) & 0 < q < \frac{1}{2} \end{cases} \tag{4.12}$$

The Jacobian of the transformation is easily shown to be equal to one. Therefore the mapping U is area preserving.

The effect of the Baker's transformation is to stretch an initial area element into filaments throughout the unit square (cf. Fig. 5)

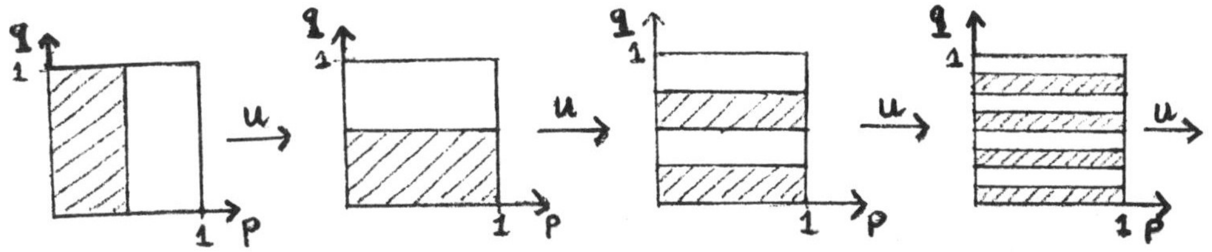

Fig.. 5: Baker's transformation stretches an initial area element into filaments.

much as a baker does in kneading dough. We note that whenever $S_0=0$ the point corresponding to the sequence S will lie to the left of $p = \frac{1}{2}$, and when $S_0 = 1$ it will lie to the right of $p = \frac{1}{2}$. Therefore, points corresponding to sequences with 0 and 1 distributed at random in the positions S_k, will be shifted to the right or left of $p = \frac{1}{2}$ by U at random.

Following Penrose[4], let us now look at how a probability density evolves under the action of the shift U. Let $\rho_0(p,q)$ be the initial probability density and let us assume that it is continuous and smooth. After we allow U to act n times, we obtain

$$\rho_n(p,q) = U^n \rho_0(pq) = \rho_0(U^n p,\ U^n q) \tag{4.13}$$

More explicitly,

$$\rho_{n+1}(p,q) = \begin{cases} \rho_n(\frac{1}{2}p,\ 2q) & 0 \le q \le \frac{1}{2} \\ \rho_n(\frac{1}{2}p + \frac{1}{2},\ 2q-1) & \frac{1}{2} \le q \le 1 \end{cases} \tag{4.14}$$

We do not expect that the probability density $\rho_n(p,q)$ will become uniform as $n \to \pm\infty$. However, a coarse grained probability density will. As we have seen, there is an element of randomness in the position of a point in the p-direction. Let us therefore look at the reduced probability density $\phi(p)$. The quantity $\phi(p)dp$ is the probability of finding a point in the interval $p \to p + dp$. It is defined

$$\phi_n(p) \equiv \int_0^1 dq\ \rho_n(p,q). \tag{4.15}$$

Using Eq. (4.14), $\phi_{n+1}(p)$ becomes

$$\phi_{n+1}(p) = \frac{1}{2}\,\phi_n(\tfrac{1}{2}p) + \frac{1}{2}\,\phi_n\ (\tfrac{1}{2}\,p+1) \tag{4.16}$$

We will now show[16] that $\lim_{n\to\infty} \phi_n(p) = 1$ and therefore that the reduced or coarse grained probability density approaches the microcanonical distribution. If we iterate Eq. (4.16) we obtain

$$\phi_n(p) = \frac{1}{2^n}\sum_{k=0}^{2^n-1} \phi_0\left(\frac{1}{2^n}\,p + \frac{k}{2^n}\right). \tag{4.17}$$

For a continuous and smooth, function $\phi_o(p)$, we obtain

$$\lim_{n\to\infty} \phi_n(p) = \lim_{n\to\infty} \int_0^{1-1/2n} dy\ \phi_o\left(\frac{p}{2n} + y\right) = \int_0^1 dy\ \phi_o(y) = 1 \qquad (4.18)$$

where we have let $y = k2^{-n}$. Notice that for $n\to-\infty$, $\phi_n(p)$ will not approach the microcanonical ensemble. However, a reduced probability density defined in terms of the variable q does. Therefore, the Bakers' transformation exhibits irreversibility in a coarse grained sense.

The mixing condition implies ergodicity[6]. Let A be an invariant set $U_t A=A$. Then

$$U_t A \cap A = A$$

But from the mixing condition

$$\lim_{t\to\infty} \mu(U_t A \cap A) = \mu(A)\mu(A) = \mu(A)$$

which implies that $\mu(A) = 0$ or 1. Therefore, the invariant set must have measure zero or must cover the available phase space. This means that the surface is metrically transitive and therefore ergodic.

Moser[15] discusses two very interesting examples of applications of the Bernoulli shift to problems in celestial mechanics. In one example, he uses the Bernoulli shift to show that, in the case of the restricted three body problem, successive periods of oscillation of the massless particle can take on random values. He also shows that for anharmonic systems, (cf. the lecture notes of G. Contopoulos in this volume) for which the eigencurves of the hyperbolic fixed points join to form homoclinic or heteroclinic points, there are no invariants of the motion in the neighborhood of the homoclinic or heteroclinic points.

5. SPECTRUM OF THE TIME EVOLUTION OPERATOR

The behavior of the probability density, as it evolves in time, is determined by the spectrum of the time evolution operator U_t. For example, a system will be ergodic[9,17] if and only if 1 is a simple eigenvalue of U_t. If it is ergodic then the absolute value of every eigenfunction of U_t is constant, every eigenvalue is simple and the set of all eigenvalues of U_t is a subgroup of the circle group. The minimum condition[9,17] for a system to exhibit mixing behavior is that U_t have a continuous spectrum or a discrete

Lesbesque spectrum (and have a simple eigenvalue of, 1, to ensure a decay to a uniform equilibrium distribution).

A major difficulty with most systems is to actually find the spectrum of their time evolution operators. There are a few systems where this may be done. The time evolution operator of quantum systems, and harmonically coupled classical systems, which have a finite number of degrees of freedom are known to have a discrete spectrum. In both cases in order to obtain irreversible behavior one must go to the thermodynamic limit (let the number of degrees of freedom and size of the system become infinite in such a way that the density remains constant). Then the spectrum becomes continuous and one gets irreversible behavior.

Because the thermodynamic limit plays such an important role in statistical mechanics, we would like to include here a simple example of how irreversible behavior can arise in a non ergodic classical system, if we take the thermodynamic limit.

The type of irreversible behavior that we would like to reproduce with a classical harmonic oscillator system is that of a large mass M, attached to a sping with force constant k and immersed in a pot of oil (cf. Fig. 6).

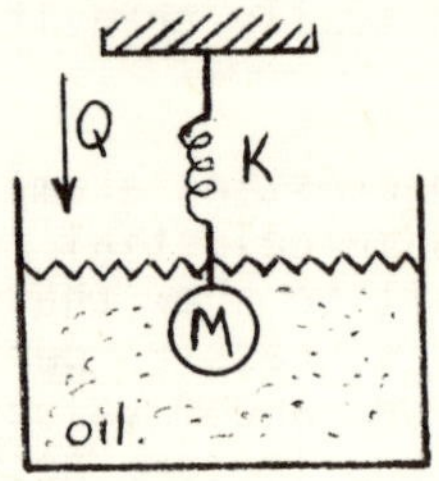

Fig. 6: System with irreversible behavior.

If we displace the mass from its equilibrium position, viscous forces between the mass and the oil will damp the motion of mass and eventually it will come to rest. We can write Newton's law for the mass in the form

$$M \frac{d^2Q}{dt^2} + C \frac{dQ}{dt} + KQ = 0 \tag{5.1}$$

where $C \frac{dQ}{dt}$ is the frictional force between the mass and the oil. The constant C is proportional to the coefficient of viscosity for the oil. If we let $Q(0) = Q_o$ and chose $C^2 = 4KM$, then we get pure exponential decay

$$Q(t) = e^{-\frac{C}{M}t} Q_o \tag{5.2}$$

of the mass to its equilibrium position.

We shall now reproduce this result with a classical harmonically coupled system by taking the thermodynamic limit.[5,18-21] Let us consider a mass M which coupled to a chain of $2N$ lighter masses $m < M$ (cf. Fig. 7).

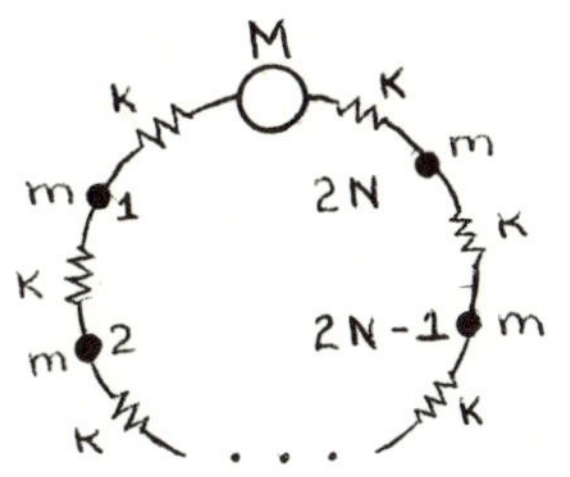

Fig. 7: Heavy mass, M, coupled to chain of lighter masses, m , via harmonic nearest neighbor interactions.

We assume that all masses are coupled to their nearest neighbors by harmonic forces with force constant k. The Hamiltonian for the system is

$$H = \frac{1}{2} \sum_{\ell=0}^{2N} \frac{P_\ell^2}{m_\ell} + \frac{1}{2}k \sum_{\ell=0}^{2N} (q_\ell - q_{\ell+1})^2 . \tag{5.3}$$

The coordinates of the heavy mass are given by p_o, q_o, and $m_o = M$. Those of the ℓth light mass on the chain are given by p_ℓ, q_ℓ and $m_\ell = m$, for $\ell = 1,2,\ldots 2N$.

The Hamiltonian in Eq. (5.3) can be diagonalized, and therefore we can write an exact expression for the position of the large mass as a function of time. It is convenient to diagonalize the Hamiltonian in two steps. We first diagonalize the Hamiltonian for the chain, and then couple the mass M to the normal modes of the chain. We then can easily diagonalize the total Hamiltonian for the system. As a result, the total Hamiltonian may be written in the form

$$H = \frac{1}{2} \sum_{\nu=0}^{2N} (\pi_\nu^2 + \Omega_\nu^2 \eta_\nu^2) \tag{5.4}$$

where π_ν and η_ν are the normal mode coordinates of the entire system and the frequencies Ω_ν are given by solutions of the secular equation

$$(\lambda_o^2-\Omega^2) - \sum_{\ell=1}^{2N} \frac{\varepsilon_\ell^2}{(\lambda_\ell^2-\Omega^2)} = 0 \tag{5.5}$$

In Eq. (5.5) we have

$$\lambda_o^2 = \frac{2k}{M} \tag{5.6}$$

$$\lambda_\ell^2 = \frac{4k}{m} \sin^2 \frac{\ell\pi}{2(2N+1} \tag{5.7}$$

and

$$\varepsilon_\ell = \begin{cases} \frac{-2K}{M^{1/2}m^{1/2}} \sqrt{\frac{2}{2N+1}} \sin \left(\frac{\ell\pi}{2N+1}\right), & \ell \text{ odd} \\ 0 & \ell \text{ even} \end{cases} \tag{5.8}$$

The values of Ω^2 can be found graphically. In figure 8, we have sketched[18,21] $(\lambda_o^2-\Omega^2)$ and $\sum_{\ell=1}^{2N} \frac{\varepsilon_\ell^2}{(\lambda_\ell^2 - \Omega^2)}$ as functions of Ω^2 for $M > m$. The points at which they intersect give us values of Ω_ν for ν odd. For ν even we see from Eq. (5.13) that $\Omega_\nu = \lambda_\ell^\nu$ for $\ell = \nu$ even.

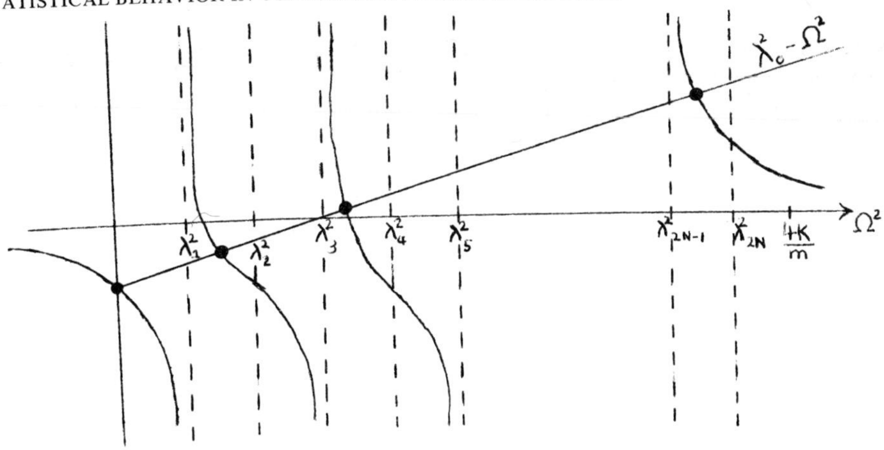

Fig. 8: Plot of $(\lambda_0^2-\Omega^2)$ vs $\sum_{\ell=1}^{2N} \frac{\varepsilon_\ell^2}{(\lambda_\ell^2-\Omega^2)}$ Intersections of the two curves give allowed values of Ω^2 .

The eigenvalues are positive and distinct and the smallest eigenvalue is $\Omega^2=0$ while the largest is bounded by $\Omega^2<\frac{4k}{m}$ when $m < M$.

The equation of motion for $Q(t)$ is found by first finding the equation of motion of $\eta(t)$ and then writing $Q(t)$ in terms of $\eta(t)$. For simplicity let us assume that initially all masses on the chain are at rest and that only the mass M is displaced from its equilibrium position: $P(0) = 0$, $Q(0) = Q_0$, $p_\ell(0)=q_\ell(0)=0$. We then obtain the following exact equation of motion for the position of the heavy mass

$$Q(t) = \sum_{\nu=0}^{2N} \frac{\cos(\Omega_\nu t)\, Q_0}{\left(1 + \sum_{\ell=1}^{2N} \frac{\varepsilon_\ell^2}{(\lambda_\ell^2-\Omega_\nu^2)^2}\right)} \tag{5.9}$$

We note that there is no irreversibility in Eq. (5.20). We could go forward or backward in time and expect to see the same type of behavior. It is a quasi periodic equation for the time evolution of the heavy mass, and describes a situation in which the energy initially given to the heavy mass will spread to the other masses, but can eventually return and be concentrated again in the heavy mass (at least most of it). This is very far from the exponential decay we wish to have. However, as we shall now show, if we take the 'thermodynamic limit', we can obtain the desired expontential behavior. Let us write the secular equation (Eq. 5.5) in the following form

$$G(z) \equiv z - \lambda_0^2 - \sum_{\ell=1}^{2N} \frac{\varepsilon_\ell^2}{(z-\lambda_\ell^2)} = 0 \tag{5.10}$$

Values of z for which $G(z)=0$ give the frequencies Ω_ν^2. We can then write the time evolution $Q(t)$ in terms of a contour integral

$$Q(t) = \sum_{\nu=0}^{2N} \frac{\cos(\Omega_\nu t)\, Q_o}{\left.\frac{dG(z)}{dz}\right|_{z=\Omega_\nu^2}} = \frac{1}{2\pi i} \oint \frac{\cos\sqrt{z}t \; dt}{G(z)} Q_o \tag{5.11}$$

The contour C in Eq. (5.11) surrounds all poles of $G^{-1}(z)$ on the real axis (cf. Fig. 9). The poles are distinct and discrete. In order to have irreversibility the poles must form a cut. This will happen in the 'thermodynamic limit'.

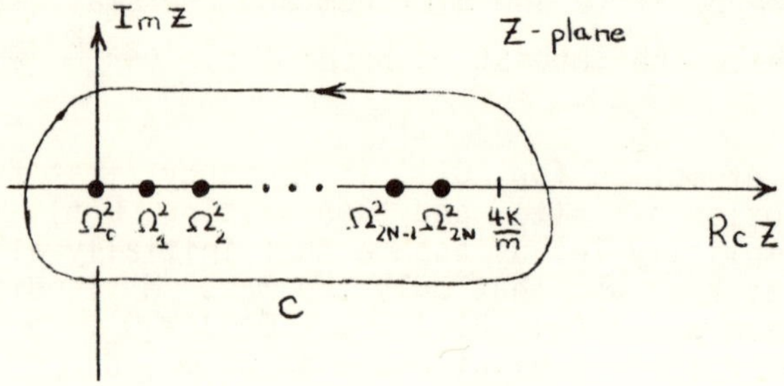

Fig. 9: Poles of $G(z)^{-1}$ lie on the Re z axis

Let us first note that the quantity ε_ℓ is only non-zero for ℓ odd (cf. Eq. (5.8)). We therefore write it in the form

$$\varepsilon(n) = \frac{-2k}{M^{1/2} m^{1/2}} \sqrt{\frac{2}{2N+1}} \sin\left[\frac{(2n-1)\pi}{2N+1}\right], \quad n = 1,2,\ldots N \tag{5.12}$$

We also write λ_ℓ^2 in the form

$$\lambda^2(n) = \frac{4k}{m} \sin^2 \frac{(2n-1)\pi}{2(2N+1)} \qquad n = 1,\ldots N. \tag{5.13}$$

When we take the limit $N \to \infty$, the change in $\varepsilon(n)$ and $\lambda^2(n)$

for each integer change in n becomes infinitesimal and we can replace the summation over n by an integration over λ.

$$G(z) = z - \frac{2k}{M} - \int_0^{\lambda_0} \frac{\varepsilon^2(\lambda)\left(\frac{dn}{d\lambda}\right)d\lambda}{z - \lambda^2} \tag{5.14}$$

We must now find the density of states $\left(\frac{dn}{d\lambda}\right)$ and an expression for $\varepsilon(\lambda)$. The density of states we can find by differentiating Eq. (5.13) with respect to λ. We obtain

$$\left(\frac{dn}{d\lambda}\right) = \frac{(2N+1)}{\pi\left(\frac{4k}{m} - \lambda^2\right)^{1/2}} \tag{5.15}$$

Integration over λ yields the total number of states

$$\int_0^{\sqrt{\frac{4k}{m}}} \left(\frac{dn}{d\lambda}\right)d\lambda = \frac{(2N+1)}{2} \tag{5.16}$$

Furthermore if we combine Eqs. (5.12) and (5.13), we may show that

$$\varepsilon^2(\lambda) = \frac{m}{M}\frac{2}{(2N+1)}\lambda^2\left(\frac{4k}{m} - \lambda^2\right)^{1/2} \tag{5.17}$$

From the secular equation we have the condition

$$\int_0^{\sqrt{\frac{4k}{m}}} \frac{\varepsilon^2(\lambda)}{\lambda^2}\left(\frac{dn}{d\lambda}\right) d\lambda = \frac{2k}{M} \tag{5.18}$$

since $\Omega=0$ is an allowed frequency. Using Eqs. (5.16) and (5.17) we can easily show that Eq. (5.18) is verified.

Let us now use the above results to obtain an expression for $G(z)$. If we substitute Eq. (5.18) into Eq. (5.14) and rearrange terms, we obtain

$$G(z) = z\left[1 - \frac{2m}{M}\int_0^{\sqrt{\frac{4k}{m}}} d\lambda \frac{\left(\frac{4k}{m} - \lambda^2\right)^{1/2}}{z - \lambda^2}\right] \tag{5.19}$$

We will now consider the limit $k \to \infty$, $M \to \infty$ in such a way that $\frac{k^{1/2}}{M} \equiv \alpha$ = constant. (Our expressions are still exact. We are only considering various limiting cases). We then obtain

$$G(z) = z \left(1 - \frac{4\alpha}{m^{1/2}\pi} \int_0^\infty d\lambda \, \frac{1}{z-\lambda^2}\right) \tag{5.20}$$

Let us now look at $Q(t)$. In the limit $N \to \infty$, the poles of $G^{-1}(z)$ form a cut from $z=0$ to $z = \frac{4k}{m}$. In the limit $k \to \infty$ the cut is extended along the entire real axis. To evaluate $Q(t)$ we first contract the contour along the cut. We then obtain

$$Q(t) = \frac{1}{2\pi i} \int_0^\infty \left\{ \frac{1}{G(x)} - \frac{1}{G_t(x)} \right\} \cos \sqrt{x}\, t \, dx \tag{5.21}$$

where we have let $z_\pm = x \pm i\varepsilon$ and

$$G_\pm(x) = \lim_{\varepsilon \to 0} G(x \pm i\varepsilon) = x \left(1 + \frac{4\alpha}{m^{1/2}\pi} \, p \int_0^\infty d\lambda \left(\frac{1}{\lambda^2 - x}\right) \pm \frac{2\alpha i}{m^{1/2}\sqrt{x}} \right) \tag{5.22}$$

The principle part which appears in Eq. (5.33) is identically zero as one can show by direct calculation. Therefore, we obtain

$$G_\pm(x) = x \left(1 \pm \frac{2\alpha i}{\sqrt{x}\, m^{1/2}}\right). \tag{5.23}$$

We can calculate $Q(t)$:

$$Q(t) = \frac{1}{2\pi i} \int_0^\infty dx \left[\frac{1}{1 - \frac{2\alpha i}{m^{1/2}\sqrt{x}}} - \frac{1}{1 + \frac{2\alpha i}{m^{1/2}\sqrt{x}}} \right] \frac{\cos\sqrt{x}\, t}{x} \, Q_o$$

$$= \frac{1}{2\pi i} \int_0^\infty dx \left[\frac{\frac{i\, 4\alpha}{m^{1/2}\sqrt{x}}}{1 + \frac{4\alpha^2}{xm}} \right] \frac{\cos\sqrt{x}\, t}{x} \, Q_o$$

It is convenient to set $S = \sqrt{x}$. Then

$$Q(t) = \frac{4\alpha}{m^{1/2}\pi} \int_0^\infty \frac{\cos st \, ds}{s^2 + 4\frac{\alpha^2}{m}} \, Q_o \tag{5.24}$$

Equ. (5.36) may be evaluated simply to obtain:

$$Q(t) = e^{\frac{-4\alpha t}{m^{1/2}}} Q_0 \tag{5.25}$$

for $t > 0$.

We therefore obtain simple exponential decay for the large oscillator. This is clearly an irreversible equation for $Q(t)$.

6. ANHARMONIC OSCILLATOR SYSTEMS

In Section 5, we found that in order to obtain irreversible behavior for a harmonic oscillator system, it was necessary to let the number of degrees of freedom become infinite so we could obtain a continuous spectrum. Anharmonic oscillator systems usually can't be diagonalized and therefore we can't say much about their spectrum However, the unperturbed frequencies of anharmonic systems are continuous and therefore weakly coupled anharmonic systems with a few degrees of freedom can exhibit a rich variety of stable and unstable (mixing) behavior.

Anharmonic oscillator systems were discussed in great detail by Contopoulos in his lectures at this conference (see also Ref. 22) and therefore we will only discuss them briefly here.

The study of anharmonic oscillator systems has long been important in statistical mechanics in connection with the theory of heat transport in solids. One would like to know the mechanism for heat conduction in finite solids. It was recognized very early[5] that for a finite harmonic lattice there is no mechanism for heat conduction. If heat is added to the lattice and divided unequally between the normal modes, there is no way for the system to reach equilibrium because the normal modes do not interact. However, if slight anharmonicities exist in the lattice, it was expected that equipartition of energy would occur and the system would this reach equilibuium.

In 1955, Fermi, Pasta, and Ulam[23] conducted a computer experiment intending to show this. They studied a system of 64 oscillators with cubic and broken linear coupling. They found that when energy was added to a few of the lower modes there was <u>no</u> tendency for the energy to spread to the other modes. This result of Fermi, Pasta, and Ulam is now understood in terms of the K.A.M. theorem (Kolmogorov[24], Arnold[25], Moser[26], and has been studied by a number of authors[27-30]. For a system with weak anharmonic coupling, which satisfies the conditions of the K.A.M. theorem;[30,31] one expects to see stable non-ergodic behavior. Most of the phase space remains filled with invariant tori which are only slight distortions of the invariant tori which existed in the unperturbed system. The anharmonic lattice therefore behaves like a harmonic lattice. As the coupling is increased however, the invariant regions of phase space breakdown and at some point one expects to see a sharp transition[27-30] to ergodic behavior and equipartitions of energy between the modes.

There is now a variety of nonlinear oscillator systems which have been studied and which exhibit a transition from stable to ergodic behavior as certain parameters are changed.[32-34] Applications include plasma physics, celestial mechanics and molecular dissociation.[35,36]

7. CONCLUSION

The relation between irreversible behavior of finite classical systems and the 'infinite' classical systems remains an unsolved problem. Lebowitz[6,12] has shown that transport coefficients (Kubo formulas) do not exist for finite mixing systems.

Finally, let us conclude with a positive note. Ergodic theory may at last have found a means of justifying Boltzmann's assumption of randomness in deriving his kinetic equation. Y. G. Sinai[37] has proved (the proofs have not been published) that N hard spheres $N \geq 2$ in a box is a mixing system.

ACKNOWLEDGEMENTS

The author wishes to thank Professor I. Prigogine for his hospitality in Brussels during the summer of 1975, where much of this manuscript was prepared, and NATO Research Grant 644 for a travel grant.

REFERENCES

1. E. G. D. Cohen and W. Thirring, editors, The Boltzmann Equation (Springer-Verlag, Wien 1973).
2. H. Goldstein, Classical Mechanics, (Addison-Wesley Pub. Co., London, 1950).
3. I. E. Farquhar, Ergodic Theory in Statistical Mechanics, (Interscience Pub., New York, 1964).
4. O. Penrose, Foundations of Statistical Mechanics, (Pergamon Press, Oxford, 1970).
5. H. Wergeland in Irreversibility in the Many-Body Problem, Edited by J. Biel and J. Rae (Plenum Press 1972).

6. I. E. Farquhar in Irreversibility in the Many-Body Problem, Edited by J. Biel and J. Rae (Plenum Press, 1972).
7. N. Van Kampen, Physica 53, 98 (1971).
8. J. L. Lebowitz and O. Penrose, Physics Today, Feb. 1973, p. 23.
9. V. I. Arnold and A. Avez, Ergodic Problems of Classical Mechanics, (W. A. Benjamin, Inc., New York, 1968).
10. P. R. Halmos, Lectures on Ergodic Theory,(Chelsea Pub. Co., New York, 1956).
11. A. S. Wightman in Statistical Mechanics at the Turn of the Decade, Edited by E. G. D. Cohen (North Holland, Amsterdam, 1908).
12. J. L. Lebowitz in Lecture Notes in Mathematics 322: Nonlinear Problems in the Physical Sciences and Biology, Edited by I. Stakgold (Springer-Verlag, Berlin, 1973).
13. A. E. Taylor, Introduction to Functional Analysis (John Wiley and Sons, New York, 1958).
14. W. Rudin, Real and Complex Analysis, (McGraw-Hill, Inc., New York, 1966).
15. J. Moser, Stable and Random Motions in Dynamical Systems, (Princeton University Press, Princeton, N. J., 1973).
16. This proof is due to Professor M. Kac - private communication.
17. E. Hopf, J. Math. Phys. 13, 51, (1934).
18. P. Ulhersma, Physica 32, 27 (1966); 32, 74 (1966).
19. P. Mazur and E. Braun, Physica 30, 1973 (1964).
20. G. W. Ford, M. Kac, and P. Mazur, J. Math. Phys. 6, 504 (1965).
21. R. I. Cukier and P. Mazur, Physica 53, 157, (1971).
22. G. Contopoulos in Dynamical Structure and Evolution of Stellar Systems, Swiss Society of Astronomy and Astrophysics - 3rd Advanced Course, (Geneva Observatory, CH - 1290, Sauverny, 1973).
23. E. Fermi: Collected Papers, Vol. II (Univ. of Chicago Press, Chicago, 1965), p. 978.
24. A. N. Kolmogorov in R. Abraham, Foundations of Mechanics, (W. A. Benjamin, Inc., New York, 1967), Appendix D.
25. V. I. Arnold, Russian Math. Surveys 18, 9, (1963); 18, 85 (1963).
26. J. Moser, Nachr. Akad. Wiss. Göttingen, II Math. Physik Kl. 1 (1962).
27. F. M. Israilev and B. V. Chirikov, Sov. Phys. (Dklady) 11, 30, (1966).
28. N. J. Zabusky and G. J. Deem, J. Computational Phys. 2. 126, (1967).
29. H. Hirooka and N. Saito, J. Phys. Soc. (Japan), 26, 624, (1969).
30. G. H. Walker and J. Ford: Phys. Rev. 188, 416, (1969).
31. J. Ford and G. H. Lunsford, Phys. Rev. A1, 59, (1970).
32. J. Ford in Fundamental Problems in Statistical Mechanics, Vol. 3, Edited by E. D. G. Cohen (North-Holland Pub. Co., Amsterdam, 1975).

33. J. Ford, Adv. Chem. Phys. 24, 155, (1973).
34. G. M. Zaslovskii and B. V. Chirikov, Sov. Phys.(Uspekhi) 14, 549, (1972).
35. E. Thiele and D. J. Wilson, J. Chem. Phys. 35, 1256, (1961).
36. D. Bunker, J. Chem. Phys. 37. 393, (1962).
37. For a discussion of some of the ideas involved see Y. G. Sinai in Ref. 1.

GLOBAL STABILITY OF AREA-PRESERVING MAPPINGS

James H. Bartlett

Department of Physics and Astronomy, The University of Alabama, Tuscaloosa, Alabama.

ABSTRACT

The area-preserving mapping

$$x' = x + a(y - y^3)$$

$$y' = y - a(x' - x'^3)$$

has been examined for long-time stability, with attention concentrated on the region surrounding the origin. For $a = 1.7$, the fixed points of order 21 are somewhat closer to the origin than the fixed points of order 16. Whereas some points on the eigencurves from fixed points of order 16 can readily escape to large distances, those on the eigencurves from fixed points of order 21 can escape with extreme difficulty.

The long-time stability of a Hamiltonian system, for motion in one dimension, is essentially determined by the circumstance that area in the phase plane is preserved. For a harmonic oscillation, the system is linear, the trajectories in the phase plane are ellipses, and the system is stable. One also obtains closed trajectories if the system is non-linear and autonomous. However, if the restoring force is spatially non-linear and periodic in time, one must study the area-preserving mappings through multiples

V. Szebehely and B.D. Tapley (eds.), Long-Time Predictions in Dynamics, 99–110. *All Rights Reserved. Copyright* © 1976 *by D. Reidel Publishing Company, Dordrecht-Holland.*

of the basic period, and the system, as we shall show, will not be completely stable.

The purpose of the present communication is to discuss the escape probability from suitably chosen finite regions and to show how a designer can find regions such that this probability may be made less than some pre-assigned amount. This consideration is entirely independent of the existence or location of invariant curves.

Studies were first made (Bartlett, 1968; Bartlett and Wagner, 1970) of the differential equation

$$\frac{d^2x}{dt^2} + p(t)\, x^3 = 0 \tag{1}$$

where $p(t)$ is a periodic square wave function with period τ and

$$p = p_o \quad \text{for} \quad -\tau/4 < t \leq \tau/4 \ , \quad p_o > 0$$

$$p = -p_o \quad \text{for} \quad \tau/4 < t \leq 3\tau/4 \ .$$

High accuracy, limited only by that of the computer, was possible because we could construct and store a table of elliptic functions.

After one period, i.e. at $t = \tau$, a point $P(x, \dot{x})$ in the phase plane will, if subject to (1), be transformed or mapped into another point $P' = TP$, where T is symbolic for the transformation, which is area-preserving. A fixed point of order n is one such that $T^nP = P$, and the origin is a fixed point for (1). Fixed points of order n will encircle the origin, with elliptic and hyperbolic F.P.'s alternating. If H_1 be a hyperbolic fixed point, then two eigencurves issuing from H_1 will straddle a neighboring elliptic fixed point and each one will intersect a corresponding eigencurve from the next hyperbolic point H_2 (at a point called homoclinic). The intersection angles will be nonzero for the general nonlinear mapping, but can become extremely small as the order of the fixed point is made greater and the origin is approached.

Figure 1 illustrates the situation near the elliptic fixed point E_1 $(N = 11)$, with associated hyperbolic points H_1 and H_2, eigencurves H_1G and $H_2\overline{G}$, and homoclinic points Y and W. Particles can escape from the cell $H_1GY\overline{G}H_2FW\overline{F}H_1$ via two routes: $A \rightarrow B \rightarrow C \rightarrow D$ and $5 \rightarrow 4 \rightarrow 3 \rightarrow 2 \rightarrow 1$. Fixed points for $n = 33$ are denoted by e_1, e_2, e_3, h_1, h_2, and h_3. The intersections of the $n = 33$ eigencurve h_2R with the eigencurve ($n = 11$) bounding regions 7 and 4 are called heteroclinic points. Particles can leave the vicinity of h_2, enter regions 7 and 4, and be conveyed outward to escape.

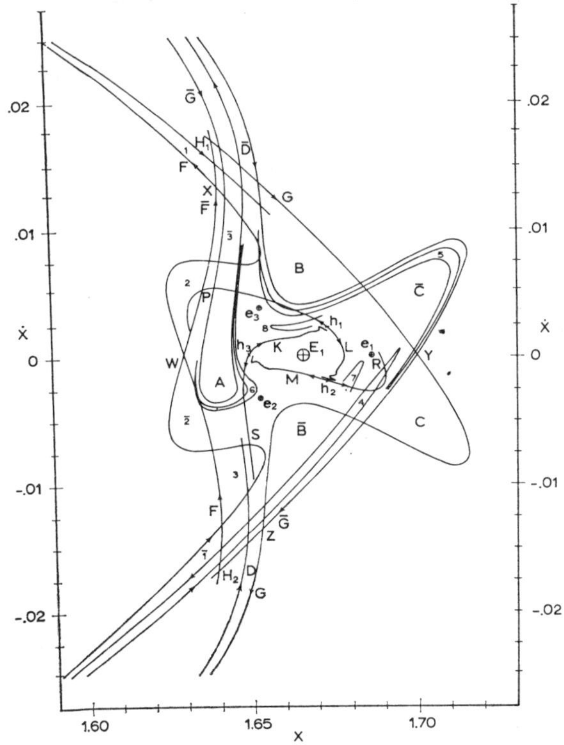

Fig. 1. Hyperbolic invariant curves of the mapping T^{11} and their relations to those of the mapping T^{33}.

The converse process, that of entry of particles from the outside, is shown in Figure 2. Regions $\bar{4}$, $\bar{7}$ . . . are mirror images, in the x-axis, of regions 4, 7 . . ., and the eigencurve $h_1\bar{R}$, along which motion is inward toward h_1, is the mirror image of h_2R (Figure 1). The regions $\bar{4}$, $\bar{7}$, . . . are compressed against the inner eigencurve h_1L. In general, in order that there be motion inward from an inner eigencurve of order n_1 toward an inner eigencurve of higher order $n_2 > n_1$, it is necessary that the outer eigencurve of order n_2 intersect the (extended) inner eigencurve of order n_1. In other language, there must be "interaction of the two resonances n_1 and n_2" as evidenced by the existence of heteroclinic points belonging to the pair (n_1, n_2). If, for a given order n_1, there are no heteroclinic points for

$n_2 > n_1$ which are closer to the origin than the fixed points of order n_1, then entry appears to be impossible, so that an invariant curve should exist. If, however, such heteroclinic points do exist, then these provide the mechanism for entry (or escape). Resolution of this question is necessary to show whether or not there are closed invariant curves, third integrals, and boundaries of eternal stability.

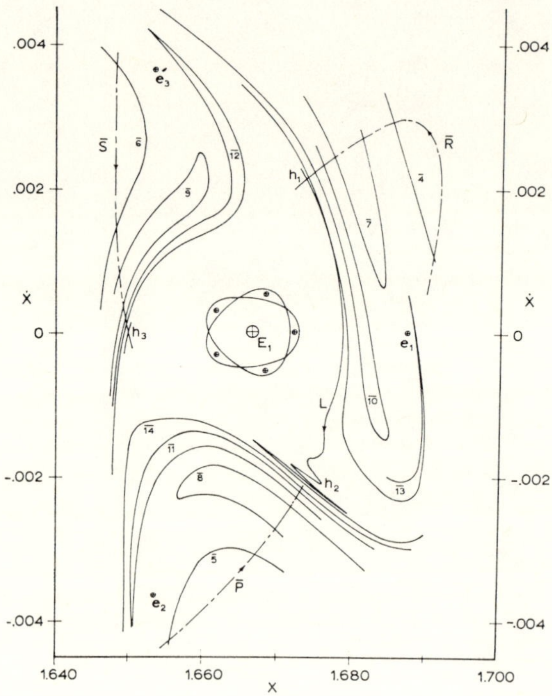

Fig. 2. Hyperbolic invariant curves of the mapping T^{33} and their relations to those of the mappings T^{11} and T^{55}.

Figure 3 shows the inner eigencurves for $n = 33$ and the eigencurves for $n = 55$, with fixed points for the latter denoted by $e(i)$ and $h(i)$. The area of an oscillation loop, a measure of escape or entry probability, was 5×10^{-5} for $n = 11$, and is 1.4×10^{-7} for $n = 33$, and $< 10^{-16}$ for $n = 55$. The intersection angle of inner eigencurves for $n = 55$ at the homoclinic point on the x-axis is vanishingly small, but probably not zero (on the basis of evidence to be displayed below, see Figure 6). No attempt was made to find heteroclinic points for $n_2 > 33$, $n_1 = 33$. The eigencurves for $n = 55$ appear to form one continuous closed curve, to the accuracy of the computations, but this can be deceptive.

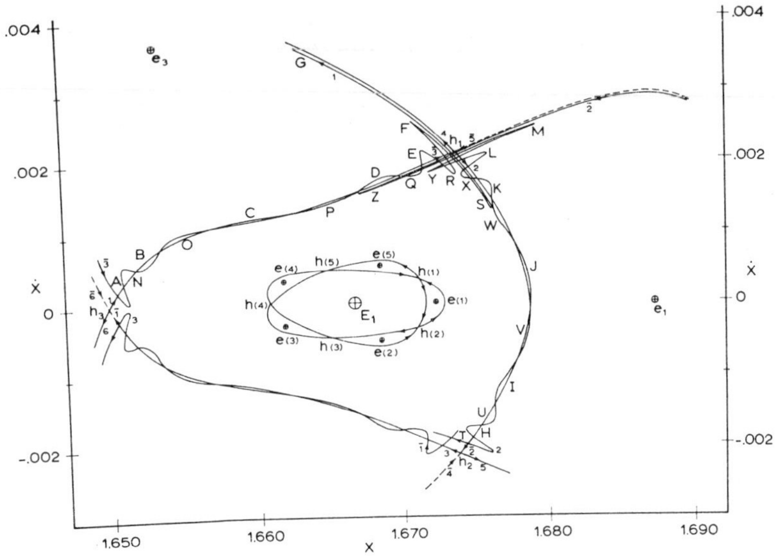

Fig. 3. Hyperbolic invariant curves of the mapping T^{33} and their relations to those of the mapping T^{55}.

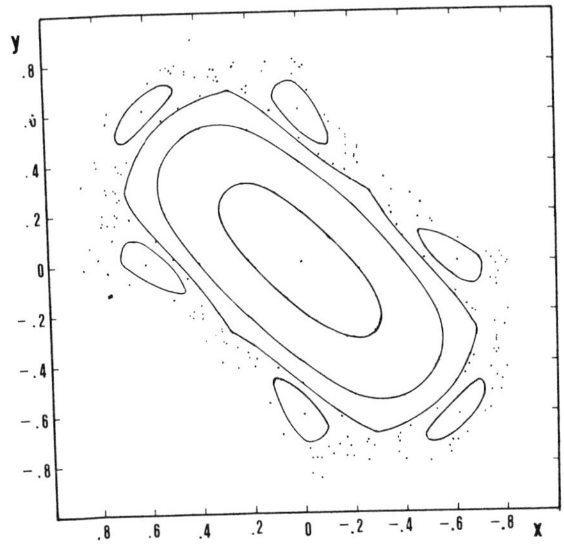

Fig. 4. The iterated mapping, T, for a = 1.6.

Since Equation (1) has limitations because of the use of elliptic functions and the consequent complicated program, and since the results are essentially due to the property of area preservation, Jenkins and Bartlett (1972) investigated the algebraic area-preserving mapping of Hénon and Heiles (1964)

$$x' = x + a(y - y^3)$$
$$y' = y - a(x' - x'^3) \quad (2)$$

where the numerical operations are readily performed. The fixed

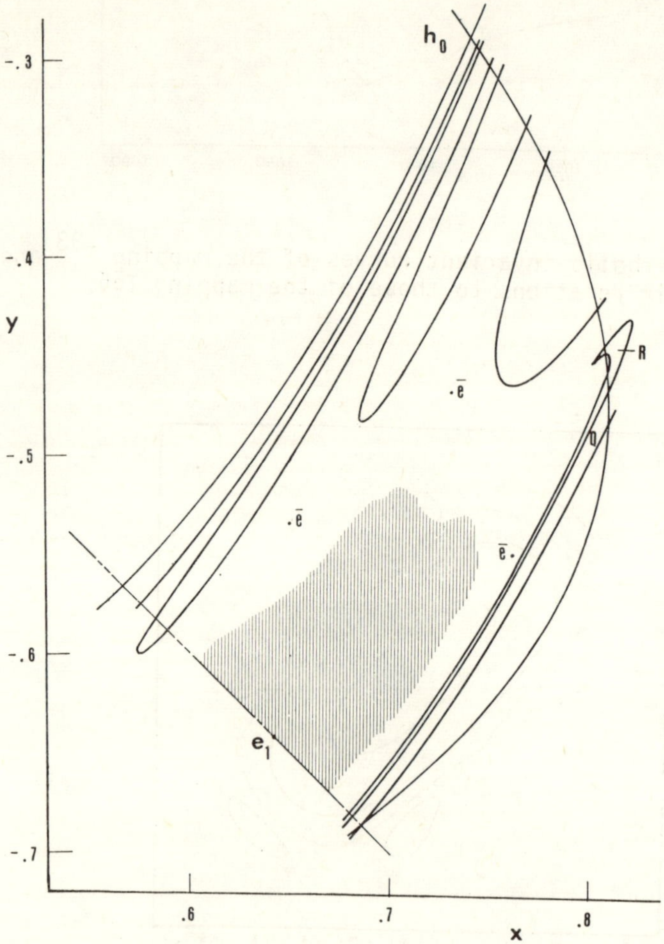

Fig. 5. One further mapping of the eigencurve oscillations of Fig. 4 (points $\bar{e}$ are eighteenth-order satellites of e_1)

points lie on the lines $x = \pm y$, $x = 0$, $y = 0$, or are maps thereof, and are thus easily obtained.

Figure 4 (due to Henon and Heiles) shows apparent closed invariant curves (to the accuracy of the calculations) about the origin as well as closed invariant curves around the elliptic fixed points of order 6, for $a = 1.6$. In addition, the successive mappings of a single point far from the origin are shown as scattered points outside the central closed curve region, and they appear to obey no discernable pattern (due to the oscillations above noted). The hyperbolic fixed points of order 6 lie approximately halfway between the above elliptic fixed points of order 6, but are not shown in this figure.

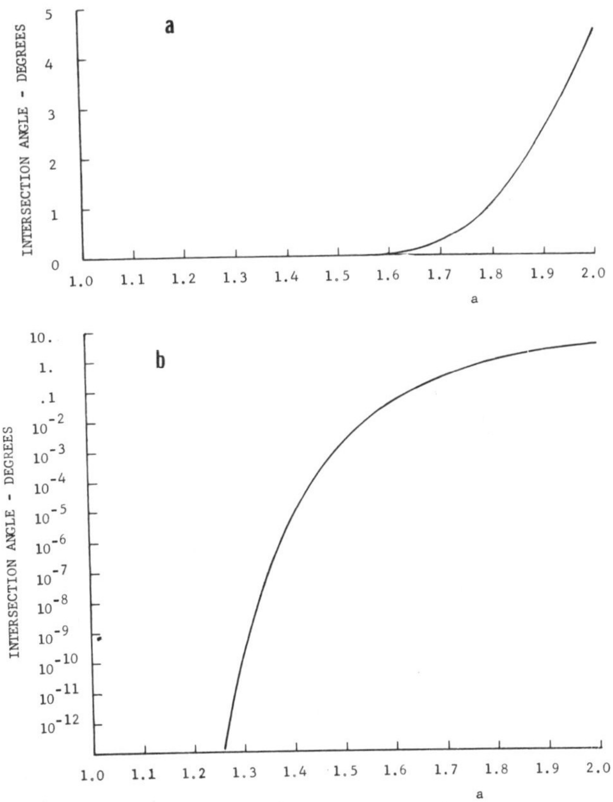

Fig. 6. Intersection angle of the incoming eigencurve h_1 with the outgoing eigencurve of h_0 at the line of symmetry, $x + y = 0$, as a function of the parameter, a.

Figure 5, for a = 1.7 , shows inner and outer eigencurves (n = 6) from the hyperbolic fixed point h_o (x = 0.73512227, y = -0.2871799) as far as the symmetry axis x = -y. The shaded region is a stable one about the elliptic fixed point e_1 (x = 0.64168895). The small oscillation near h_o is of the inner eigencurve issuing from the hyperbolic fixed point h_1 , which is h_o reflected in x = -y . The area of an oscillation loop is of order 10^{-4}.

Figure 6 shows how the intersection angle of the inner eigencurves (between h_o and h_1) at the point x = -y depends on the parameter a . It is seen that there is a very high order of contact with the a-axis, which is made more explicit by the logarithmic plot 6b.

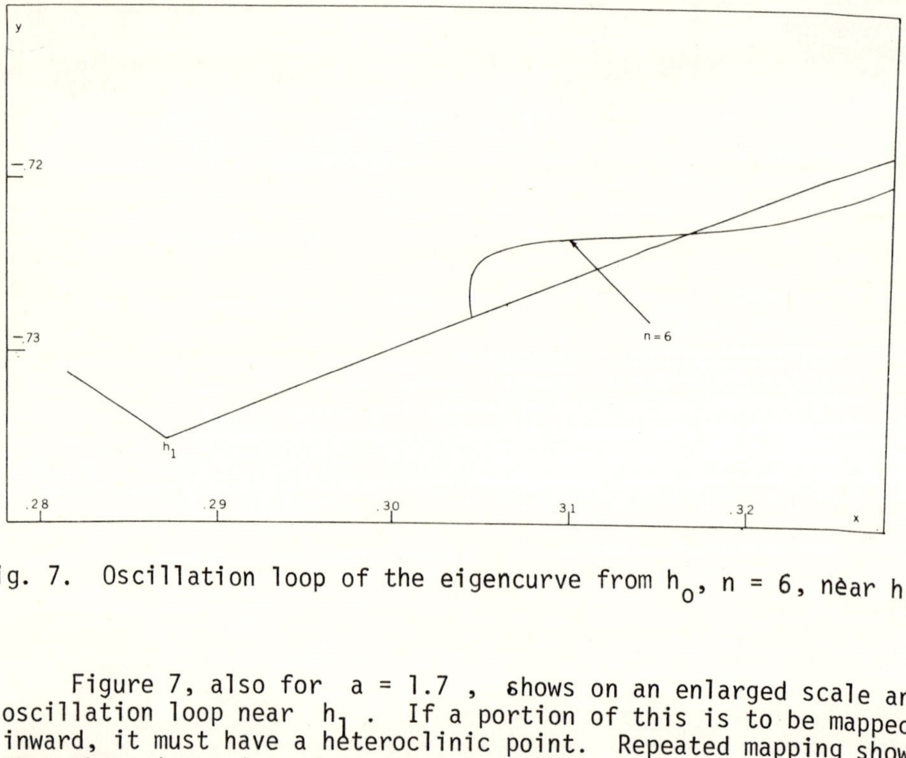

Fig. 7. Oscillation loop of the eigencurve from h_o, n = 6, near h_1.

Figure 7, also for a = 1.7 , shows on an enlarged scale an oscillation loop near h_1 . If a portion of this is to be mapped inward, it must have a heteroclinic point. Repeated mapping shows that there is an inward movement toward the n = 54 eigencurves, and Figure 8 shows the map of a very small part of the above loop under n = 54 . It is labelled "Extended eigencurve for n = 6 ". Figure 8 also shows inner and outer eigencurves for n = 16 , as well as the (outer) eigencurve for n = 21 . To the accuracy of this figure, this eigencurve for n = 21 goes from one hyperbolic point to the next, but this is because the detail is not fine enough (see below).

Figure 9 shows regions which are apparently stable for 10,000 mappings, when a coarse mesh (.01 x .01) is used. One region (one-fourth of the total) extends from the origin almost to h_1 , and another near the elliptic point e_2 is shown. All the mesh points appear to be stable, but when the mesh size is made finer (.0001 x .0001) many points now turn out to be unstable. This is seen in Figure 10 where the points unstable after 10,000 mappings are indicated by unshaded lacunae in the diagram. The upper ones fall

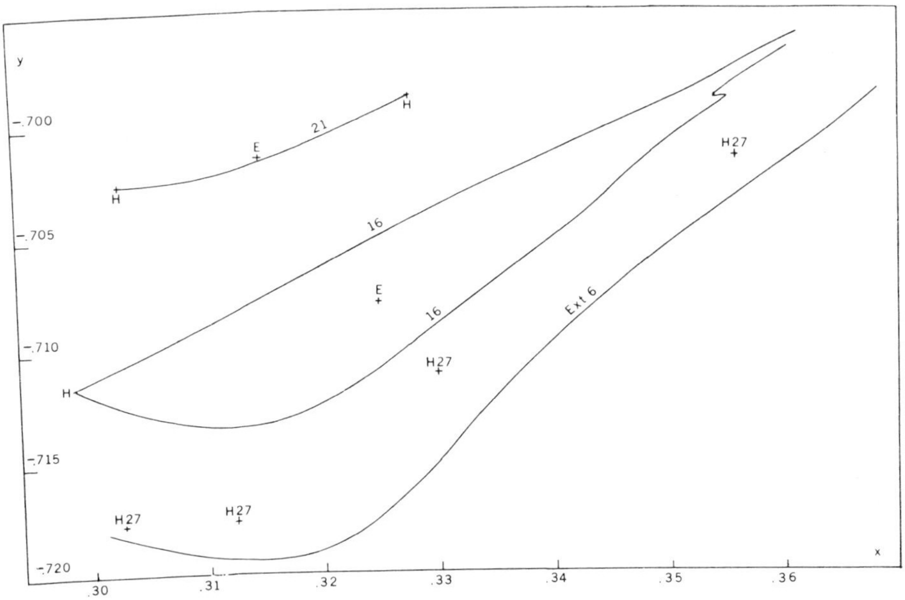

Fig. 8. Extended eigencurve for n=6 (labelled "Ext. 6") and for n=16 and n=21, and hyperbolic fixed points (H 27) and n=27.

close to the inner eigencurve for $n = 16$, and some of the lower ones close to the outer eigencurve for $n = 16$. The instability increases rapidly as one goes outward.

The eigencurves for $n = 21$ intersect at a very small angle, which creates the illusion of a smooth curve from one hyperbolic point to the next. That this is not really so is demonstrated by Figure 11, which shows an oscillation loop between forward and backward eigencurves, plotted on an expended scale. (The first six figures for both ordinate and abscissa have been subtracted off, and the results plotted in units of 10^{-8}.) The area of the outer part of the loop is of order 10^{-14} . For one circuit of the origin, about 50,000 mappings are required, and there are 42 hyperbolic points, so that the escape probability is roughly 42×10^{-14}

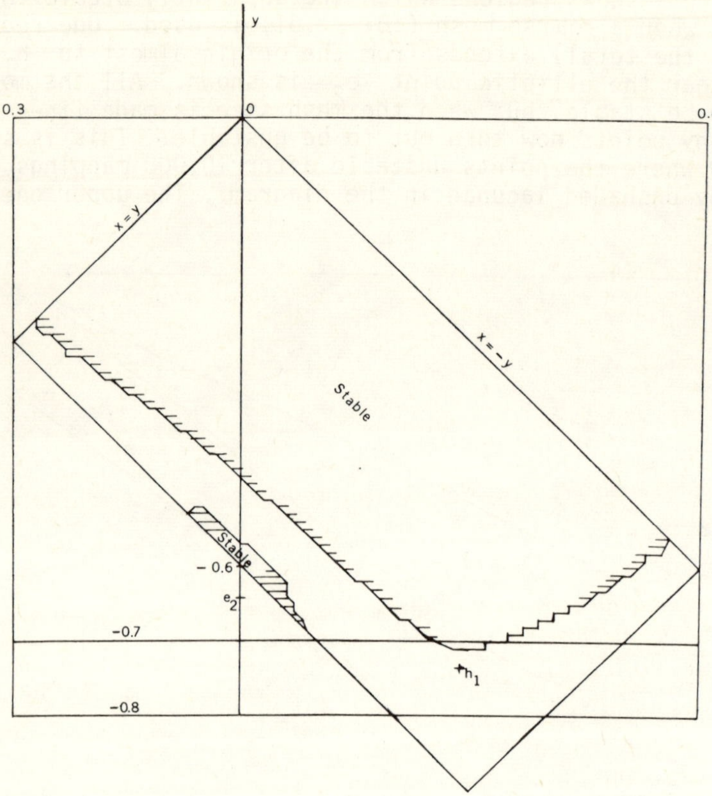

Fig. 9. Stable (shaded) regions, under 10,000 mappings, a = 1.7, with coarse mesh (.01 x .01). This covers one fourth of the domain of Fig. 6.

for 50,000 mappings for particles moving outward from the cell bounded by the n = 21 eigencurves.

Even though we do not as yet have theoretical relations for the dependence of intersection angle or loop area on the order of the fixed point, this is not a handicap for design purposes. We need only specify a maximum escape probability which can be tolerated, and then find a mapping of order high enough so that the calculated loop area multiplied by twice the order is below this maximum value. The escape to which we refer is from the inside of the cell to the outside, but not necessarily to infinity. Complete escape would mean that appropriate heteroclinic points would be present (which would mediate "interaction with resonances of lower

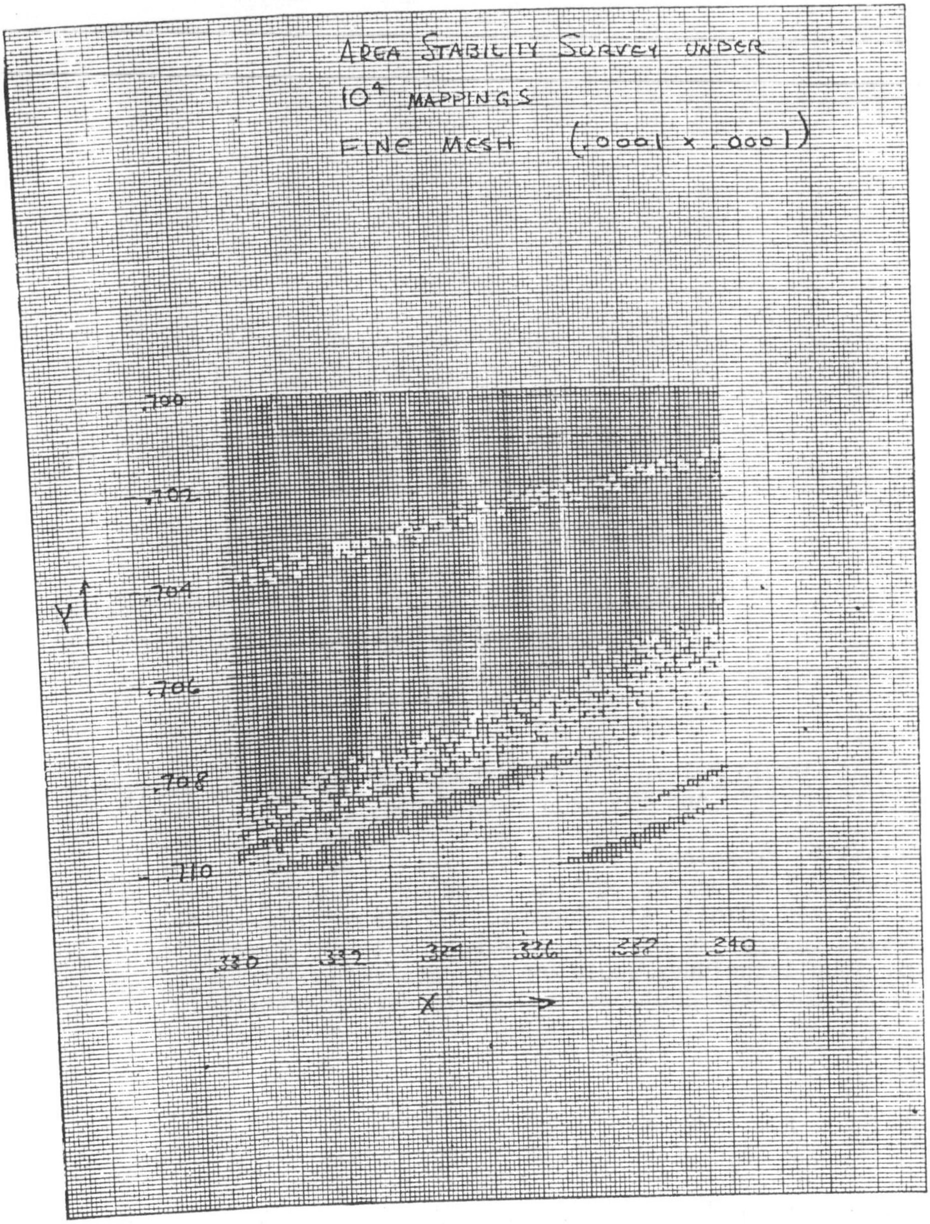

Fig. 10. Stable (shaded) regions for a small part of Fig. 9, under 10,000 mappings and fine mesh size (.0001, .0001). The upper unshaded lacunae fall close to the inner eigencurve for n=16. (a=1.7).

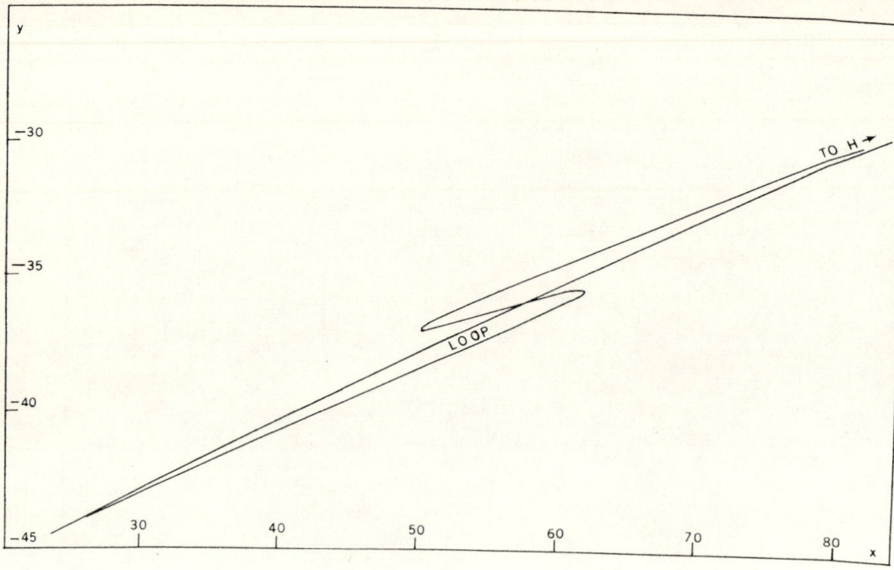

Fig. 11. Oscillation loop for n=21, a=1.7, ordinates and abscissae in units of 10^{-8}, relative to the point x=.327603, y=-.698737. Area of loop is of order 10^{-14}.

order") and would rule out the existence of invariant curves enclosing the origin and the cell under study.

Finally, dynamical calculations should be made, insofar as possible, to the accuracy of the computer, if one is to find accurate fixed points, eigencurves, and escape probabilities. This means that great care is necessary with integration techniques, because such techniques could be accurate enough for orbit determination, but not for the appraisal of long-time stability.

REFERENCES

Bartlett, J. H.: 1968, Mat. Fys. Medd. Dan. Vid. Selsk. 36, No. 11.
Bartlett, J. H. and Wagner, C. A.: 1970, Celest. Mech. 2, 228.
Henon, M. and Heiles, C.: 1964, Astron. J. 69, 73.
Jenkins, B. Z. and Bartlett, J. H.: 1972, Celest. Mech. 5, 407.

ON THE METHOD OF AVERAGING

Urs Kirchgraber

ETH
Zürich, Switzerland

ABSTRACT. In this paper we present several new aspects of the method of averages: first we describe some formal properties of the method, second we apply it in order to reprove Hopf's bifurcation theorem (and obtain a direction of bifurcation formula which is similar to that of Hsü and Kazarinoff), thirdly we offer a theorem concerning error bounds. Since the details will appear elsewhere we only describe the results.

1. FORMAL PROPERTIES OF THE METHOD OF AVERAGES

Many systems of ordinary differential equations describing oscillatory phenomena can be put, after a suitable transformation of the variables, in the following form:

$$\begin{cases} \dot{\phi} = \omega(a) + \varepsilon\, R(\phi,\Omega,a) \\ \dot{\Omega} = \qquad\qquad \varepsilon\, S(\phi,\Omega,a) \\ \dot{a} = \qquad\qquad \varepsilon\, T(\phi,\Omega,a) \end{cases} \tag{1}$$

where R, S, T are 2π -periodic with respect to the components of ϕ and Ω; actually we assume that the Fourier representation of these functions with respect to ϕ and Ω are finite.

Furthermore $\omega(a)$ is assumed to be linearly independent over the integers (i.e. $\sum_{i=1}^{n} m_i\omega_i(a) = 0$ in any open region, $m_i \in Z$ if, and only if $m_1 = m_2 = \ldots. = m_n = 0$, n denotes the dimension of ϕ). Let us introduce the averaged system (with respect to ϕ) corre-

V. Szebehely and B.D. Tapley (eds.), Long-Time Predictions in Dynamics, 111–117. All Rights Reserved.
Copyright © 1976 by D. Reidel Publishing Company, Dordrecht-Holland.

sponding to (1):

$$\begin{cases} \dot{\bar{\phi}} = \omega(\bar{a}) + \varepsilon\, \bar{R}(\bar{\Omega},\bar{a}) & \text{where } \bar{R}(\bar{\Omega},\bar{a}) = \dfrac{1}{(2\pi)^n} \int_0^{2\pi}\!\!\dots\int_0^{2\pi} R(\phi,\bar{\Omega},\bar{a})\,d\phi_1\dots d\phi_n \\ \dot{\bar{\Omega}} = \varepsilon\, \bar{S}(\bar{\Omega},\bar{a}) & \text{etc.} \\ \dot{\bar{a}} = \varepsilon\, \bar{T}(\bar{\Omega},\bar{a}) \end{cases} \qquad (2)$$

We are now able to state our first theorem.

Theorem 1

Assumptions: a) Let the following symmetry conditions hold true:

$$R(-\phi,-\Omega,a) = R(\phi,\Omega,a),\ S(-\phi,-\Omega,a) = S(\phi,\Omega,a);$$

$$T(-\phi,-\Omega,a) = -T(\phi,\Omega,a)$$

b) $\bar{S}$ is independent of $\bar{\Omega}$: $\bar{S} = \nu(\bar{a})$, $\nu(\bar{a})$ is linearly independent over the integers; $\bar{T}$ vanishes: $\bar{T} = 0$;

Conclusion: The solutions of (1) are formally quasiperiodic i.e. there exists a formal near identical transformation

$$\begin{cases} \phi = \tilde{\phi} + \varepsilon r(\tilde{\phi},\tilde{\Omega},\tilde{a},\varepsilon) \\ \Omega = \tilde{\Omega} + \varepsilon s(\tilde{\phi},\tilde{\Omega},\tilde{a},\varepsilon) \\ a = \tilde{a} + \varepsilon t(\tilde{\phi},\tilde{\Omega},\tilde{a},\varepsilon) \end{cases},$$

such that the transformed system reads as follows:

$$\begin{cases} \dot{\tilde{\phi}} = \omega(\tilde{a}) + \varepsilon\tilde{R}(\tilde{\Omega},\tilde{a},\varepsilon) \\ \dot{\tilde{\Omega}} = \varepsilon\tilde{S}(\tilde{a}) \\ \dot{\tilde{a}} = 0 \end{cases}.$$

The proof of this theorem as well as other versions of it are given in a joint paper by the author and M. Vitins [1].

The crucial condition of Theorem 1 is condition b). Of course we can compute $\bar{S}$, $\bar{T}$ and check whether this condition holds or not. Remember however that system (2) is obtained by a change of variables and in addition requires the averaging procedure with respect to ϕ. Moreover there may exist several transformations that put a physical system into the form (1). One or the other choice may lead to a system (2) that satisfies condition b).

Unfortunately we are not able to offer a sufficient condition for assumption b) to hold true. Nevertheless we can propose a necessary condition by means of which transformations not satisfying condition b) can be eliminated. The following theorem is due to M. Vitins and is proved in [1].

Theorem 2

Let p be a first integral of (1), which is independent of ϕ; but depends on Ω. Then condition b) does not hold.

A supplement of this theorem and applications are given in [1].

2. ON E. HOPF'S BIFURCATION THEOREM

Consider an autonomous system of differential equations depending on a parameter μ:

$$\dot{\xi} = g(\xi,\mu) \tag{3}$$

Assume that we are given a μ-dependent family of equilibrium solutions $\xi(\mu)$. Introduce y by $\xi = \xi(\mu) + y$:

$$\dot{y} = g(\xi(\mu) + y,\mu) = A(\mu)y + \frac{1}{2}\sum_{i,j} F^{ij}(\mu)y_i y_j + \frac{1}{3!}\sum_{i,j,k} F^{ijk}(\mu)y_i y_j y_k + \dots \tag{4}$$

$A(\mu)$ is a nxn-Matrix ($\dot{y} = A(\mu)y$ are the equations of first variations of (3) attached to the equilibrium solution $\xi(\mu)$), $F^{ij}(\mu)$, $F^{ijk}(\mu)$ are n-vectors. Let us introduce the following abbreviations:

$$A = A(0), \quad A^1 = \left.\frac{dA(\mu)}{d\mu}\right|_{\mu=0}, \quad F^{ij} = F^{ij}(0), \quad F^{ijk} = F^{ijk}(0)$$

As to the eigenvalues of A we make the assumption: Precisely two of them are purely imaginary, while the remaining ones have negative real parts. We assume therefore, without loss of generality:

$$A = \left(\begin{array}{cc|c} 0 & 1 & \mathbf{0} \\ -1 & 0 & \\ \hline \multicolumn{2}{c|}{\mathbf{0}} & \mathbf{D} \end{array}\right)$$

where D is a (n-2)x(n-2) - Matrix with all eigenvalues having negative real parts.

In order to state our result we need the following abbreviations:

$$\underline{F}^{ij} = \begin{pmatrix} F_3^{ij} \\ F_4^{ij} \\ \vdots \\ F_n^{ij} \end{pmatrix}, \quad \underline{F}_j^{i} = \begin{pmatrix} F_j^{i3} \\ F_j^{i4} \\ \vdots \\ F_j^{in} \end{pmatrix}, \quad G = (E + \tfrac{1}{4} D^2)^{-1}$$

The quantity χ, to be defined now, is crucial for our theorem

$$\chi = \frac{1}{16} [F_2^{11}F_1^{11} + F_2^{11}F_2^{12} + F_2^{22}F_2^{12} - F_1^{11}F_1^{12} - F_1^{22}F_1^{12} - F_1^{22}F_2^{22}] +$$

$$+ \frac{1}{16} [F_1^{111} + F_1^{122} + F_2^{112} + F_2^{222}] +$$

$$+ \frac{1}{32} [-4(\underline{F}_1^{1^T} + \underline{F}_2^{2^T})D^{-1}(\underline{F}^{11}+\underline{F}^{22}) + (\underline{F}_1^{1^T} - \underline{F}^{2^T})(-\tfrac{1}{2}GD[\underline{F}^{11}-\underline{F}^{22}] + 2G\underline{F}^{12})$$

$$+ (\underline{F}_1^{2^T} + \underline{F}_2^{1^T})(-G[\underline{F}^{11}-\underline{F}^{22}] - GD\underline{F}^{12})]$$

<u>Theorem 3</u>

If μ is sufficiently small and sign μ = -sign $\chi/(A'_{11}+A'_{22})$ then (3) possesses a periodic solution (different from $\xi(\mu)$. If, in addition, sign χ = -1 this periodic solution is asymptotically orbitally stable with asymptotic phase.

This theorem is principally due to E. Hopf [2]. His conditions, however, are stated in a fairly implicit manner. More recently Hsü and Kazarinoff [3] have made them much more explicit. The conditions given above are even slightly simpler than those of Hsü and Kazarinoff. Moreover our approach does not make reference to Hopf's result, but is based on the method of averaging.

<u>Outline of the proof</u>.

Consider Eq. (4). Introduce new variables μ_1,x by $\mu = \varepsilon^4\mu_1$, $x = \varepsilon^2 y$, ε being a new parameter. The new system is of the form:

$$\dot{x} = Ax + \varepsilon^2 f^1(x) + \varepsilon^4 f^2(x) + O(\varepsilon^6)$$

Now perform a near-identical change of variables

$$x \to \bar{x} + \varepsilon^2 u^1(\bar{x}) + \varepsilon^4 u^2(\bar{x})$$

Denote the transformed system as follows.

$$\dot{\bar{x}} = A\bar{x} + \varepsilon^2 \bar{f}^1(\bar{x}) + \varepsilon^4 \bar{f}^2(\bar{x}) + O(\varepsilon^6)$$

Next we perform the following change of variables.

$$\bar{x} \to \begin{pmatrix} a \cos \phi \\ -a \sin \phi \\ \underline{z} \end{pmatrix} \longrightarrow \quad (\underline{z} \text{ denotes a } (n-2)\text{-vector}$$

the corresponding system reads as follows:

$$\begin{cases} \dot{\phi} = 1 + \varepsilon^2 g_1^1(\phi,a,\underline{z}) + \varepsilon^4 g_1^2(\phi,a,\underline{z}) + O(\varepsilon^6) \\ \dot{a} = \varepsilon^2 g_2^1(\phi,a,\underline{z}) + \varepsilon^4 g_2^2(\phi,a,\underline{z}) + O(\varepsilon^6) \\ \dot{\underline{z}} = D\underline{z} + \varepsilon^2 \underline{g}^1(\phi,a,\underline{z}) + \varepsilon^4 \underline{g}^2(\phi,a,\underline{z}) + O(\varepsilon^6) \end{cases}.$$

The following lemma is the essential step of the proof.

Lemma

u^1, u^2 can be chosen such that:

$$g_2^1(\phi,a,\underline{0}) = 0 \;\; ; \;\; g_2^2(\phi,a,\underline{0}) = a[\mu_1 \frac{A'_{11}+A'_{22}}{2} + \chi a^2] \stackrel{\text{Def}}{=} g_2^2(a)$$

$$\underline{g}^1(\phi,a,\underline{0}) = \underline{0} \;\; ; \;\; \underline{g}^2(\phi,a,\underline{0}) = \underline{0}$$

Next we introduce ϕ instead of t as independent variable, Δa, Δz instead of a, z by $a = a^0 + \varepsilon\Delta a$, $z = \varepsilon^4 \Delta z$, a^0 being defined by $g_2^2(a^0) = 0$.

The resulting system is

$$\begin{cases} \dfrac{d\Delta a}{d\phi} = \varepsilon^4 \dfrac{\partial g_2^2}{\partial a}(a^0)\Delta a + O(\varepsilon^5) \\ \dfrac{d\Delta z}{d\phi} = D\Delta z + O(\varepsilon^2) \end{cases}.$$

By studying the corresponding Poincaré-map and applying the contraction mapping principle the desired result is obtained. The details are given in [4].

3. ERROR BOUNDS FOR PERTURBED SYSTEMS

The problem of error bounds of the method of averaging can be put, after some preliminary steps, in the following form.

Consider the following system of differential equations,

$$\dot{y} = g(y) + h(y) \tag{5}$$

and consider h as a small perturbation of the corresponding unperturbed system

$$\dot{z} = g(z) \quad . \tag{6}$$

Let us denote the general solutions of (5), (6) by $y(t,y^0)$, $z(t,z^0)$ respectively. While $z(t,z^0)$, is assumed to be available, $y(t,y^0)$ is not known in general. Actually $z(t,\xi)$ is hoped to be an approximation to $y(t,\xi)$ and an estimate of the difference

$$y(t,\xi) - z(t,\xi)$$

is desired. A reasonable result is obtained if a scalar function $\omega(t)$ is introduced as follows. Assume that $z(t,\xi)$ admits a function $\omega(t)$ such that

$$|z(t,\xi^1) - z(t,\xi^2)| \leq \omega(t)\ |\xi^1-\xi^2| \quad . \tag{7}$$

As to the perturbation term h we require it to be bounded by a constant M (in the region under consideration). We then have:

Theorem 4

Under the conditions mentioned before, the following inequality holds for t in the domain of definition of $y(t,\xi)$, $z(t,\xi)$:

$$|y(t,\xi) - z(t,\xi)| \leq M \int_0^t \omega(s)ds \quad .$$

The type of condition (7) was introduced in [5]. The same type of results have been established by M. Vitins [6] and A. Stokes [7]. The proof of Theorem 4 is given [8]. Let us finally mention a much more general result. Consider instead of (5), (6), the systems

$$\dot{y} = g(t,y) + h(t,y) \tag{8}$$

and

$$\dot{z} = g(t,z) \quad . \tag{9}$$

Denote the general solutions by $y(t,t^0,y^0)$, $z(t,t^0,z^0)$ respectively. Assume that we are given a function $\omega(t,t^0,x^1,x^2)$ such that

$$|z(t,t^0,x^1) - z(t,t^0,x^2 \leq \omega(t,t^0,x^1,x^2)\ |x^1-x^2|$$

We then have:

Theorem 5

The following inequality holds true:

$$|y(t,t^0,\xi)-z(t,t^0,\xi)| \leq \int_{t^0}^{t} \omega(t,s,y(s,t^0,\xi),y(s,t^0,\xi))\,|h(s,y(s,t^0,\xi))|\,ds$$

This result, of course, is much less explicit than the one given before. However it offers the opportunity to take into account the stability behavior of z as well as the growth of h.

The proof of Theorem 5 will appear elsewhere.

REFERENCES

1. U. Kirchgraber, M. Vitins, "The Structure of the Secular System in Nonlinear Oscillations", to appear in Cel. Mech.
2. E. Hopf, Ber. Verh. Sachs. Akad. Wiss. Leipzig. Math. Nat. 94 (1942), 3-22.
3. I. Hsü, N. Kazarinoff, "An Applicable Bifurcation Formula and Instability of Small Periodic Solutions of the Field-Noyes Model", to appear.
4. U. Kirchgraber, "Bifurcation and Averaging", to appear
5. U. Kirchgraber, Mech. Res. Comm. 1, (1974), 173-175.
6. M. Vitins, "Error Bounds in the Method of Averages Based on Properties of Average Motions", to appear in SIAM J. of Appl. M.
7. A. Stokes, "Comparison Theorems and Instability with Applications to Numerical Integration, Satellite Orbits and Averaging", to appear.
8. U. Kirchgraber, "Error Estimation for Perturbed Systems", to appear in J. Reine Angew. Math.

ON THE THEORY OF AVERAGING

Ferdinand Verhulst

Department of Mathematics, University of Utrecht,
Utrecht, The Netherlands

1. INTRODUCTION

The purpose of this paper is to present a number of recently obtained results in averaging, both in theory and in applications, and to show at the same time what kind of pitfalls there are in applying the theory to problems in dynamics.

In section 2 the main averaging theorems are formulated. In section 3 we discuss the intricate problem of extending the time-scale of validity of the approximations obtained by averaging. Section 4, finally, is concerned with the formulation of the averaging method for certain autonomous and non-autonomous problems in perturbed two-body motion. In a number of appendices theory and applications are illustrated by examples.

We shall adopt the conventions and definitions given in Verhulst (1975).

2. THE MAIN THEOREMS OF AVERAGING

In the theory of asymptotic approximations of the solutions of non-linear initial value problems an important part is played by the theory developed by Krylov, Bogoliubov and Mitropolsky (1961), Roseau (1966) and many others. In various applications

V. Szebehely and B.D. Tapley (eds.), Long-Time Predictions in Dynamics, 119–140. *All Rights Reserved.*
Copyright ©1976 *by D. Reidel Publishing Company, Dordrecht-Holland.*

this theory, the theory of averaging ("la methode de centrage" or "the theory of synchronization") was used much earlier.

The purpose of this section is to formulate the main theorems of averaging with conditions which are reduced as much as possible. This presentation of the theorems under weak conditions is not just for mathematical entertainment as this precise formulation will be used to understand the breakdown of the method in a number of examples derived from dynamics.

As in the initial value problem (1) we shall study approximations of functions $x(t, \varepsilon)$ defined in a domain $D \subset \mathbb{R}^n$. A useful preliminary is the <u>definition</u> of <u>interior</u> <u>subset</u> of D: $D_0 \subset D$ is an interior subset of D if the distance between the boundary of D_0 and the boundary of D is bounded from below by a positive constant, independent of ε, for all $\varepsilon \in [0, \varepsilon_0]$.

<u>Theorem 1</u> (general averaging theorem)

Consider the initial value problem

$$\frac{dx}{dt} = \varepsilon\, f(x, t, \varepsilon)\ ,\quad x(0, \varepsilon) = x_0 \tag{1}$$

with $t \in [0, \infty]$, $\varepsilon \in [0, \varepsilon_0]$; $x \in D$, an open, bounded set in $\mathbb{R}^n$;

1. f is defined in a connected set $G = D \times [0, \infty) \times [0, \varepsilon_0]$;
2. f is continuous and uniformly bounded in G;
3. f is Lipschitz-continuous with respect to x in G;
4. the limit $\lim_{\varepsilon\to 0} f(x, t, \varepsilon) = f(s, t, 0)$ exists uniformly in G;
5. the limit $\lim_{T\to\infty} \frac{1}{T}\int_0^T f(y, \tau, 0)\, d\tau = f_0(y)$ exists in G;

Consider the associated initial value problem

$$\frac{dy}{dt} = \varepsilon f_0(y)\ ,\quad y(0, \varepsilon) = x_0$$

and assume

6. $y(t, \varepsilon)$ exists on the time-scale $1/\varepsilon$ and $y(t, \varepsilon) \in D_0$, an interior subset of D,

then

$$x(t, \varepsilon) - y(t, \varepsilon) = o(1) \text{ on the time-scale } 1/\varepsilon\ .$$

<u>Some Comments</u>

a. The conditions 1-3 are mostly the classical conditions to ensure existence and uniqueness of the solution $x(t, \varepsilon)$; uniform boundedness of f on the whole time-axis has been assumed to ensure existence of solutions on a sufficiently large time interval. This, however, is unsatisfactory, as in most cases this condition presupposes apriori knowledge about the behavior of the solution on the whole time-axis. If one looks carefully at the lemmas which need this condition, in for instance Eckhaus's (1975) proof, one finds that the condition can be weakened to uniform boundedness of f on the time-scale $1/\varepsilon$. That such a restriction is useful is illustrated in appendix 2.

b. Condition 4 is needed to use $f(x, t, 0)$ as an asymptotic approximation to $f(x, t, \varepsilon)$.

c. The theorem contains the result that $y(t, \varepsilon)$ is an asymptotic approximation to $x(t, \varepsilon)$ on the time-scale $1/\varepsilon$; a more precise estimate cannot be given in this general formulation but more precision is possible if we put somewhat heavier requirements on f, for instance periodicity in t. We then have,

<u>Theorem 2</u> (averaging in the periodic case)

Consider the initial value problem of theorem 1 and assume that conditions 1-4 are satisfied; assume moreover

7. f posesses a bounded derivative with respect to ε in G;
8. f is periodic in t,

then condition 5 is satisfied and if we assume that condition 6 holds good we have

$$x(t, \varepsilon) - y(t, \varepsilon) = O(\varepsilon) \text{ on the time-scale } 1/\varepsilon .$$

<u>Some Comments</u>

a. Proofs of theorems 1 and 2 under these weak conditions were given by Besjes (1969) and Eckhaus (1975). Other proofs of theorem 2 were given by Roseau (1966) and Van der Burgh (1974). In the last reference the relation between theorem 2 and results obtained by two-variable expansion procedures is made clear.

b. A number of confusing remarks can be found in the general literature on asymptotic approximations concerning the time interval of validity of the results of theorems 1 and 2. Some authors claim uniform validity on the whole time-axis; others ignore the whole question while stating their results. We stress here that the approximations obtained by averaging are valid only on the bounded interval $0 \leq t \leq c/\varepsilon$, where c is a constant independent of ε; it is made clear by the work of Perko (1969) and Van der Burgh (1974) that generally no extension of the time-scale is obtained by applying two-variable (multiple time-scale) methods. A trivial example to illustrate the boundedness of the interval can be found in appendix 1. Some special cases in which the time-scale can be extended are discussed in section 3.

c. If we put somewhat heavier requirements on the perturbation problem of theorem 2, especially involving the differentiability properties of f, the approximation $y(t, \varepsilon)$ can be shown to be the first term in an asymptotic series approximating $x(t, \varepsilon)$. The asymptotic expansion is in integral powers of ε and can be used to obtain higher accuracy. There are certain limitations as to the accuracy improvement as the expansion series obtained is asymptotic and need not be convergent. Some questions are still unanswered here, even in the theory of approximation of periodic solutions. In calculating higher approximations one does not extend the time-scale of validity beyond $1/\varepsilon$ except in certain special cases. It is not clear how to obtain higher order approximations in the general non-periodic case (theorem 1). It will be shown in the subsequent section that one can obtain higher order approximations in the non-periodic case if one constructs solutions in the attraction domain of an asymptotically stable critical point.

d. It can be useful to compare approximate solutions obtained by averaging and by Poincaré expansion procedures. In the non-periodic case the two methods clearly do not produce the same results owing to the fact that Poincaré expansion generally involves a smaller time-scale of validity. In approximating periodic solutions by Poincaré expansion the same argument concerning the time-scale of validity holds but the analytical expressions obtained by the two methods turn out to be the same in a number of cases. In this respect there are still a number of problems to be solved.

e. The perturbation problems formulated in theorems 1 and 2 will be called perturbation problems in the standard form. Not all perturbation problems can be reduced to the standard form (1) but if they can, this reduction is usually performed by applying the "variation of constants" method of Lagrange. A more extensive discussion of this reduction problem can be found in Verhulst (1975); examples of reduction to the standard form can be found in the subsequent section and in several appendices. There are examples in the literature of averaging without first reducing to the standard form. This procedure might be called "crude averaging" and generally leads to incorrect results; an example is given in appendix 3.

3. EXTENSION OF THE TIME-SCALE OF VALIDITY

The extension of the time-scale of validity of the approximations obtained with theorems 1 and 2 is a notorious problem which is far from being solved. There are two types of problems with which serious progress has been made. The first type occurs whenever averaging leads to

$$f_0(y) \equiv 0$$

This type of phenomenon is directly related to resonance problems and will be the subject of A. H. P. Van der Burgh's lectures. We will only remark here that in these cases the time-scale of validity is extended to a higher but still finite order (e.g., $1/\varepsilon^2$ or $1/\varepsilon^3$).

The second type of problem concerns extension of the time-scale of validity to the whole time-axis (asymptotic validity on all time-scales $1/\delta(\varepsilon)$ or uniform asymptotic validity) and is, not unexpectedly, intimately connected with the theory of stability. The results obtained here are not restricted to periodic solutions but we consider this case as an introduction.

We already know from appendix 1, that generally periodic solutions are not approximated uniformly on the whole time-axis by averaging but the approximation is uniform in the following well-known case (see for example Roseau (1966) or Hale (1969)):

Theorem 3 (bifurcation theorem)

Supposing that the associated system obtained in theorem 2

$$\frac{dy}{dt} = \varepsilon f_0(y) , \quad y(0, \varepsilon) = x_0$$

has a critical point (corresponding to a stationary solution) η which is asymptotically stable in linear approximation, then the original problem (1) has a periodic solution $y(t, \varepsilon)$ with the property:

$$y(t, \varepsilon) - \eta = o(1) \quad \text{uniformly in } t .$$

One expects that an obvious application of theorem 3 would be uniform validity of asymptotic approximations of the periodic solutions of the equation

$$\ddot{x} + x = \varepsilon f(x, \dot{x}) . \tag{2}$$

The application, however, is not straightforward; it will be shown in appendix 4 that one can prove in a number of cases uniform asymptotic validity for the approximation of the amplitude of the periodic solutions. An example will be found in the discussion of the Van der Pol equation.

a. Approximations with the averaged equations containing an attractor

In the following theorem uniform asymptotic validity is stated for the approximations of all solutions starting in the attraction domain of a critical point.

Theorem 4 (Eckhaus, 1975)

Consider the initial value problem

$$\frac{dx}{dt} = \varepsilon f(x, t, \varepsilon) , \quad x(0, \varepsilon) = x_0 , \tag{1}$$

and assume that the conditions of theorem 1 are satisfied. Suppose that

a. $y = \eta$ is a critical point of the associated system

$$\frac{dy}{dt} = \varepsilon f_0(y) , \quad y(0, \varepsilon) = x_0$$

$y = \eta$ is asymptotically stable in the linear approximation;

b. x_0 belongs to the domain of attraction of η ;

c. $y(t, \varepsilon) \in D_0$ for $0 \leq t < \infty$,

then

$$x(t, \varepsilon) - y(t, \varepsilon) = o(1)$$

uniformly on $0 \leq t < \infty$.

Some comments

a. Theorem 4 provides us with the global result that if we start in the domain of attraction of the critical point, the approximation is uniformly valid. Local results like this, i.e. by putting the initial value x_0 in the neighborhood of the critical point, were already known for a long time.
b. A result similar to theorem 4 was obtained by C. Banfi (1967) and C. Banfi and D. Graffi (1969), who replaced condition a. by requiring $y = \eta$ to be an asymptotically stable solution of the associated system, uniformly in ε . Unfortunately the result did not attract much attention.
c. We shall give a proof of theorem 4 which differs from the Eckhaus (1975) proof. From the theory on the behavior of solutions near a critical point (Roseau, 1966), we easily derive:

Theorem 5

Consider the equation

$$\frac{dx}{dt} = (A + B(t))x + f(x, t, \varepsilon) , \tag{3}$$

in which $x \in D \subset R^n$, $t \in [0, \infty)$, $\varepsilon \in [0, \varepsilon_0]$; the eigenvalues of the constant matrix have only negative real parts; $B(t)$ is a continuous matrix with the property $\lim_{t\to\infty} |B| = 0$;

$f(x, t, \varepsilon)$ is continuous in t, Lipschitz-continuous in x and has the property $||f|| = o||x||$ as $||x|| \to 0$ uniformly in ε. Then we have:
$x = 0$ is an asymptotically stable critical point of the equation; there exists a neighborhood $D = \{x \mid ||x|| \leq \sigma\}$ with the property that each solution x which starts at $t = t_0$ in D remains in D for $t \geq t_0$ and for each pair of such solutions x_1 and x_2 we have in D

$$|x_1 - x_2|_{t = t_0 + t_1} \leq k|x_1 - x_2|_{t = t_0} ,$$

with t_0 and t_1 (arbitrary) positive constants and $k < 1$.

Theorem 5 contains a local result which can help us to prove asymptotic validity of approximations which start in the vicinity of $x = 0$. To obtain global results we need the following:

<u>Proof of Theorem 4</u> (Eckhaus and Sanchez-Palencia, 1975).

Theorem 1 yields $||x(t, \varepsilon) - y(t, \varepsilon)|| \leq \delta(\varepsilon)$ for $0 \leq t \leq c/\varepsilon$ with c a constant independent of ε; $\delta(\varepsilon) = o(1)$. Consider the associated system in the form $(\tau = \varepsilon\tau)$

$$\frac{dy}{dt} = f_0(y) .$$

$y = \eta$ is asymptotically stable critical point of this equation; for each $\sigma > 0$ we can find a $T(\sigma)$ with: $\tau \geq T(\sigma)$, $|y(0)-\eta| \leq \sigma \longrightarrow ||y(\tau) - \eta|| \leq \sigma$ (theorem 5) where T is independent of ε (the parameter has disappeared from the equation). Consider the following partition of the time-axis $(n = 1, 2, \ldots)$

$$[0, \frac{T}{\varepsilon}] \cup [\frac{T}{\varepsilon}, \frac{2T}{\varepsilon}] \cup \ldots \cup [\frac{nT}{\varepsilon}, \frac{(n + 1)T}{\varepsilon}] \cup \ldots .$$

From theorem 1 we have for all finite n

$$||x - y_n||_{I_n} \leq \delta(\varepsilon) ,$$

in which $y_n(t, \varepsilon)$ is obtained from the associated system on $I_n = [\frac{nT}{\varepsilon}, \frac{(n + 1)T}{\varepsilon}]$ with initial value $y_n(\frac{nT}{\varepsilon}, \varepsilon) = x(\frac{nT}{\varepsilon}, \varepsilon)$.
It is easy to show that $\delta(\varepsilon)$ is independent of n.
Theorem 5, on the other hand, produces the estimate $(k < 1)$

$$||y - y_n||_{I_n} \leq k||y - y_n||_{I_{n-1}} .$$

So we find with Schwarz's inequality from these two estimates,

$$||x - y||_{I_n} \leq \delta(\varepsilon) + k||x - y||_{I_{n-1}} \quad .$$

Using this recursion relation we find

$$||x - y||_{I_n} \leq \delta(\varepsilon)(1 + k + k^2 + \ldots + k^n) \quad .$$

Taking the limit for $n \to \infty$ finally yields that for $t \to \infty$

$$||x(t, \varepsilon) - y(t, \varepsilon)|| \leq \frac{\delta(\varepsilon)}{1 - k} \quad .$$

An application to second-order autonomous equations can be found in appendix 4.

b. Approximations with the original equation (1) containing an attractor

In theorem 5, eq. (3), we made a number of assumptions on the non-linear part of $f(x, t, \varepsilon)$ to guarantee asymptotic stability of the critical point $x = 0$ $(x, f \in \mathbb{R}^n)$. To obtain asymptotic approximations of the solutions, we have to assume an asymptotic series expansion of $f(x, t, \varepsilon)$ with respect to ε .

For the construction of the first order approximation we only keep the zero and first order terms.

$$\frac{dx}{dt} = (A + B(t))x + f(x, t) + \varepsilon g(x, t) \ ,$$

$$x(0, \varepsilon) = x_0 \quad . \tag{4}$$

Later on, we shall return to this procedure.

We assume that the conditions of theorem 5 are satisfied; so $x=0$ is an asymptotically stable critical point of the equation in the linear approximation; $D \subset \mathbb{R}^n$ is the domain of attraction, bounded by a constant independent of ε . These assumptions imply that $g(x, t)$ is uniformly bounded on $D \times [0, \infty)$.

We introduce the associated equation

$$\frac{dy}{dt} = (A + B(t))y \quad ,$$

and associated initial value problem

$$\frac{dz}{dt} = (A + B(t))z + f(z, t) \ , \qquad z(0) = x_0 \ . \tag{4a}$$

The assumption is that $y(t)$ and $z(t)$ are known; $\phi(t)$ is the principal fundamental matrix of the equation for $y(\phi(0) = E)$ with the property

$$c_0 e^{-\mu t} < |\phi(t)| < c_1 e^{-\nu t}$$

for $t \in [0, \infty)$; c_0 , c_1 , μ , and ν are positive constants (see for instance Roseau, 1966). If the real parts of the eigenvalues of the matrix A are $-\lambda_1$, . . . , $-\lambda_n$, we have

$$0 < \nu < \lambda_1 < \dots < \lambda_n < \mu \ .$$

We introduce the transformation (variation of constants)

$$x(t) = \phi(t)u(t) + \phi(t) \int_0^t \phi^{-1}(s)\, f(x(s), s)\, ds \qquad (5)$$

to obtain from eq. (4) the equation

$$\frac{du}{dt} = \varepsilon\phi^{-1}(t)\, g(x, t), \qquad u(0) = x_0 \ ,$$

in which x still has to be eliminated by eq. (5). We introduce as a new time-like variable

$$\tau = e^{\mu t} \ ,$$

and find after averaging over τ and by using the properties of the fundamental matrix $\phi(t)$ and g the system

$$\frac{d\tilde{u}}{d\tau} = 0 \ , \qquad \tilde{u}(1) = x_0 \ ,$$

and clearly

$$u\left(\frac{\ln\tau}{\mu}\right) - x_0 = o(1) \quad \text{on the } \tau \text{ time-scale } 1/\varepsilon \ .$$

It follows that in the original variable t

$$u(t) - x_0 = o(1) \quad \text{on the time-scale } 1/\ln(1/\varepsilon) \ .$$

It will be clear from transformation (5) and the equation for z that

$$x(t) - z(t) = o(1) \text{ on the time-scale } 1/\ln(1/\varepsilon) \ .$$

From this point on we can use exactly the same method as in the Eckhaus and Sanchez-Palencia version of the proof of theorem 4 to obtain uniform asymptotic validity; the difference being that here we use the partition of the time-axis

$$[0, \frac{T}{\ln(1/\varepsilon)}] \cup [\frac{T}{\ln(1/\varepsilon)}, \frac{T}{\ln(1/\varepsilon)}] \cup \ldots .$$

We summarize in

Theorem 6

We consider the initial value problem (4) and the associated problem (4a). The conditions of theorem 5 are satisfied for eq. (4); $D \subset R^n$ is the domain of attraction of the asymptotically stable critical point of eq. (4); then we have if $x_0 \in D$,

$$|x(t, \varepsilon) - z(t)| = o(1) \text{ uniformly on } 0 \leq t < \infty .$$

Comment

The procedure outlined in the proof of theorem 6 can be used to obtain higher order approximations. If we carry out the calculations, we need more terms of the asymptotic expansion of the original $f(x, t, \varepsilon)$ in eq. (3). If we require $f(x, t, \varepsilon)$ to be analytic with respect to ε and $(x - z(t))$, the approximation $z(t)$ of $x(t, \varepsilon)$ becomes the first term of a Poincaré expansion with respect to the small parameter ε.

Apart from the result of uniform asymptotic validity of the approximation, theorem 6 is useful because of the rather weak conditions imposed on the right-hand side of eq. (4).

4. THE PERTURBED TWO-BODY PROBLEM

It turns out that in the case of autonomous (time-independent) perturbations averaging can be performed in the case of special three-dimensional perturbations. There are still many problems left here and we restrict ourselves to mentioning the interesting discussions by Struble (1960), Kyner (1965), and Eckstein, Shi, and Kevorkian (1964).

If we restrict ourselves to the planar case (perturbations only acting in the plane of revolution of the unperturbed problem), the formulation of the averaging procedures can be given in a relatively simple form. A related application can be found in the restricted three-body problem; see for instance Sergysels-Lamy (1975).

In the case of non-autonomous perturbations, which are so much more important in astrophysics, there are only a few applications. One of them is by Morrison (1965), who considers periodic planar perturbations. Another application is by Verhulst (1975), who considers secular, non-periodic perturbations in systems with variable mass.

a. Planar autonomous perturbations

Consider the two-body problem with perturbations $\beta(r, \dot{r}, \phi)$ and $\gamma(r, \dot{r}, \phi)$ in the respectiv ly radial and tangential directions respectively ($\underline{r}$ indicates a vector in R^2, r and ϕ are polar coordinates):

$$\ddot{\underline{r}} = -\frac{Gm_0}{r^3}\underline{r} + \beta(r, \dot{r}, \phi)\underline{r} + \gamma(r, \dot{r}, \phi)\dot{\underline{r}} \quad .$$

Writing the equation in scalar form one finds

$$\ddot{r} = -\frac{Gm_0}{r^2} + \frac{J^2}{r^3} + r\beta(r, \dot{r}, \phi) + \dot{r}\gamma(r, \dot{r}, \phi) \ , \qquad (6a)$$

$$J = r^2\dot{\phi} = ce^{\int_0^t \gamma(r(\tau), \dot{r}(\tau), \phi(\tau))\, d\tau} \qquad (6b)$$

c is the angular momentum per unit mass at time $t = 0$; the variable J gives us this quantity as it varies with time t . The procedure will be to transform $r = 1/\rho$ and use ϕ as a new time-like variable. This can be done as ϕ is a monotonically increasing function of time t (assuming $c > 0$). If in this way we find periodic solutions of ϕ , we still have to prove that, for the particular perturbation at hand, the transformation $t \to \phi$ corresponds with a one-to-one mapping of the time-axis into itself. If this is the case, we have found periodic solutions in t . The transformations which we introduced imply (' indicates differentiation with respect to ϕ)

$$\dot{r} = -J\rho' \quad ,$$

and we find from eqs. (6ab)

$$\rho''(\phi) + \rho(\phi) = \frac{Gm_0}{J^2} - \frac{\beta}{J^2\rho^3} \quad ; \qquad (7a)$$

$$J'(\phi) = \frac{\gamma}{\rho^2} \qquad (7b)$$

in which $\beta = \beta(1/\rho, -J\rho', \phi)$, $\gamma = \gamma(1/\rho, -J\rho', \phi)$.
Eqs. (7ab) constitute a system of the third-order which, together with appropriate initial values, enables us to construct approximate solutions by averaging. To do this we assume β and γ

to be small in a sense to be specified in each particular perturbation problem and to transform the equations to the standard form (1). This transformation can be carried out using the fundamental matrix of the unperturbed system in a suitable form, for instance the phase-amplitude form as given in appendix 1.

We put

$$\rho(\phi) = \frac{Gm_0}{J^2(\phi)} + k(\phi) \cos(\phi + \psi(\phi) - \phi_0) \quad ;$$
$$\rho'(\phi) = - k(\phi) \sin(\phi + \psi(\phi) - \phi_0) \quad . \tag{8}$$

With eqs. (7ab), the system in the standard form becomes

$$\frac{dk}{d\phi} = 2Gm_0 \frac{\gamma}{J^3\rho^2} \cos(\phi + \psi - \phi_0) + $$
$$+ \frac{\beta}{J^2\rho^3} \sin(\phi + \psi - \phi_0) \quad ; \tag{9a}$$

$$\frac{d\psi}{d\phi} = -2Gm_0 \frac{\gamma}{kJ^3\rho^2} \sin(\phi + \psi - \phi_0) + $$
$$+ \frac{\beta}{kJ^2\rho^3} \cos(\phi + \psi - \phi_0) \quad ; \tag{9b}$$

$$\frac{dJ}{d\phi} = \frac{\gamma}{\rho^2} \quad , \tag{9c}$$

in which ρ and ρ' should still be replaced by using eq. (8). One of the advantages of the standard form (9) is that certain properties of the approximate solutions may become immediately clear. In appendix 5 for instance, we show that in a simple atmospheric drag problem we immediately find that the phase angle ψ is constant in the first asymptotic approximation. This is a phenomenon analogous to what Bogoliubov and Mitropolsky (1961, ch. 1.3) called quasi-synchronous motion in non-linear friction problems.

In appendix 6 we consider a problem in which no tangential perturbation is present. As could be expected, one can prove in the case $\beta = \beta(r)$ that the amplitude k is constant in the first approximation. Only a small perturbation in the frequency (period) occurs and it is clear as in appendix 1 that we have here an approximation on the time-scale $1/\varepsilon$ and no better.

b. Non-autonomous perturbations

An extension to planar, non-autonomous perturbations of two-body systems was given by Morrison (1965). We shall briefly sketch his approach, at the same time somewhat generalizing the results. We suppose that the perturbations weakly depend on time t. Introducing the small parameter ε, the equations of motion become

$$\ddot{\underline{r}} = -\frac{Gm_0}{r^3}\underline{r} + \varepsilon\mu\underline{r} + \varepsilon\nu\dot{\underline{r}} \quad .$$

The perturbations μ,ν depend on r, $\dot{r},\phi$ and εt. We find the scalar equations

$$\ddot{r} = -\frac{Gm_0}{r^2} + \frac{J^2}{r^3} + \varepsilon\mu(r,\dot{r},\phi,\varepsilon t)r + \varepsilon\nu(r,\dot{r},\phi,\varepsilon t)\dot{r} \ , \qquad (10a)$$

$$J = r^2\dot{\phi} = ce^{\varepsilon\int_0^t \nu(r(\tau),\dot{r}(\tau),\phi(\tau),\varepsilon\tau)\, d\tau} \quad . \qquad (10b)$$

Again we transform $r = 1/\rho$ and use ϕ as a new time-like variable. Moreover we introduce a new dependent variable

$$\lambda = \varepsilon t \ .$$

We find again (' indicates differentiation with respect to ϕ)

$$\dot{r} = -J\rho' \quad .$$

The perturbation equations become

$$\rho''(\phi) + \rho(\phi) = \frac{Gm_0}{J^2} - \frac{\varepsilon\mu}{J^2\rho^3} \quad ; \qquad (11a)$$

$$J'(\phi) = \varepsilon\frac{\nu}{\rho^2} \quad ; \qquad (11b)$$

$$\lambda'(\phi) = \frac{\varepsilon}{J\rho^2} \quad ; \qquad (11c)$$

in which $\mu = \mu(1/\rho, -J\rho', \phi, \lambda)$; $\nu = \nu(1/\rho, -J\rho', \phi, \lambda)$.
Eqs. (11a-c) constitute a system of the fourth-order, one order more than eqs. (7ab), owing to the explicit occurrence of time t in eqs. (10ab). Again, we can obtain equations in the standard form (1) by applying transformation (10).
The resulting equations are 2π-periodic in ϕ and assuming some regularity conditions for μ and ν (theorem 2) we may average to obtain an $o(\varepsilon)$ approximation on the ϕ time-scale $1/\varepsilon$.
This is a remarkably strong result as we did not even assume the perturbation terms μ and ν to be periodic in t.
The application by Morrison is concerned with the case that μ and ν depend on r and $(\phi-\varepsilon t)$ and are 2π-periodic in $(\phi-\varepsilon t)$; as an example planar lunar orbits are considered.

A different type of non-autonomous perturbations, as those represented by eqs. (10ab), is found in studying two-body problems with variable mass. We shall demonstrate here the averaging approach to these problems as given in Verhulst (1975). We consider the following model, formulated by Mesherskii and later by Jeans and others. In a system of two bodies, acting upon each other by Newtonian force fields, mass is ejected isotropically from the system at very high velocities and is lost to the system. One finds (see Hadjidemetriou, 1967) that angular momentum is conserved; the orbits are planar. The equations of motion in polar coordinates are

$$\ddot{r} = -\frac{\mu(t)}{r^2} + \frac{c^2}{r^3} \quad ;$$

$$r^2\dot{\phi} = c \quad ,$$

in which $\mu(t)$ ($\mu(t) = Gm$) is a monotonically decreasing function of time t.

The model attracted attention in various contexts: the theory of evolution of binary systems, the theory of the veolcity distribution of stars in the galaxy, cometary orbit theory and cosmological theories involving a decreasing gravitational 'constant' G. We shall restrict our considerations to the case in which $\mu(t)$ obeys a modified Jeans-Eddington relation of the form

$$\dot{\mu} = -\alpha(\mu-\mu_r)^n \ , \quad \mu(0) = \mu_0$$

where α, μ_r and n are positive constants with $0 \leq \mu_r < m_0$, $n \geq 1$. If G is a constant, we call μ_r/G the rest mass of the two-body system.

Again we introduce $\rho = 1/r$ and ϕ as a new time-like variable. We obtain the third-order initial value problem

$$\rho'' + \rho = \frac{\mu_r}{c^2} + u \quad ;$$

$$u' = -\alpha c^{2n-3} \frac{u^n}{\rho^2} \quad , \tag{12}$$

with $\rho(\phi_0)$, $\rho'(\phi_0)$ determined by $r(0)$, $\dot{r}(0)$; $u(\phi_0) = \frac{\mu_0 - \mu_r}{c^2}$.

We assume that α is small and put $\alpha c^{2n-3} = \varepsilon$. The third-order system (12) can be brought into the standard form for averaging by transforming

$$\rho(\phi) = \frac{\mu_r}{c^2} + u(\phi) + A(\phi)\cos(\phi-\phi_0) + B(\phi)\sin(\phi-\phi_0);$$

$$\rho'(\phi) = - A(\phi)\sin(\phi-\phi_0) + B(\phi)\cos(\phi-\phi_0).$$

The results vary according to the values of the parameters μ_r and n. As an illustrative example we look at the case $\mu_r = 0$, $n = 3$. We find after averaging over ϕ for the approximations of A, B and u, integration yields

$$\frac{\tilde{A}(\phi)}{A(\phi_0)} = \frac{\tilde{B}(\phi)}{B(\phi_0)} = \frac{\tilde{u}(\phi)}{u(\phi_0)} ,$$

and

$$\tilde{A}(\phi) = A(\phi\ )\exp.\ (-\varepsilon \frac{u^3(\phi_0)}{\Gamma^{3/2}(\phi_0)} (\phi-\phi_0)) ,$$

with $\Gamma = \tilde{u}^2 - \tilde{A}^2 - \tilde{B}^2$.

The energy of the two-body system is given by

$$E = \frac{1}{2}\dot{r}^2 + \frac{1}{2}\frac{c^2}{r^2} - G\frac{m}{r} .$$

APPENDIX 1: a trivial example showing the boundedness of the time-interval of validity.

Consider the equation $\ddot{x} + x = \varepsilon g(x,\dot{x},t)$ with initial values $x(0) = A$, $\dot{x}(0) = 0$. To obtain the standard form for averaging we substitute

$$x = k(t)\cos(t + \phi(t)) \quad ;$$

$$\dot{x} = -k(t)\sin(t + \phi(t)) \quad ,$$

to find

$$\dot{k} = -\varepsilon\sin(t + \phi)g(k\cos(t + \phi),-k\sin(t + \phi),t) \quad ,k(0) = A \;;$$

$$\dot{\phi} = -\varepsilon\frac{\cos(t + \phi)}{k}\, g(k\cos(t + \phi),-k\sin(t + \phi),t) \quad ,\phi(0) = 0 \;.$$

We choose $g(x,\dot{x},t) = x$; the right hand side of the equations for $k(t)$ and $\phi(t)$ is 2π-periodic in t and averaging yields the associated system

$$\dot{\tilde{k}} = 0; \quad \dot{\tilde{\phi}} = -\frac{1}{2}\varepsilon \;,$$

which means that

$$\tilde{k}(t) = A; \quad \tilde{\phi}(t) = -\frac{1}{2}\varepsilon t \quad ,$$

and we have the approximate solutions $\tilde{x}(t) = A\cos((1 - \frac{1}{2}\varepsilon)t)$;

$$\dot{\tilde{x}}(t) = -A\sin((1 - \tfrac{1}{2}\varepsilon)t) \;.$$

The exact solution is $x(t) = A\cos(\sqrt{1-\varepsilon}\,t)$;

$$x(t) = -A\sin(\sqrt{1-\varepsilon}\,t) \quad .$$

It is easy to see that on the time-scale $1/\varepsilon$ the error is $o(\varepsilon)$, on the time-scale $1/\varepsilon^2$ the error is not even $o(1)$.

APPENDIX 2: an example showing the need to restrict the time-scale.

Consider the equation $\ddot{x} + x = 8\varepsilon\dot{x}^2\cos t$; initial values $x(0)=A$, $\dot{x}(0)=0$. Reduction to the standard form (appendix 1) and averaging yields

$$k\cos^3\tilde{\phi} = A \quad \text{and} \quad \frac{1}{4}\sin(2\tilde{\phi}) + \frac{1}{2}\tilde{\phi} = -\varepsilon At$$

$\tilde{\phi}(t)$ decreases monotonically and equals $-\frac{\pi}{2}$ when $\varepsilon At = \frac{\pi}{4}$.

The approximate solution for $k(t)$ is bounded on the time-scale $1/\varepsilon$ by an ε-independent constant if one puts, for instance, $\varepsilon At < \pi/6$.

APPENDIX 3: a counter-example for the validity of 'crude averaging'.

Consider the equation $\ddot{x} + x = -4\varepsilon\dot{x}\cos^2 t$ with given initial values $x(0) = 0$, $\dot{x}(0) = A$. "Crude averaging' over the right hand side of the equation yields

$$\ddot{z} + z = -2\varepsilon\dot{z}$$

in which $z(t)$ is supposed to approximate $x(t)$ on some time-interval.

The solution becomes

$$z(t) = \frac{A}{\sqrt{1-\varepsilon^2}} e^{-\varepsilon t} \sin(\sqrt{1-\varepsilon^2}\, t) \quad .$$

Averaging after reduction to the standard form as in appendix 1 yields that

$$\tilde{x}(t) = Ae^{-3/2\varepsilon t}\sin t$$

is the asymptotic solution with the property

$$\tilde{x}(t)-x(t) = o(\varepsilon), \text{ on the time-scale } 1/\varepsilon \ .$$

On that time-scale we have

$$\begin{aligned} x(t)-z(t) &= x(t)-\tilde{x}(t)+\tilde{x}(t)-z(t); \\ &= o(\varepsilon)+o(1) \quad . \end{aligned}$$

APPENDIX 4: Uniform asymptotic validity of approximations of periodic solutions of $\ddot{x} + x = f(x,\dot{x})$ (Urabe, 1967 and Eckhaus and Sanchez-Palencia, 1975).

We consider periodic solutions of equation

$$\ddot{x} + x = f(x,\dot{x}) \quad . \tag{2}$$

It is easy to prove that the right hand side of the averaged equations only depends on $\tilde{k}$:

$$\dot{\tilde{k}} = \varepsilon f_1(\tilde{k}); \quad \dot{\tilde{\phi}} = f_2(\tilde{k}).$$

Solutions of $f_1(\tilde{k}) = 0$ are critical points of the equation for the amplitude $\tilde{k}$ and they correspond with (approximate) periodic solutions of the original equation (2). An example of such a case is found in studying the van der Pol equation. In that case

$$f(x,\dot{x}) = (1-x^2)\dot{x} \quad ;$$

$$f_1(\tilde{k}) = \frac{1}{2}\tilde{k}(1-(\frac{\tilde{k}}{2})^2) \ , \quad f_2(\tilde{k}) = 0 \quad .$$

However, no asymptotically stable critical point of the averaged equations exists, so we cannot apply theorem 4 to this system.

Progress can be made by limiting our attention to the orbital curves in the phase-plane. We transform

$$x = r \cos \theta \ ; \quad \dot{x} = r \sin \theta \quad .$$

We find for the equation describing the orbital curves:

$$\frac{dr}{d\theta} = \varepsilon \frac{r \sin \theta f(r \cos \theta, r \sin \theta)}{\varepsilon \cos \theta f(r \cos \theta, r \sin \theta)-r} \quad .$$

The right hand side of the equation is 2π-periodic in the time-like variable θ. Excluding a neighborhood of the origin and putting some mild regularity conditions on f (theorem 2) we find after averaging:

$$\frac{d\tilde{r}}{d\theta} = - \varepsilon f_1(\tilde{r}).$$

Application of theorem 2 yields

$$r(\theta)-\tilde{r}(\theta) = o(\varepsilon) \text{, on the } \theta \text{ time-scale } 1/\varepsilon .$$

If $\tilde{r}(\theta) = r_0$ is an asymptotically stable critical point in the linear approximation of the equation for $\tilde{r}$, theorem 3 asserts that the original equation for $r(\theta)$ contains a periodic solution $r_p(\theta)$ with the property

$$r_p(\theta)-r_0 = o(1) \text{ uniformly on } 0 \leq \theta < \infty \quad .$$

Theorem 4, moreover, extends theorem 2 by asserting that for all solutions $r(\theta)$ atracted by r_0 we have

$$r(\theta)-\tilde{r}(\theta) = o(1) \text{ uniformly on } 0 \leq \theta < \infty .$$

An example is again found in the van der Pol equation in which

$$\frac{d\tilde{r}}{d\theta} = -\frac{\varepsilon}{2}\,\tilde{r}\,(1-(\frac{\tilde{r}}{2})^2).$$

Here $\tilde{r} = 2$ is an asymptotically stable critical point in the linear approximation so that we have a uniform asymptotic approximation for the amplitude $\tilde{r} = 2$ of the periodic solution (Urabe, 1967, ch. 8.4) and for the orbital curves attracted to the periodic solution in the time-like variable θ.

APPENDIX 5: Autonomous perturbations with a tangential force only.

We consider some simple atmospheric drag problems by putting

$$\beta \equiv 0; \quad \gamma = -\varepsilon(\dot{x})^{2p}f(r) , \quad p = 0,1,2,...$$

in which $f(r)$ is constant or decreases as r increases. Eqs. (12a-c) become, if we choose $\phi_0 = 0$,

$$\frac{dk}{d\phi} = -2\varepsilon Gm_0 J^{2p-3}k^{2p}\frac{f(\frac{1}{\rho})}{\rho^2}\cos(\phi+\psi)\sin^{2p}(\phi+\psi) \quad ;$$

$$\frac{d\psi}{d\phi} = 2\varepsilon Gm_0 J^{2p-3}k^{2p-1}\frac{f(\frac{1}{\rho})}{\rho^2}\sin^{2p+1}(\phi+\psi) \quad ;$$

$$\frac{dJ}{d\phi} = -\varepsilon J^{2p}k^{2p}\frac{f(\frac{1}{\rho})}{\rho^2}\sin^{2p}(\phi+\psi) ,$$

with $\rho = \frac{Gm_0}{J^2} + k\cos(\phi+\psi)$ and appropriate initial values. It is easy to show that averaging always produces for the approximation $\tilde{\psi}$ of ψ,

$\frac{d\tilde{\psi}}{d\phi} = 0$, so that $\tilde{\psi} = \psi(0)$ in the first approximation. To study the behavior of $\tilde{k}$ and $\tilde{J}$ we consider two simple cases:

$p = 0; \quad f(r) \equiv 1 \quad (\gamma \text{ is a constant})$.

By returning to eqs. (7ab) and comparing with the results in section 4b for the two-body problem with decreasing mass, we find that the solutions of this problem can be found from the solutions of the decreasing mass problem with $\mu_r = 0$ and Jeans-Eddington index $n = 3/2$.

For the case $f(r) = r^{-2}$, $p = 1,2,\ldots$,

we find $\frac{d\tilde{k}}{d\phi} = 0$.

J can be found from $\frac{d\tilde{J}}{d\phi} = -\varepsilon\lambda\tilde{J}^{2p}$ with $\lambda = \frac{k_{(0)}^{2p}}{2\pi}\int_0^{2\pi}\sin^{2p}(\phi+\psi)\,d\phi$.

The solutions cannot be continued on an interval beyond the time-scale $1/\varepsilon$ for all values of p.

APPENDIX 6: An example in which no tangential perturbation is present.

If $\gamma \equiv 0$, the perturbation problem is simplified considerably. Angular momentum is conserved, $J(\phi) = c$, and eqs. (9ab) become

$$\frac{dk}{d\phi} = \frac{\beta}{c^2\rho^2(\phi)}\sin(\phi+\psi-\phi_0) ,$$

$$\frac{d\psi}{d\phi} = \frac{\beta}{c^2k\rho^2(\phi)}\cos(\phi+\psi-\phi_0) .$$

We choose initial values $k(\phi_0) = k_0 > 0$, $\psi(\phi_0) = 0$; ρ has still to be transformed by eq. (8). The solution becomes very simple if β only depends on ρ(or r). Putting $\beta = \varepsilon f(r)$ one easily proves that averaging yields

$$\frac{d\tilde{k}}{d\phi} = 0 \quad ;$$

$$\frac{d\tilde{\psi}}{d\phi} = \varepsilon\lambda \quad .$$

in which the constant λ depends on c and k_0 and is given by

$$\lambda = \frac{1}{2\pi c^2 k_0} \int_0^{2\pi} \frac{f(1/\rho(\phi))\cos(\phi+\psi-\phi_0)}{\rho^2(\phi)}\, d\phi \quad ,$$

where $\rho = \frac{Gm_0}{c^2} + k_0\cos(\phi+\psi-\phi_0)$.

The solutions can be continued on an interval beyond the time-scale $1/\varepsilon$ but the approximations cannot be expected to be asymptotically valid on a longer time-scale (cf. appendix 1).

ACKNOWLEDGEMENT

The author is indebted to Prof. W. Eckhaus who provided him with the contents of appendix 4 and the alternative proof of theorem 4 and to Dr. A.H.P. van der Burgh who suggested a number of improvements to the manuscript.

REFERENCES

1. Banfi, C., (1967), "Sull' approssimazione di processi non stationari in meccanica non lineare", Boll. Un. Mat. Italiana, 22, 442-450.
2. Banfi, C., and D. Graffi, (1969), "Sur les méthodes approchées de la mécanique non lineaire", Actes Coll. Eq. Diff. Non Lin., Mons, 33-41.
3. Besjes, J. G., (1969), "On the asymptotic methods for non-linear differential equations", J. de Mecanique 8, 357-372.
4. Boboliubov, N.N., and Y. A. Mitropolsky, (1961), Asymptotic Methods in the Theory of Non-linear Oscillations, Gordon and Breach Publ., New York.
5. Eckhaus, W., (1975), "New approach to the asymptotic theory of non-linear oscillations and wave-propagation", J. Math. An. Appl. 49, 575-611.
6. Eckhaus, W., and Sanchez-Palencia, (1975), Private communication.
7. Eckstein, M., Y. Shi, and J. Kevorkian, (1964), "Satellite motion for all inclinations around an oblate planet", Proc. I.A.U. vol. 25.
8. Hadjidemetriou, J.D., (1967), "Secular variation of mass and the evolution of binary systems", Adv. Astr. and Astroph. 5, 131-188.
9. Hale, J.K., (1969), Ordinary Differential Equations; Wiley-Interscience, New York.
10. Kyner, W.T., (1965), "A mathematical theory of the orbits about an oblate planet", J. Soc. Industr. Appl. Math. 13. 136-171.

11. Morrison, J.A., (1965),"Application of the method of averaging to planar orbit problems", J. Soc. Industr. Appl. Math. 13. 96-117.
12. Perko, L. M., (1969),"Higher order averaging and related methods for perturbed periodic and quasi-periodic systems", Siam J. Appl. Math. 17, 698-724.
13. Roseau, M., (1966), Vibrations nonlinéaires et théorie de la stabilite, Springer-Verlag.
14. Sanchez-Palencia, E., (1975),"Méthode de centrage et comportement des trajectoires dans l'espace des phases", Compt. Rend. Acad. Sci. Paris, Ser. A, 280, 105-107; (an extended version will appear in the Intern. J. Nonlinear Mech.).
15. Sergysels-Lamy,A., (1975),"Existence of periodic orbits of the second kind in the elliptic restricted problem of three bodies", Celestial Mech. 11, 43-51.
16. Struble, R.A., (1960),"A geometrical derivation of the satellite equations", J. Math. An. Appl. 1, 300-307.
17. Urabe, M., (1967), Nonlinear Autonomous Oscillations; Ac. Press, New York.
18. Van der Burgh, A.H.P., (1974)"Studies in the asymptotic theory of nonlinear resonance", diss. Technical University Delft.
19. Van der Laan, L., and F. Verhulst, (1972),"The transition from elliptic to hyperbolic orbits in the two-body problem by slow loss of mass", Celestial Mechanics 6, 343-351.
20. Verhulst, F., (1975),"Asymptotic expansions in the perturbed two-body problem with application to systems with variable mass", Celestial Mechanics 11, 95-129.

QUALITATIVE INVESTIGATION OF ALMOST SEPARABLE HAMILTONIAN SYSTEM OF TWO DEGREES OF FREEDOM

N. Sigrist

Seminar fur angewandte Mathematik der ETH, Zurich, Switzerland.

ABSTRACT. The present paper is a condensed survey of some qualitative methods to investigate the orbits of Hamiltonian dynamical systems. In particular we consider autonomous problems of two degrees of freedom which are globally almost separable or admit a periodic solution (local almost separability) and which can be reduced to a non-autonomous, periodic system of one degree of freedom (by isoenergetic reduction). In general the a priori given separable or unperturbed problem is not structurally stable at all, or is not an interesting subset of the phase space (as e.g., regions of resonance). To deal with this situation we propose to perform a finite perturbation procedure based on the averaging principle that defines a new separable problem which is structurally stable on a larger set than on the original one. The content of this paper consists in exploring this idea from several points of view. Some applications in celestial mechanics are included. More details will be published elsewhere.

1. THE POINCARÉ-MAPPING

Consider a non-autonomous dynamical system of one degree of freedom with Hamiltonian $H(y_1,y_2,t)$, H being 2π-periodic with respect to the independent variable t. The evolution of the system is governed by the canonical equations of motion

$$\dot{y}_1 = -\frac{\partial H}{\partial y_2} \; ; \; \dot{y}_2 = \frac{\partial H}{\partial y_1} \qquad \cdot = \frac{d}{dt} \; . \tag{1}$$

V. Szebehely and B.D. Tapley (eds.), Long-Time Predictions in Dynamics, 141–150. *All Rights Reserved. Copyright ©1976 by D. Reidel Publishing Company, Dordrecht-Holland.*

Let $y^T = (y_1,y_2)$ and assume that the solutions of (1) given by the function $y = \Psi(t,y^o)$ (initial time $t^o = 0$, initial condition $\Psi(0,y^o) = y^o = (y_1^o,y_2^o)$) exist at least for $0 \le t \le 2\pi$. Then we may define the Poincaré map P: $y^o \to y = P(y^o) = \Psi(2\pi,y^o)$. By virtue of the periodicity of H we have the basic relation $\Psi(t+2\pi\ k,y^o) = P^k(\Psi(t,y^o))$ for $t \in R$ and $k \in Z$ as far as the solutions exist. Thus we may hope to obtain information about the long-time behaviour of the solutions of (1) by investigating P, in particular by looking for specific invariant sets of P, as e.g. fixed-points corresponding to 2π-periodic solutions or closed curves, giving rise to quasi-periodic motions and invariant tori in the extended phase space with coordinates y_1, y_2, $t(\mathrm{mod}\ 2\pi)$. It will turn out that these invariant sets can be obtained qualitatively by merely discussing the level lines of a function of two variables.

2. ALMOST SEPARABLE PROBLEMS. THE TWIST-THEOREM

Consider an almost separable problem with Hamiltonian $H(\alpha,\beta,t,\varepsilon) = H_o(\alpha) + \varepsilon H_1(\alpha,\beta,t) + O(\varepsilon^2)$ where $\varepsilon \ge 0$ is a small (perturbation) parameter and the pair (α,β) are action-angular variables for the unperturbed problem corresponding to $\varepsilon = 0$. H is assumed to be 2π-periodic in β and t. Upon introducing $y_1 = \sqrt{2\alpha}\cos\beta$, $y_2 = \sqrt{2\alpha}\sin\beta$ the Poincaré-map $P_\varepsilon: y^o \to y$ is defined for ε sufficiently small and it is of the form $P_\varepsilon = P_o + O(\varepsilon)$. The twist-theorem [1], p. 225, assures the existence of a set of closed invariant curves with positive measure for ε small enough provided P_o is a twist-map, i.e. $H_o''(\alpha) \neq 0$ (twist-condition). This implies existence and boundedness of all solutions for all real values of t (in terms of y_1,y_2), furthermore the variation $\alpha(t) - \alpha(0)$ is small for all $t \in R$.

3. RESONANCE

Let a Hamiltonian H be given as in section 2 and assume that for $\alpha = \alpha^*$ we have $\omega(\alpha^*) = H_o'(\alpha^*) = p/q$, p and q being relatively prime integers. The ordinary perturbation procedure based on the averaging principle would then suffer from vanishing denominators steming from the so-called critical arguments $m\beta + nt$ with $mp + nq = 0$. This phenomenon is referred to as resonance. If the twist-condition is satisfied there follows from the twist-theorem the existence of a positive constant K such that the annulus $|\alpha - \alpha^*| < K\sqrt{\varepsilon}$ contains at least two closed invariant curves C_1, C_2 of P_ε for ε small enough such that the circle $\alpha = \alpha^*$ is in between C_1 and C_2. Furthermore the circle $\alpha = \alpha^*$ consists of fixed-points of P_o^q and the fixed-point theorem of Poincaré-Birkhoff assures the existence of 2qn fixed-points of P_ε for ε sufficiently small (n a positive integer) corresponding to periodic solutions of period $2\pi q$.

The existence and approximate position of these fixed-points can also be established by a perturbation procedure as follows (compare [2], [3] p. 52): Introduce the new independent variable $s = t/q$ and new dependent variables $(\tilde{\alpha},\tilde{\beta})$ by

$$\alpha = \tilde{\alpha} \qquad \beta = ps + \tilde{\beta} \tag{2}$$

to isolate the critical terms, in fact the argument $m\beta + nt$ is transformed into $(mp + nq)s + m\tilde{\beta}$ and this is independent of s if and only if the original argument is critical. For the new hamiltonian $H(\tilde{\alpha},\tilde{\beta},s,\varepsilon) = H_o(\tilde{\alpha}) + O(\varepsilon)$ we find $H_o'(\alpha^*) = 0$ and $|H_o'(\tilde{\alpha})|<n\sqrt{\varepsilon}$ for $|\tilde{\alpha} - \overset{*}{\alpha}|<c\sqrt{\varepsilon}$. A function $f(\tilde{\alpha},\tilde{\beta},s,\varepsilon)$ is said to be $\theta(\varepsilon^r)$ if it has a factor $\varepsilon^k(H_o')^\ell$ with $k + \ell/2 = r$. An almost identical canonical transformation $\phi_S:(\gamma,\sigma) \to (\tilde{\alpha},\tilde{\beta})$ generated by $S = S_1 + S_{3/2}$ $(S_k = \theta(\varepsilon^k))$ yields the new hamiltonian $K(\gamma,\delta, s,\varepsilon) = K_o(\gamma) + \varepsilon K_1(\gamma,\delta) + \theta(\varepsilon^2)$. The new action variable γ varies on an interval of length $\theta(\sqrt{\varepsilon})$ that contains the number α^* in its interior for $\varepsilon>0$ small enough since $|\gamma - \tilde{\alpha}| = O(\varepsilon)$. Furthermore

$$K_1(\gamma,\delta) = \frac{1}{2\pi}\int_0^{2\pi} H_1(\gamma,ps + \delta,qs)\, ds$$

has period $2\pi/q$ in δ.

Consider first the level-lines and in particular the stationary points of the truncated Hamiltonian $K^{(1)} = K_o + \varepsilon K_1$ which represents an autonomous system of one degree of freedom. From $K_o'(\alpha^*) = 0$, $K_o''(\gamma) \neq 0$ and the vanishing of the mean value of $\frac{\partial K_1}{\partial \delta}$ with respect to δ there follows the existence of 2qn stationary points of $K^{(1)}$. These points give rise to fixed-points of the Poincaré-map attached to the complete Hamiltonian K provided the Hessian of $K^{(1)}$ is non-degenerate for $\varepsilon \neq 0$ but small. This can be seen by applying an appropriate version of the implicit function theorem. The corresponding periodic solutions can be expected to be pairwise stable/not stable as will be explained in section 5 example 2 in more detail. By going back to the original coordinates y_1,y_2 as introduced in section 2 we obtain the well-known picture of the invariant curves of the Poincaré-map P_ε in the region of resonance. See fig. 1.

4. STABILITY OF A PERIODIC SOLUTION

Assume that a Hamiltonian system as introduced in section 1 admits a 2π-periodic solution $y = \Psi(t)$. The canonical transformation $x = y - \Psi(t)$ where $x^T = (x_1,x_2)$ carries the periodic solution into the equilibrium $x = 0$ and the new Hamiltonian $H(x_1,x_2,t)$ is still

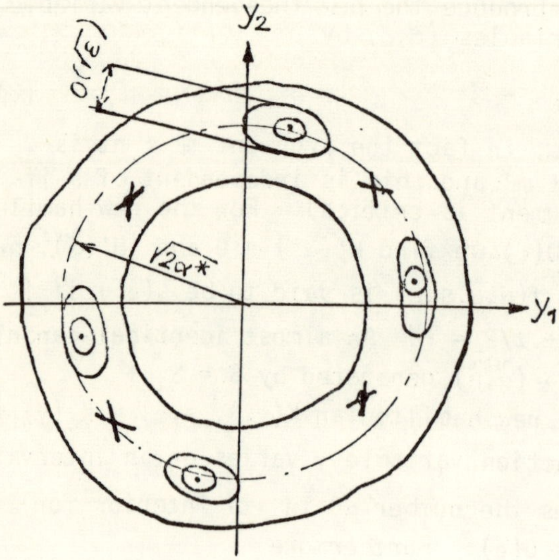

Fig. 1: Typical invariant curves of the Poincaré-map for a a resonance.

2 π-periodic in t. The origin x = 0 is a fixed-point of the Poincaré-map P attached to H and stability of the equilibrium solution x = 0 is equivalent to the following notion of stability (of a fixed-point of P): The point x = 0 is called stable if every neighbourhood N_1 of 0 contains a neighbourhood N_2 of 0 such that the iterates $(P|N_2)^k$ of the restriction of P to N_2 exist for all $k \in Z$ and $P^k(N_2) \subset N_1$ for all $k \in Z$.

In the linear stability the map P is linearized at x = 0, i.e. P(x) = Mx + ..., M being the so-called monodromy matrix. The matrix M is determined by solving the equations of variation $\dot{x} = A(t)x$ attached to the periodic solution under consideration. The matrix A is 2π-periodic in t and its trace vanishes. This implies det(M) = 1, thus the eigenvalues $\lambda_{1,2}$ of M (the so-called Floquet-multipliers) satisfy $\lambda_1 \cdot \lambda_2 = 1$. From the last relation there follows the well-known classification into (i) hyperbolic case ($\lambda \in R, \gamma \neq \pm 1$), (ii) parabolic case ($\lambda = \pm 1$) and (iii) elliptic case ($\lambda \notin R$, $\lambda_{1,2} = e^{\pm 2\pi i \gamma}$, γ = characteristic exponent). In case (i) the origin is not stable even if the nonlinear terms are added whereas is case (iii) we have linear stability. To decide the influence of the nonlinear terms we make use of Birkhoff's normal form and the twist-theorem, see [1], p. 243.

In the elliptic case Floquet's theory can be condensed in the following theorem: There exists a real, symplectic, 2 π-periodic matrix $\theta(t)$ such that the canonical transformation $x = \theta(t)z$ ($z^T = (z_1,z_2)$) carries the equations of variation into the harmonic oscillator

$$\dot{z} = \Gamma_z , \qquad \Gamma = \begin{pmatrix} 0 & \gamma \\ -\gamma & 0 \end{pmatrix} .$$

Action-angular variables (I,w) are now immediately introduced by $x_1 = \sqrt{2I}\cos w$, $x_2 = \sqrt{2I}\sin w$. The new Hamiltonian $F(I,w,t) = \sum_{k\geq 2} F_k(I,w,t)$ is of the form $F_2 = \gamma I$,

$$F_k(I,w,t) = \sqrt{I}^k \sum_{m,n} c_{mn}^{(k)} e^{i(mw+nt)} \quad , \quad k \geq 3$$

where $|m| \leq k$ and m-k is even.

The usual averaging procedure ($\sqrt{I}$ being the small quantity) defines an almost identical, canonical transformation $\phi_S:(\bar{I},\bar{w})\rightarrow(I,w)$ generated by a function $S = S_2 + S_3$ where S_k has the factor $\sqrt{I}^k$. The new hamiltonian $\bar{F} = \sum_{k\geq 2} \bar{F}_k$ is related to the old one by

$$\begin{aligned} \bar{F}_2 &= F_2 = \gamma \bar{I} \\ \bar{F}_3 &= F_3 + \frac{\partial S_3}{\partial w} + \frac{\partial S_3}{\partial t} = 0 \qquad (3) \\ \bar{F}_4 &= \frac{1}{4\pi^2} \int_0^{2\pi} [F_4 + \tfrac{1}{2}\{F_3,S_3\}]\, dw\, dt \end{aligned}$$

$\{.,.\}$ being the Poisson-bracket, provided $\lambda^3,\lambda^4 \neq 1$. Then $\bar{F}$ is found to be

$$\bar{F} = \bar{F}^{(4)} + 0(\bar{I}^{5/2}) \quad , \quad \bar{F}^{(4)} = \gamma\bar{I} + c\bar{I}^2 \qquad (4)$$

c being a constant. This result corresponds to the normal-form of Birkhoff. The Poincaré-map attached to $\bar{F}^{(4)}$ is a twist-map provided $c \neq 0$ and the origin is stable in this case. Thus in general the difference between the linear and the complete stability theory consists in the additional exceptions $\lambda^3,\lambda^4 = 1$ which, however, form a discrete set in the real numbers only.

5. APPLICATIONS

Example 1: Consider an almost separable problem with Hamiltonian $H(y_1,y_2,t,\varepsilon) = H_o(y_1,y_2) + O(\varepsilon)$ where $H_o(y_1,y_2) = f(y_1^2+y_2^2) = \frac{\omega}{2}(y_1^2+y_2^2) + \ldots$. The origin $y_1 = y_2 = 0$ is an equilibrium for $\varepsilon=0$. The continuation method gives a 2π-periodic solution $y = \Psi(t)$ with $||\Psi(t)|| = O(\varepsilon)$ of the perturbed problem for ε small enough provided $\omega \notin Z$. We are in the elliptic case if $\omega \neq n/2$ for all $n \in Z$ and ε small enough (note $\gamma = \omega + O(\varepsilon)$). The quantity c in eq. (4) is an analytic function of ε and $c(0) \neq 0$ implies stability provided $\omega \neq n/3, n/4$ for all $n \in Z$ and ε small enough. Note that $f''(0) \neq 0$ implies $c(0) \neq 0$. As examples for this situation we mention the perturbed simple pendulum and the periodic orbits of the first kind in the restricted (planar and circular) problem of three bodies. The latter problem is brought into the desired form by first performing the isoenergetic reduction followed by the introduction of Poincaré variables y_1,y_2 that remove the geometrical singularity for the circular orbits, see [6] p. 219. The exceptions $\lambda^3,\lambda^4 = 1$ can be related to the Kirkwood gaps in the asteroid belt [1] p. 249.

In the case of a perturbed harmonic oscillator we have $f'' = 0$ and $c(0) = 0$. But if the relevant part of the perturbation is taken into account we find $c(\varepsilon) = \varepsilon\bar{c}(\varepsilon)$ with $\bar{c}(0) \neq 0$ in general and stability still holds for $\omega \neq n/3, n/4$ for all $n \in Z$ and ε small enough. A typical picture of the invariant curves of the Poincaré-map is shown in fig. 2, see also [3] p. 47.

Example 2 ([3] p. 52 and p. 74): Let us go back to the resonant periodic orbits established in section 3. Let $(\bar{\gamma},\bar{\delta})$ be a stationary point of the truncated hamiltonian $K^{(1)}(\gamma,\delta,\varepsilon) = K_o(\gamma) + K_1(\gamma,\delta)$.

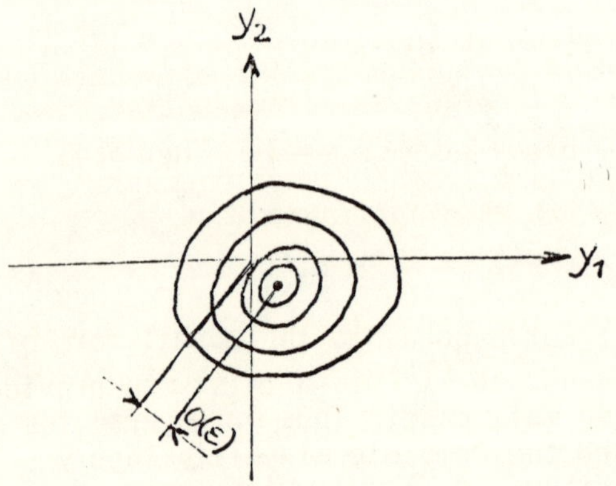

Fig. 2: Invariant curves surrounding a stable orbit of the first kind ($\lambda^3,\lambda^4 \neq 1$)

The corresponding Hessian matrix h is

$$h = \begin{pmatrix} a + O(\varepsilon) & O(\varepsilon) \\ O(\varepsilon) & \varepsilon b \end{pmatrix}$$

where $a = k_o''(\gamma)$ and $b = \frac{\partial^2 K_1}{\partial\delta^2}(\bar{\gamma},\bar{\delta})$. We have $a \neq 0$ by assumption and since $b \neq 0$ in general, the Hessian is non-degenerate for $\varepsilon \neq 0$ small. The real eigenvalues of h are $\mu_1 = a + O(\varepsilon)$, $\mu_2 = \varepsilon b + O(\varepsilon^2)$ and the stationary points are pairwise elliptic/hyperbolic according to the change in the sign of b as δ increases. Let us concentrate ourselves to an elliptic point (i.e. $\varepsilon ab > 0$). The Floquet-multipliers are found to be $\lambda_{1,2} = \exp(\pm 2\pi i\omega) + \ldots$ where $\omega = \sqrt{\varepsilon ab}$ and the theory of section 4 applies. The verification of the condition $c \neq 0$, however, is rather complicated because we are close to the degenerate parabolic case (the function S_3 in (3) contains denominators having $\sqrt{\varepsilon}$ as a factor). As an example we mention the periodic orbits of the second kind in the restricted problem of three bodies.

Example 3: As an illustration of the exceptional cases λ^3, $\lambda^4 = 1$ let us first consider the well-known spring-pendulum system in the plane [4] for small deviations from the stable equilibrium. The equations pertinent to this problem are after truncation

$$\ddot{q}_1 + q_1 = \varepsilon q_2^2 \quad , \qquad \ddot{q}_2 + \nu^2 q_2 = 2\varepsilon q_1 q_2 \ . \tag{5}$$

Here the q_1-axis is in the direction of the gravitational force and the q_2-axis is perpendicular to the q_1-axis. For $q_2 = 0$ we have a family of harmonic motions and we want to investigate the stability of these periodic solutions for $\nu \approx 1/2$. Eq. (5) are equivalent to canonical equations attached to the Hamiltonian $H(p_1,p_2,q_1,q_2,\ ) = \frac{1}{2}(p_1^2 + p_2^2 + q_1^2 + \nu^2 q_2^2) - \varepsilon q_1 q_2^2$. After the introduction of oscillator-elements (α_i,β_i) by $p_1 + iq_1 = \sqrt{2\alpha_1}\exp(i\beta_1)$, $p_2 + iq_2 = \sqrt{2\alpha_2/\nu}\exp(i\beta_2)$ and the isoenergetic reduction (solve $H(\alpha_1,\alpha_2,\beta_1,\beta_2,\varepsilon) = h$ for α_1) we arrive at a Hamiltonian $H(\alpha,\beta,s,\varepsilon)$ where $\alpha = \alpha_2$, $\beta = \beta_2$, $s = \beta_1$ being the new independent variable as introduced in section 2. The non-critical terms are eliminated by the usual averaging procedure and the application of the transformation (2) yields the Hamiltonian

$$K(\tilde{\alpha},\tilde{\beta},s,\varepsilon) = K_o(\tilde{\alpha}) + \varepsilon K_1(\tilde{\alpha},\tilde{\beta}) + O(\varepsilon\ ), \quad K_o = (2\nu - 1)\tilde{\alpha}. \tag{6}$$

The orbits of the truncated hamiltonian $K^{(1)} = K_o + \varepsilon K_1$ can be discussed by investigating the level-lines of the function

$$\phi(r,\tilde{\beta}) = r^2(c + \sqrt{1 - r^2}\,\sin 2\tilde{\beta}) \;, \quad 0 \le r \le 1$$

where $r = \sqrt{\nu\tilde{\alpha}/h}$ and the qualitative behaviour is characterized by a single resonance-parameter $c = \nu(1-2\nu)/(\sqrt{2h}\,\varepsilon)$.

The level-lines of ϕ are plotted for $c = 0$ (exact resonance), $0<c<1$ (close to resonance) and $c>1$ (far from resonance) in fig. 3. The origin $r = 0$ corresponds to the periodic solutions mentioned above. Clearly the resonance causes the origin to become not stable. The neglected terms $O(\varepsilon^2)$ in (6) do not alter this behaviour as can be seen by an application of an implicit function theorem (for the stationary points) and the twist-theorem (for the closed curves).

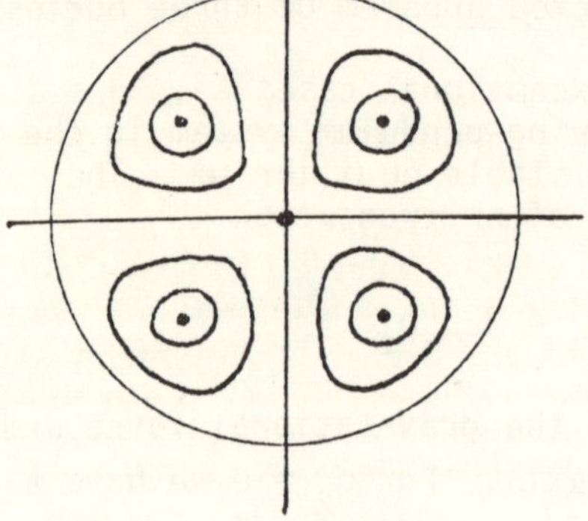

c = 0

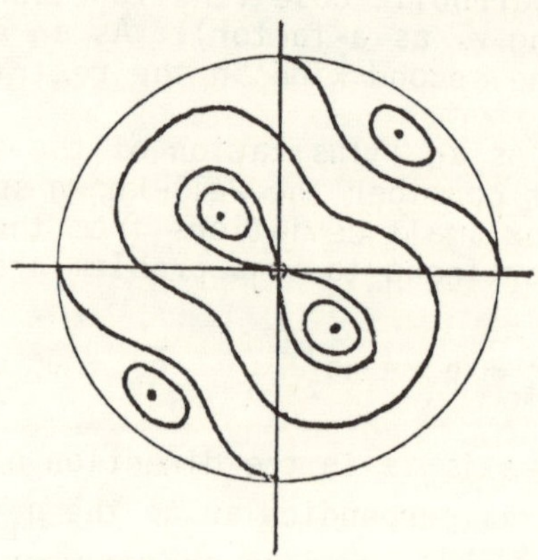

0<c<1

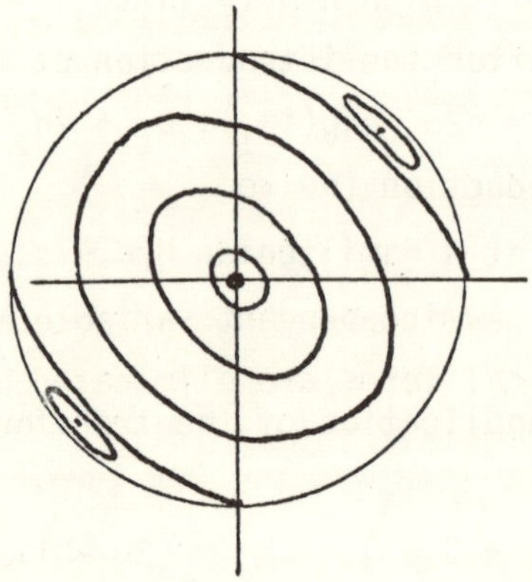

c>1

Fig. 3: Invariant curves for the spring-pendulum system. $\lambda^2 \approx 1$. Polar coordinates $r, \tilde{\beta}$.

A similar study of orbits with small eccentricity and λ^3, $\lambda^4 = 1$ for $\varepsilon = 0$ in the restricted problem of three bodies led W. Jefferys in [2] to the function

$$\phi(r,\beta) = (r^2 - r_o^2)^2 + \varepsilon a\, r^q \cos q\beta$$

where $q = 1,2,3,4$, $r^2 - r_o^2 = O(\sqrt{\varepsilon})$, r_o and a are constants (depending on the value of the Jacobian integral).

As a last example we mention the almost circular orbits in the vicinity of the critical inclination in the problem of the artificial satellite (without drag, zonal harmonics of the gravitational potential of the oblate earth are taken into account only). The reader is referred to the work of S. Aoki in [5] and to [3] p. 85.

6. CONCLUSIONS

From the preceding considerations we may learn that for almost separable Hamiltonian systems of two degrees of freedom it is possible to define new separable problems that are stable with respect to additional perturbations (structural stability) on subsets of the phase space which are larger than those of the originally given unperturbed problem. Distinguished invariant sets of the Poincaré-map (fixed-points, closed curves) are qualitatively well represented by the level-lines of a function of two variables that can be obtained by a comparatively simple averaging procedure. The same concept proves to be very useful also for noncononical systems.

Some, but not all, of the foregoing ideas can be extended to systems of several degrees of freedom. In particular the appearance of several critical arguments prevents the reduction to a discussion in the plane. Statements concerning stability, boundedness of the orbits are weakened by the (possible) occurence of Arnol'd's diffusion.

REFERENCES

[1] Siegel, C. L., and J. K. Moser: Lectures on celestial mechanics, 1971.
[2] Jefferys, W.: Some dynamical systems of two degrees of freedom in celestial mechanics, Astr. J. 71, 1966, p. 306.
[3] Sigrist, N.: Investigation of Hamiltonian systems with resonances and geometrical singularities, Thesis ETH Zürich, 1975.
[4] Broucke, R., and P. A. Baxa: Periodic solutions of a spring-pendulum system, Cel. M. 8, 1973, p. 261.

[5] Aoki, S.: Contribution to the theory of critical inclination of close earth satellites, Astr. J. 68, 1963, p. 365.
[6] Abraham, R.: Foundations of mechanics, 1967.

PART II

NUMERICAL AND STATISTICAL ANALYSIS

STABILIZATION, MANIPULATION AND ANALYTIC STEP ADAPTION

J. Baumgarte

Lehrstuhl für Mechanik, Technische Universität,
Braunschweig, Pockelsstrasse 4, West Germany

ABSTRACT. In order to reduce the error propagation in the numerical calculation of orbits two methods of stabilization of the differential equations of the perturbed Keplerian motion are offered: (1) a nonconservative method by asymptotical stabilization of the energy relation, (2) a conservative method by manipulation of the Hamiltonian. In both methods the stabilization is combined with the introduction of a new independent variable in order to achieve an analytic step size adaption.

1. INTRODUCTION

In this paper we consider the implications of stabilization of the differential equations of motion with respect to the application to the numerical calculation of orbits. It is obvious, and also shown by numerical experiments, that Liapunov - stable differential equations develop fewer unavoidable numerical errors during a numerical integration than Liapunov - unstable ones. The reason is that a numerical error during the nth step produces erroneous initial values for the next step. Thus the subsequent integration of unstable differential equations deteriorates in a much more unstable manner than does the integration of stable differential equations.

It is well known that the classical differential equations of the Keplerian motion are unstable in the sense of Liapunov. In order to improve the accuracy and efficiency during the numerical

V. Szebehely and B.D. Tapley (eds.), Long-Time Predictions in Dynamics, 153–163. *All Rights Reserved.*
Copyright ©1976 *by D. Reidel Publishing Company, Dordrecht-Holland.*

calculation of elliptic orbits we propose in this paper two methods of stabilization of the Keplerian problem.

The first method is nonconservative, the other method is purely conservative with respect to its physical character. In both stabilization methods the following facts are valid in general.

1. In Keplerian motion the period T_τ and therefore also the frequency $2\pi/T_\tau$ depend on the initial conditions. The stabilization is performed by modifying the differential equations of the motion in such a way that the period T_τ, or the frequency $2\pi/T_\tau$ respectively, is independent of the initial conditions.

2. This modification is performed by introducing an element p_o, which is by definition the negative total energy. The introduction of p_o is the presupposition for the stabilization of the differential equations of motion. The element p_o is a constant in the unperturbed motion, or when the perturbation is conservative. The value of p_o is fed once and for all into the computer. In nonconservative cases the differential system is supplemented by a differential equation for the slowly varying element p_o. Such an element differential equation is affected only by a very small error production.

3. The strict Liapunov - stability only exists in the case of the pure elliptic Keplerian motion. In all perturbation cases we have lost in principle the strict stability. But the remaining instability, which is of the order of the perturbation parameter ε, is very small compared with the removed instability. Therefore we will speak about ε-stability. Stabilization is at first the stabilization of the differential equations of the pure (elliptic) Keplerian problem, under the presupposition that the initial value of the element p_o will not be varied. Then the applications to a perturbed problem is always easily handled [1, 2, 4, 13].

4. The stabilization is directly related to the error propagation . It appears from the results of the numerical experiments that the positive effects of the stabilization of the pure Keplerian motion carry over to the perturbed problems [9, 14]. The experiments show that the error propagation in the stabilized equations is more favorable in contrast to that in the unstabilized ones. Namely, a linear error growth occurs in stabilized differential equations and a quadratic (superlinear) error growth occurs in unstabilized equations of motion.

5. In the case of highly eccentric orbits or close approaches to third bodies, it is practical to use an indepenent variable s different from the time t in order to perform an analytic step adaption (often called analytical step size control) during the numerical integration. This procedure is called time transformation, and s is called fictitious time. With respect to analytic step size adaption the following remark is in order. Let

$$\dot{\underline{y}} = \underline{f}\,(t,\underline{y}) \qquad (1)$$

be a differential system for the unknown vectorial function $\underline{y}(t)$. An analytical step adaption is performed by introducing a new independent variable s by means of a relation

$$dt = \mu(\underline{y},t)ds$$

where $\mu(\underline{y},t) > 0$ is an adopted scaling function such that an integration step of size ds produces a step $dt = \mu\, ds$ with respect to t. There results the differential system

$$\underline{y}' = \frac{d\underline{y}}{ds} = \mu(\underline{y},t)\;\underline{f}(\underline{y},t)$$
$$t' = \mu \quad , \quad ' = \frac{d}{ds} \qquad (2)$$

for the unknown $\underline{y}(s)$ and $t(s)$.
In this paper we always combine the stabilization with the step size adaption.

6. Let us assume that the system (1) is Liapunov - stable, than the set (2) may be L apunov - unstable. But this assumption is harmless for the following reasion. We are only interested in the relation between $\underline{y}$ and t, which is obtained by elimination of s in the solution $\underline{y}(s)$, t(s) of the system (2). It is obvious that this relation does not suffer from instability.

As a corollary it follows: let us assume that the differential system (1) was stabilized by adding, for instance, a control term [1,2,3,4,7]. Then this stabilization is not spoiled by an analytic step adaption, because the stability of $\underline{y}$ as a function of t is preserved.

2. NONCONSERVATIVE STABILIZATION

In a nonconservative stabilization the frequency is fixed by asymptotic stabilization of the energy relation. This method is often described in the literature [1,2,3,4,5,8,12]. The stabilizing effect is obtained by adding the control term to the original system of differential equations of motion. A control term, or a control vector respectively, is an expression which vanishes at the exact solution and which products the stability by its presence in the erroneous numerical integration, the stability by its presence.

Let be

$$H = T(p_i, q_i) + U(q_i, t) \tag{3}$$

be the Hamiltonian expressed as the sum of the kinetic energy T $T(p_i, q_i)$ and the potential energy $U(q_i, t)$, where

$$U(q_i, t) = -\frac{K^2}{r} + \varepsilon\, V(q_i, t) \quad ; \quad r = r(q_i) \tag{4}$$

(K^2: gravitational parameter, r: distance, ε: perturbation parameter, mass: m = 1) and where $\varepsilon P_i(q_k, p_k, t)$ are the nonconservative forces.

We add to the original system (with time transformation) a stabilizing force (control vector), which has the direction of p_i [2], in such a way that the modified system obeys the differential relation

$$[H(p_i, q_i) + p_o]' + \gamma[H(p_i, q_i) + p_o] = 0 \quad ; \quad \gamma > 0 \tag{5a}$$

This requirement leads to the following stabilized equations of motion:

$$p_i' = \{-\frac{\partial H}{\partial q_i} + \varepsilon P_i\}\,\mu - \frac{\gamma}{\sum\limits_{(k)} p_k \frac{\partial H}{\partial p_k}}\,[H + p_o]\,p_i \tag{6a}$$

$$q_i' = \frac{\partial H}{\partial p_i}\,\mu = \frac{\partial T}{\partial p_i}\,\mu \tag{6b}$$

$$p_o' = -\{\frac{\partial H}{\partial t} + \varepsilon \sum_{(k)} P_k \frac{\partial H}{\partial p_k}\}\,\mu = -\varepsilon\,\{\frac{\partial V}{\partial t} + \sum_{(k)} P_h \frac{\partial T}{\partial p_k}\}\,\mu \tag{6c}$$

$$t' = \mu \; ; \quad \mu = \mu(p_i, q_i, p_o, t) > 0 \tag{6d}$$

The expression with the scalar factor $\gamma > 0$ in eqn. (5a) is the control term, or the control-vector respectively, which represents the stabilizing force. The factor $[H+p_o]$ is the energy relation which vanishes in the exact solution but not in the computed solution.

The addition of the control vector (manipulation of the differential system) performs an asymptotic stabilization of the energy relation. We see, that by multiplying eqn. (5a) by $[H+p_o]$ we obtain

$$\frac{1}{2}\frac{d}{ds}[H + p_o]^2 = -\gamma[H + p_o]^2 \leq 0 \quad , \quad \gamma > 0 \tag{5b}$$

which shows the asymptotic stabilization of the energy relation, provided $\gamma>0$, where γ may be otherwise arbitrarily chosen. It is advantageous to choose γ as a constant. Numerical experiments show that the choice of the constant γ is not very critical [9]. A good value is

$$\gamma \leq \frac{2\pi}{S_r} \quad , \tag{7}$$

where S_r is the period with respect to the fictitious time s.

The case $\mu = 1$, $ds = dt$, leads to the perturbed, but stabilized problem with the physical time t as independent variable. It is shown in other publications [2,4,8] that the pure Keplerian problem ($\varepsilon = 0$, $p_o = \text{const}$) in the form (6) with t as independent variable is <u>Liapunov</u> - stable, since

$$\frac{d}{dt}[H + p_o]^2 \leq 0 \quad , \tag{8}$$

which also implies asymptotic stability for the energy relation. Therefore in the system (6a-d) the ε-stability for the dependent variables q_i, p_i, p_o as functions of the physical time t is guaranteed. This is sufficient, because we only need the ε-stability of $q_i(t)$, $p_i(t)$, $p_o(t)$ and no more.

<u>Remark</u>: The usual choice of μ is $\mu = \mu(q_i)$, especially

$$\mu = c\, r^n \quad , \tag{9}$$

(r;distance) with c as a constant numerical scaling factor and the constant n varing form 0 to 2. In this case we can obtain, after elimination of the p_i, a system for the accelerations q_i'') which may be useful somtimes during numerical integration.

3. CONSERVATIVE STABILIZATION BY MANIPULATION OF THE HAMILTONIAN

It is characteristic of all conservative stabilization methods, that they make the frequency (or frequencies) independent of the initial conditions [6,7], without introduction of a nonconservative control vector.

The fact, that the frequencies of any multidimensional harmonic oscillator are apriori constants and are, therefore independent of the initial conditions is exactly the reason for the oscillator's Liapunov-stability. The aim, of making the frequency of the Keplerian motion an apriori constant, is therefore equivalent to replacing the given original Hamiltonian $H(p_i,q_i)$ by a new (manipulated) Hamiltonian H^* which is the Hamiltonian of a "quasi-oscillator", leading to Liapunov - stable differential equations of motion [6,7]. The direct way to find H^* uses the theory of transformation into action and angle variables. The relationship between the problem of stability and the action-angle transformation was found by Sigrist [12].

The canonical transformation into action variables α_k and angle variables β_k

$$q_i = q_i(\alpha_k, \beta_k) \; ; \quad p_i = p_i(\alpha_k,\beta_k) \tag{10}$$

has the following well known properties:

1. The functions (10) are 2π-periodic with respect to the angle variables β_k.
2. The tranformed Hamiltonian depends only on the action variables α_k,

$$H = H(\alpha_k) \; . \tag{11}$$

The corresponding equations of motion are:

$$\dot{\alpha} = - \frac{\partial H}{\partial \beta_k} = 0; \quad \dot{\beta}_k = \frac{\partial H}{\partial \alpha_k} = \omega_k(\alpha_r) \; ; \quad \dot{} = \frac{d}{dt} \tag{12}$$

It is seen that the action variables α_k are constants during the actual motion. It is most important to observe that the numerical values of these constants depend, in general, on the initial conditions. The right-hand sides of the $\dot{\beta}_k$-equations (12) are called "frequencies" $\omega_k(\alpha_r)$, which depend also, in general, on the initial conditions.

In the case of Keplerian motion there are canonical elements such that two frequencies vanish and the third one is essentially the reciprocal of the period multiplied by 2π. The Keplerian motion is called, therefore, completely degenerated.

With respect to our stabilizing procedure we need only some facts of the theory of action-angle transformation. But we use, in the stabilized equations of motion, the old variables p_i and q_i.

With respect to this stabilization method it is sufficient to know only two facts:

1. In our special case of complete degeneration, the original Hamiltonian H depends, after the action-angle transformation, only on a single action variable, the intrinsic action variable, which we will call α. This leads to

$$H = H(\alpha) \cdot \tag{13}$$

It follows

$$\omega = \frac{dH}{d\alpha} \tag{14a}$$

and

$$\omega = \omega(\alpha) = \omega(\alpha(H)) = \omega(H) \ . \tag{14b}$$

This is the only knowledge we need of the action-angle transformation.

2. We have to know, that the relation between frequency and energy is established in the elementary theory of elliptic Keplerian motion by Kepler's third law. With this knowledge we obtain for $\omega = \omega\ (\gamma_0)$, or $\omega = \omega(H)$ respectively:

$$\omega(p_0) = \frac{(2p_0)^{3/2}}{K^2} \ ; \quad \omega(H) = \frac{(-2\ H)^{3/2}}{K^2} \tag{15a,b}$$

With these preparations we go over to the manipulated (stabilized) Hamiltonian H^* for the unperturbed case, and we write:

$$H^* = - \frac{1}{\frac{df(p_0)}{dp_0}} f(-H) \ ; \qquad H = H\ (p_i, q_i) \ , \tag{16}$$

where the function f will be discussed later. The manipulated Hamiltonian H^* leads to the following manipulated system of differential equations of motion:

$$\dot{p}_i = -\frac{\partial H^*}{\partial q_i} = -\lambda\frac{\partial H}{\partial q_i} \quad ; \quad \dot{q}_i = \frac{\partial H^*}{\partial p_i} = \lambda\frac{\partial H}{\partial p_i}$$

$$\lambda = -\frac{\dfrac{df(-H)}{dH}}{\dfrac{df(p_o)}{dp_o}} \tag{17a,b,c}$$

We see that in the exact analytical solution the factor λ is equal to 1 Therefore the system (17) is equivalent, with respect to the exact analytical solution, to the original equations of motion

$$\dot{p}_i = -\frac{\partial H}{\partial q_i} \quad ; \quad \dot{q}_i = \frac{\partial H}{\partial p_i} , \tag{18}$$

produced by the original Hamiltonian $H(p_i,q_i)$.

But the important fact is that the behavior of stability of system (17) may be different from system (18), resulting from the special choice of the function f. This implies that the erroneous computed solutions of the sets (17) and (18) may show a much different error propagation. Therefore we choose the function f in such a way that in eqn. (16), H^* is the Hamiltonian of a "quasi-oscillator" which means, that the frequency is an apriori constant This implies, that H^* has to be linear in its intrinsic action variable α. Thus,

$$H^* = \Omega\alpha , \tag{19}$$

where Ω is given apriori and $\alpha = \alpha(H)$. These conditions are fulfilled if we choose

$$\Omega = -\frac{1}{\dfrac{df(p_o)}{dp_o}} = \omega(p_o) = \frac{(2p_o)^{3/2}}{K^2} , \tag{20a}$$

or

$$\frac{df(p_o)}{dp_o} = -K^2(2p_o)^{-3/2} , \tag{20b}$$

and

$$\alpha(H) = f(-H) \ . \tag{21}$$

By integration of eqn. (20b) we find the function f, from which we obtain the stabilized Hamiltonian H^*[6,7]:

$$H^* = \frac{(2p_o)^{3/2}}{\sqrt{-2H}} \ , \tag{22}$$

producing the stabilized equations of motion

$$\dot{p}_i = -\lambda \frac{\partial H}{\partial q_i} \ ; \quad \dot{q}_i = \lambda \frac{\partial H}{\partial p_i} \quad ; \quad \lambda = \left(\frac{p_o}{-H}\right)^{3/2} \ , \tag{23}$$

where $H = H(p_i, q_i)$ is the original Hamiltonian.

We now perform a time transformation by introducing a new independent variable s. We define s by the differential relation

$$\frac{dt}{ds} = t' = \tilde{\mu}(q_i, p_i, t, p_o) > 0 \ . \tag{24}$$

The differential system (23) is transformed into the new one

$$p_i' = -\tilde{u}\lambda \frac{\partial H}{\partial q_i} ; \quad q_i' = \tilde{\mu}\lambda \frac{\partial H}{\partial p_i} \quad ; \quad t' = \tilde{\mu} \ ; \lambda = \left(\frac{p_o}{-H}\right)^{3/2} . \tag{25}$$

Let us adopt the following practical choice of $\tilde{\mu}$:

$$\tilde{\mu} = \mu \frac{1}{\lambda} \ , \tag{26}$$

where $\mu = \mu(q_i, p_i, t, p_o) > 0$ is the prescribed scaling function for the time transformation (e.g., $\mu = c\, r^n$). Thus system (25) is reformulated as:

$$p_i' = -\frac{\partial H}{\partial q_i} \mu \ ; \quad q_i' = \frac{\partial H}{\partial p_i} \mu \ ; \ t' = \frac{1}{\lambda}\mu ; \quad \frac{1}{\lambda} = \left(\frac{-H}{p_o}\right)^{3/2} \tag{27}$$

The advantage of this set of equations is that only the t' - equation in (27) is supplemented by a control factor. This stabilization procedure modifes the original, but time transformed system as little as possible.

We now list the equations of the perturbed problems:

$$p_i' = \{-\frac{\partial H}{\partial q_i} + \varepsilon P_i\} \mu \tag{28a}$$

$$q_i' = \frac{\partial H}{\partial p_i} \mu = \frac{\partial T}{\partial p_i} \mu \tag{28b}$$

$$p_o' = - \{\frac{\partial H}{\partial t} + \varepsilon \sum_{(j)} P_j \frac{\partial H}{\partial p_j} \}\mu = - \varepsilon\{\frac{\partial V}{\partial t} + \sum_{(j)} P_j \frac{\partial T}{\partial p_i} \}\mu \tag{28c}$$

$$t' = (\frac{-H}{p_o})^{3/2} \mu \tag{28d}$$

Even in this set only the t'-equation (28d) is supplemented by a control factor in order to stabilize the differential equations.

REMARKS

1. After elimination of s the solutions $q_i(t)$, $p_i(t)$, $p_o(t)$ are ε-stable (cf. remark introduction).

2. In the case $\mu=cr^n$ the elimination of the p_i leads to q_i''-equations.

3. In the case $\mu=c\, r^{3/2}$ the variational equations of the eqn. (28) for a circular motion of the unperturbed Keplerian problem show Liapunov-stability.* Thus not only the solutions $q_i(t)$, $p_i(t)$ are stable but also the functions $q_i(s)$, $p_i(s)$, $t(s)$.

4. Numerical experiments show that a combination of the non conservative stabilization method (cf. eqn. (6a-d)) with this conservative method may be useful with respect to error reduction. We have only to replace the p_i' - equation (28a) by

$$p_i' = -\{-\frac{\partial H}{\partial q_i} + \varepsilon P_i\}\mu - \frac{\gamma}{\sum_{(j)} p_j \frac{\partial H}{\partial p_j}} [H + p_o] p_i \,. \tag{29}$$

By doing this it may be advantageous to choose now γ smaller than usual, that is in the case of system (6a-d), in order to affect the p_i'-equations only in a very small manner.

*W. Machaelis, unpublished.

REFERENCES

1. Baumgarte, J., "Numerical Stabilization of the Differential Equations of Keplerian Motion", Celestial Mech. 5, (1972), 490-501.

2. Baumgarte, J., "Stabilization of Constraints and Integrals of Motion in Dynamical Systems", Computer Methods in Applied Mechanics and Engineering, 1, (1972), 1-16, North-Holland Publishing Company, Amsterdam.

3. Baumgarte, J., Zeitschrift für angewandte Mathematik und Mechanik (ZAMM) 53, (1973), 701-704.

4. Baumgarte, J., "Stabilization of the Differential Equations of Keplerian Motion", in Recent Advances in Dynamical Astronomy, (1973), 38-44, B. D. Tapley and V. Szebehely (eds), D. Reidel Publishing Company, Dordrecht-Holland.

5. Baumgarte, J., "Stabilized Kepler Motion Connected with Analytic Step Adaption", to appear in Celestial Mech. (1975).

6. Baumgarte, J., and E. Stiefel, "Stabilization by Manipulation of the Hamiltonian", Celestial Mech., 10, (1974), 71-85.

7. Baumgarte, J., and E. Stiefel, Mechanics Research Communications, 1, (1974), 49-50.

8. Baumgarte, J., and E. Stiefel, "Examples of Transformations Improving the Numerical Accuracy of the Integration of Differential Equations", in D. G. Bettis (ed.) Lecture Notes in Mathematics, Springer-Verlag, Heidelberg, vol. 362, (1974), 207-236.

9. Beandet, P.R., "Dynamic Stabilization and Error Propagation in Perturbed Keplerian Motion", Prepared by Computer Sciences Corporation for Goddard Space Flight Center, (1974), Contract No. NAS 11999, Task, Assignment No. 086.

10. Hochfeld, E., (ed), "Stabilization of Computer Circuits", University of Chicago, Nov. 1957, Available from U.S. Government Printing Office, Washington D.C., as WADC TR 57-425 (AD 155740).

11. Nacozy, P.E., Astrophys. Space Sci. 14, (1971), 40-51.

12. Sigrist, N., "Restricted Liapunov Stability and Stabilization of Ordinary Differential Equations", Journal of Applied Mathematics and Physics (ZAMP), 25, (1974), 37-46.

13. Stiefel, E.L., and G. Scheifele, "Linear and Regular Celestial Mechanics, Springer-Verlag, (1971).

14. Velez, C. E., Celestial Mech. 10, (1974), 405-422.

ON THE INTERPRETATION OF LEAST SQUARES COLLOCATION*

B. D. Tapley

Department of Aerospace Engineering and Engineering Mechanics, The University of Texas at Austin, Austin, Texas.

1. INTRODUCTION

One of the central issues in current geodetic data reduction is the question of how noise-corrupted measurements of physical parameters, derived from two independent measurement processes, can be combined to obtain an optimal estimate of the parameters. The problem is complicated further when there is apriori information on a subset of the parameters to be estimated. The problem of combining surface measurements of gravitational anamolies with observations of satellites to determine the earth's geopotential coefficients is an important example of this class of problem. The difficulties with this problem lead Moritz[1] to propose the least squares collocation algorithm as a general approach to geodetic data reduction. Subsequent investigations have claimed to validate Moritz's conclusion that the least squares collocation method is a more general parameter estimation method than the classical least squares, minimum variance or maximum likelihood methods[2,3].

In Ref. (4), Tapley and Schutz demonstrated a partial equivalence between the least squares collocation and the minimum variance or least squares algorithms. In this discussion, it was pointed out that the least squares collocation algorithm can be inefficient computationally if the number of observations exceeds the number of parameters and a more efficient algorithm was suggested.

---

* This research was sponsored by NASA under grant No. NGR 44-012-283

V. Szebehely and B.D. Tapley (eds.), Long-Time Predictions in Dynamics, 165–172. *All Rights Reserved. Copyright © 1976 by D. Reidel Publishing Company, Dordrecht-Holland.*

The purpose of this presentation is to demonstrate a strict mathematical equivalence between the least squares collocation and the classical minimum variance estimates. Specifically, it is shown that the least squares collocation algorithms are a special case of the modified minimum variance estimates. Finally, the computational efficiency of several forms of the general minimum variance estimation algorithm is discussed.

2. THE GENERAL PROBLEM

For any given parameter estimation problem assume that the nx1 parameter vector x can be separated into two components; a px1 vector z for which there is no aprori information and a qx1 vector s for which there is aprori information. In this case,

$$x^T = [z^T : s^T] \tag{1}$$

Assume further that the observations and the parameters are related through the linear model

$$y = Hz + Bs + v \tag{2}$$

where s and v are random vectors whose aprori statics satisfy the conditions

$$\begin{aligned} E[s] &= \bar{s} \quad , \qquad E[(s - \bar{s})(s - \bar{s})^T] = C \\ E[v] &= 0 \quad , \qquad E[v\, v^T] = R \end{aligned} \tag{3}$$

This model assumes that there is no aprori information on the parameter subset, z, while there is aprori information for the parameters subset, s. Moreover, assume that the aprori estimates of s is s. This latter information is equivalent to the statement

$$\bar{s} = s + \eta \tag{4}$$

where the error in the aprori estimate η is assumed to be distributed with zero mean and covariance, C.

The problem considered then is that of finding the best linear unbiased estimate of z and s, such that the variance in the estimate is minimized, given the linear observation relations (1) and (4) and the aprori statistics (3).

Eqs. (1) and (4) can be combined to obtain the following set of observations.

$$\begin{aligned} y &= Hz + Bs + v \\ \bar{s} &= Is + \eta \end{aligned} \tag{5}$$

Now if the following definitions are used

$$\tilde{y} = \begin{bmatrix} y \\ \cdots \\ s \end{bmatrix}, \quad \tilde{H} = \begin{bmatrix} H & \vdots & B \\ \cdots & \vdots & \cdots \\ 0 & \vdots & I \end{bmatrix}, \quad \tilde{v} = \begin{bmatrix} v \\ \cdots \\ \eta \end{bmatrix}, \quad x = \begin{bmatrix} z \\ \cdots \\ s \end{bmatrix}, \tag{6}$$

Eqs. (5) can be expressed as

$$\tilde{y} = \tilde{H}x + \tilde{v} \tag{7}$$

where $\tilde{v}$ satisfies the aprori statics

$$E[\tilde{v}] = 0 \quad , \quad e[\tilde{v}\ \tilde{v}^T] = \tilde{R}. \tag{8}$$

The augmented observation error covariance matrix $\tilde{R}$ is defined as follows:

$$R = \begin{bmatrix} E[v\ v^T] & E[v\ \eta^T] \\ E[\eta\ v^T] & E[\eta\ \eta^T] \end{bmatrix} = \begin{bmatrix} R & 0 \\ 0 & C \end{bmatrix} \tag{9}$$

where it is assumed that the observation errors v are not correlated with the errors in the aprori estimate of the parameters s, i.e., $E[v\ \eta^T] = 0$

For the linear model represented by Eq. (7), the best linear unbiased (BLUE) estimate, $\hat{x}$, which minimizes the covariance matrix associated with the estimate is

$$\hat{x} = [\tilde{H}^T\tilde{R}^{-1}\tilde{H}]^{-1}\ \tilde{H}^T\tilde{R}^{-1}\tilde{y} \tag{10}$$

The covariance matrix, P, associated with the state estimate error is given by

$$P = [\tilde{H}^T\tilde{R}^{-1}\tilde{H}]^{-1} \tag{11}$$

Using Eqs. (6); Eq. (10) can be expanded to obtain

$$\begin{bmatrix} \hat{z} \\ \hat{s} \end{bmatrix} = \begin{bmatrix} H^TR^{-1}H & H^TR^{-1}B \\ B^TR^{-1}H & B^TR^{-1}B+C^{-1} \end{bmatrix}^{-1} \begin{bmatrix} H^TR^{-1}y \\ B^YR^{-1}y+C^{-1}\bar{s} \end{bmatrix} \tag{12}$$

Eq. (12) is the general algorithm for computing the linear unbiased minimum variance estimate when there is an aprori estimate for a subset of the parameter vector.

2.1 INVERSION BY PARTITIONING

In the event that explicit solutions for z and s are sought, Eq. (12) can be rewritten as follows.

$$\begin{bmatrix} H^T R^{-1} H & H^T R^{-1} B \\ B^T R^{-1} H & B^T R^{-1} B + C^{-1} \end{bmatrix} \begin{bmatrix} \hat{z} \\ \hat{s} \end{bmatrix} = \begin{bmatrix} H^T R^{-1} y \\ B^T R^{-1} y + C^{-1} \bar{s} \end{bmatrix} \tag{13}$$

If Eq. (13) is expanded, the following system of linear equations is obtained.

$$\begin{aligned} (H^T R^{-1} H)\, \hat{z} + (H^T R^{-1} B)\, \hat{s} &= H^T R^{-1} y, \\ (B^T R^{-1} H)\, \hat{z} + (B^T R^{-1} B + C^{-1})\, \hat{s} &= B^T R^{-1} y + C^{-1} \bar{s} , \end{aligned} \tag{14}$$

If the coefficient matrix of $\hat{s}$ in the second of Eqs. (14) is positive definite, then the second set of equations can be rearranged to yield

$$\hat{s} = (B^T R^{-1} B + C^{-1})^{-1} \{B^T R^{-1} [y - Hz] + C^{-1} \bar{s}\}^{-1} \tag{15}$$

From the fundamental identities in Appendix A, it follows that

$$\begin{aligned} (B^T R^{-1} B + C^{-1})^{-1}\, B^T R^{-1} &= C B^T (B C B^T + R)^{-1} \\ (B^T R^{-1} B + C^{-1})^{-1} &= C + C B^T (B C B^T + R)^{-1} B C \end{aligned} \tag{16}$$

and Eq. (15) can be expressed as

$$\hat{s} = \bar{s} + C B^T (B C B^T + R)^{-1}\, [y - H\hat{z} - B\bar{s}] \tag{17}$$

If $\hat{s}$, as defined by Eq. (17), is substituted into the first of Eqs. (14), the result can be rearranged algebraically to obtain the following expression.

$$[H^T (B C B^T + R)^{-1} H]\, \hat{z} = H^T (B C B^T + R)^{-1} [y - B\bar{s}]$$

Solving for $\hat{z}$ leads to the result

$$\hat{z} = [H^T (B C B^T + R)^{-1} H]^{-1}\, H^T (B C B^T + R)^{-1}\, [y - B\bar{s}] \tag{18}$$

If the assumption that $\bar{s} = 0$ is used, Eqs. (17) and (18) will reduce to the following expression.

$$\begin{aligned} \hat{z} &= [H^T (B C B^T + R)^{-1} H]\, H^T (B C B^T + R)^{-1} y \\ \hat{s} &= C B^T (B C B^T + R)^{-1}\, [y - H\hat{z}) \end{aligned} \tag{19}$$

The set of Eqs. (19) and (20) are identical to the Least Squares Collocation Algorithm given by Moritz. (See for example, Eqs. (2.35) and (2.38) in Ref. (1) where the interpretation $\overline{C} = (B^T R^{-1} B + C^{-1})^{-1}$ is appropriate). Since Eqs. (19) are only an alternate expression for Eqs. (12), it can be seen that the least square collocation algorithm is only a special case of the more general linear unbiased minimum variance estimate given by Eq. (12). Recall that Eq. (14) was derived from Eq. (12) by introducing the restrictive assumption that $E[s] = \overline{s} = 0$. In the discussion given by Moritz in Ref. (1), he refers to z as a set of parameters and s as the signal; both of which are to be estimated. However, in the general case represented by Eq. (12), such distinctions are not necessary. The complete state vector x is partitioned into two sets one for which there is no aprori information, z, and one for which there is an aprori estimate, s. The state vector may be constant parameters or values of the state of a time dependent process at a given epoch or a mixture of these quantities. Note that in Eq. (12) the restrictive assumption $\overline{s} = 0$, introduced by Moritz, is relaxed to obtain a more general algorithm.

Furthermore, if the parameters z are ignored, i.e., if there is aprori information on all of the parameters in x, the Eq. (2) reduces to

$$y = Bs + v$$

and Eqs. (19) reduce to the relation

$$\hat{s} = BC^T(BCB^T + R)^{-1}\, y \tag{20}$$

This is the expression used by Moritz and Schwarz to determine an estimate of the Spherical Geopotential Coefficients using Satellite Observations.[5] In this treatment, the satellite orbit elements are determined from an independent orbit determination process and then the rates of change in the orbit elements are used as observations to estimate corrections to the geopotential coefficients. Note that Eq. (2) will allow a simultaneous determination of the satellite orbit elements with the geopotential coefficients by assuming that z represents the satellite orbit elements while s represents the geopotential coefficients.

Note that Eq. (12) is the weighted least squares solution if the covariance matrices R and C are replaced by weighting matrices W_v^{-1} and W_η^{-1}. Finally, if the observation errors, v, and the aprori estimation error, η, have a gaussian distribution, then Eq. (12) yields the maximum likelihood estimate as well.

2. CONCLUSIONS

The previous results indicates that the least squares collocation results are obtained as a special case of the general linear unbiased minimum variance estimate. For certain geodetic application the least square collocation algorithm may be a more efficient formulation of the results from the point of view of the compuations required.

3. APPENDIX A - MATRIX IDENTITIES

In the previous discussion, the inverse of an n x n matrix was required. The matrix inverse was inverted in partitioned form. The explicit relations for the inverted matrix, in terms of the partitioned matrix, are obtained as follows.

Partition $\tilde{R}$, a symmetric positive definite matrix, to obtain:

$$\tilde{R} = \begin{bmatrix} A & B \\ B^T & C \end{bmatrix} \tag{A.1}$$

where A is an $\ell \times \ell$ positive definite matrix, D is a $q \times q$ positive definite matrix, and B is an $\ell \times q$ matrix. Define the inverse of R as follows:

$$\tilde{R}^{-1} \begin{bmatrix} D & E \\ E^T & F \end{bmatrix} \tag{A.2}$$

where D is an $\ell \times \ell$ positive definite matrix, F is a $q \times q$ positive definite matrix, and E is an $\ell \times q$ matrix. Then it follows that

$$\tilde{R}\,\tilde{R}^{-1} \begin{bmatrix} A & B \\ B^T & C \end{bmatrix} \begin{bmatrix} D & E \\ E^T & F \end{bmatrix} = \begin{bmatrix} I & 0 \\ 0 & I \end{bmatrix} \tag{A.3}$$

Carrying out the indicated multiplication leads to

$$\begin{aligned} AD + BE^T &= I \\ AE + BF &= 0 \\ B^TD + CE^T &= 0 \\ B^TE + CF &= I \end{aligned} \tag{A.4}$$

From the first and third equations, it follows that

$$D = [A - BC^{-1}B^T]^{-1}$$
$$E^T = -C^{-1}B^T[A - BD^{-1}B^T]^{-1} \tag{A.5}$$

Combining the second and fourth of Eqs. (A.4) leads to

$$E = -A^{-1} B[C - C^T A^{-1} B]^{-1}$$
$$F = [C - B^T A^{-1} B]^{-1}$$

Hence, the combined results are summarized as follows:

$$\begin{aligned} D &= [A - BC^{-1}B^T]^{-1} = A^{-1}[I - BD^T] \\ E^T &= -C^{-1}B^T[A - BC^{-1}B^T] \\ E &= -A^{-1}B[C - B^T A^{-1} B]^{-1} \\ F &= [C - B^T A^{-1} B]^{-1} = C^{-1}[I - B^T E] \end{aligned} \tag{A.6}$$

On substituting E^T into the last expression for D and E into the last expression for F and noting that $(E^T)^T = E$, the following additional important identities are obtained.

$$\begin{aligned} D &= [A - BC^{-1}B^T]^{-1} = A^{-1} + A^{-1}B(C - B^T A^{-1} B)^{-1} B^T A^{-1} \\ F &= [C - B^T A^{-1} B]^{-1} = C^{-1} + C^{-1} B^T [A - BC^{-1} B^T]^{-1} BC^{-1} \\ -E &= A^{-1}B[C - B^T A^{-1} B] = [A - BC^{-1}B^T]^{-1} BC^{-1} \end{aligned} \tag{A.7}$$

REFERENCES

1. Moritz, H., "Advanced Least-Squares Methods," Report No. 175, Department of Geodetic Science, Ohio State University, 1972.
2. Deutsch, R., Estimation Theory, Prentice Hall, 1965.
3. Tapley, B. D., "Statistical Orbit Determination Theory," Recent Advances in Dynamical Astronomy, Ed. by B. D. Tapley and V. Szebehely, D. Reidel Publishers, 1973, pp. 356-425.
4. Tapley, B. D., and B. E. Schutz, "A Comparison of Estimation Methods for the Reduction of Laser Range Observations," Proceedings of the Symposium on Earth's Gravitional Field and Secular Variations in Position, Ed. by R. S. Mather and P. V. Angus-Leffan, The University of New South Wales, Sydney, Australia, December 1973, p. 489-508.

5. Moritz, H., and K. Schwarz, "On the Computation of Spherical Harmonics from Satellite Observations," Fifth Symposium on Mathematical Geodesy, Florence, Italy, 1972.

6. Schwarz, K., "Least Squares Collocation for Large Systems," Sixth Symposium on Mathematical Geodesy, Siena, April, 1975.

A NOTE ON STABILIZATION IN THREE-BODY REGULARIZATION

S. J. Aarseth

Institute of Astronomy, Cambridge, England

ABSTRACT. The time transformation $t' = R_1R_2$ in the three-body regularization method of Aarseth and Zare is modified to include the potential energy, thereby enabling the physical time to be obtained explicitly. A new recommendation for stabilizing the equations of motion further is considered briefly. It is concluded that a third type of time transformation, $t' = R_1R_2/(R_1+R_2)^{1/2}$, appears to offer the best prospects in practical calculations unless the time itself is required very accurately.

The following discussion is based on the regularization method of the general three-body problem due to Aarseth and Zare (1974). Briefly, the relative motions of two mass-points m_1 and m_2 with respect to m_3 are described by two sets of regularized coordinates and momenta, $\underline{Q}_k$, $\underline{P}_k (k=1,2)$, each variable having four components. The physical time t is related to the fictitous time τ by the transformation

$$dt = R_1R_2\, d\tau \ , \tag{1}$$

where R_1 and R_2 are the respective distances. It is well known that the numerical solution of Equation (1) is subject to instabilities. On the other hand, the equations of motion represent two coupled

V. Szebehely and B.D. Tapley (eds.), Long-Time Predictions in Dynamics, 173–177. All Rights Reserved.
Copyright ©1976 by D. Reidel Publishing Company, Dordrecht-Holland.

Keplerian systems, each of which would have stable solutions in the absence of perturbations. This raises the possibility that a significant part of the numerical errors may be associated with the time itself. It may be remarked, however, that the exact value of the time is not always required; indeed, that integration of Equation (1) may be omitted altogether if one is only interested in the final configuration satisfying certain conditions.

As shown in the original paper, the time transformation (1) removes the singularities of two Kepler motions with respect to the mass-point m_3. Although simple, this transformation is not unique. In the following we consider the alternative expression

$$dt = -U^{-1}\, d\tau \tag{2}$$

which makes use of the potential energy, U, of the three mass-points. This time transformation gives rise to the explicit relation (Baumgarte and Stiefel, 1974).

$$t = -\frac{\tau+C}{2E} + \frac{1}{2E}\sum_{i=1}^{3} \underline{r}_i\, \underline{p}_i \tag{3}$$

Here E is the total energy when the coordinates $\underline{r}_i$ and momenta $\underline{p}_i$ are expressed in the center of mass frame; the constant C is determined at t=0 when $\tau=0$. Making use of the center of mass condition $\underline{p}_3 = -(\underline{p}_1+\underline{p}_2)$ together with the relative coordinates $\underline{q}_k = \underline{r}_k - \underline{r}_3 (k=1,2)$, the summation simplifies to

$$\sum_{i=1}^{3} \underline{r}_i^T \underline{p}_i = \sum_{k=1}^{2} \underline{q}_k^T \underline{p}_k, \tag{4}$$

where T denotes the transpose and $\underline{r}_i$, etc., are column vectors.

Regularized variables are now introduced by means of the the relations $\underline{q}_k^T = \frac{1}{2} \underline{Q}_k^T \underline{A}_k$, $\underline{p}_k = \underline{A}_k^T \underline{P}_k / 4R_k$ which, together with the transformation property $\underline{A}_k \underline{A}_k^T = 4R_k$ for the matrices $\underline{A}_k$ finally reduces to the desired regular expression

$$\sum_{i=1}^{3} \underline{r}_i^T \underline{p}_i = \frac{1}{2} \sum_{k=1}^{2} \underline{Q}_k^T \underline{P}_k . \tag{5}$$

The final form of Equation (3) is completely analogous to that of the KS regularization (Stiefel and Scheifele, 1971, p. 85) and the case of the total energy being close to zero is therefore equally troublesome in the two formulations. This analogy may be pursued further by constructing an equivalent time element for perturbed three-body motion. Although some simplification can be in this derivation, the final result is probably too complicated to be of practical value and will not be given here.

Next we turn to the modifications of the equations of motion necessitated by the new time transformation. It may be noted that in the KS regularization Equation (2) has the same functional form as the standard time transformation $t'=r$, hence, the original equations of motion still apply. In order to make full use of the existing equations of motion it is convenient to write Equation (2) in the separable form

$$dt = g_1 g_2 \, d\tau \quad , \tag{6}$$

with $g_1 = R_1 R_2$ representing the transformation (1). Consequently we have

$$g_2 = (m_1 m_3 R_2 + m_2 m_3 R_1 + m_1 m_2 R_1 R_2/R)^{-1} \quad , \tag{7}$$

where R is the distance between m_1 and m_2. Following the procedure in the original paper, we modify the Hamiltonian in the extended phase space by

$$\hat{\Gamma} = g_2 \Gamma^* \quad , \tag{8}$$

where $\Gamma^* = g_1 \Gamma$ was introduced previously. This gives rise to the new equations of motion (cf. Zare and Szebehely, 1975, where $g = g_1 g_2$)

$$\frac{d\underline{Q}_k}{d\tau} = g_2 \frac{\partial \Gamma^*}{\partial \underline{P}_k} \quad , \tag{9}$$

$$(k=1,2)$$

$$\frac{d\underline{P}_k}{d\tau} = - g_2 \frac{\partial \Gamma^*}{\partial \underline{Q}_k} - \Gamma^* \frac{\partial g_2}{\partial \underline{Q}_k} \tag{10}$$

We note that these equations are still regular for collisions in which $R_1 \to 0$ or $R_2 \to 0$. Although the additional terms are readily available, evaluation of the modified equations are naturally more time-consuming per integration step. This extra effort may be worth while if it is desirable to know the physical time as accurately as possible.

From the definition of the Hamiltonian in the extended phase space, $\Gamma^* = 0$ on the solution path. It has been found empirically, however, that the numerical solution of the unperturbed Kepler problem is improved if the second term in Equation (10) is retained

(Baumgarte, 1972). A more recent discussion of time transformations in the extended phase space (Zare and Szebehely, 1975), recommends that, in order to have asymptotic stability for $|\Gamma|$, the second minus sign in Equation (10) should be reversed if the potential energy increases in absolute value. This novel suggestion has also been tried in the present method. On the basis of limited numerical experience (i.e., examples I and II in the original paper) it appears that including the Poincare control term does not improve the solutions, irrespective of whether the sign change is carried out. In the third example (perturbed equilateral triangle) the use of Equation (2) requires somewhat fewer integration steps to reach a given time; however, this result can also be achieved for an alternative time transformation discussed below. Two reasons may be advanced for each improvement in examples I and II. First, the integration may not have covered sufficiently long times to benefit from the explicit expression (3). Secondly, the errors associated with Equation (1) may not dominate in the present formulation where each Kepler motion is strongly perturbed, although in analogy with KS regularization one would expect a growth-rate proportional to τ (Peters, 1968).

As an alternative to the transformations (1) and (2) we consider briefly a third possibility given by

$$dt = \frac{R_1 R_2}{(R_1+R_2)^{1/2}} \, d\tau \ . \tag{11}$$

This time transformation should be used together with Equations (9) and (10) where now

$$g_2 = (R_1 + R_2)^{-1/2} \tag{12}$$

Experience so far indicates that the Poincare control term may be omitted without significant loss of accuracy, thereby saving the evaluation of Γ^*. One reason why Equation (11) is such a good time-step regulator may be connected with the property that $t' \propto R^{3/2}$ near triple collisions; this represents the limiting case for convergence as $R \to 0$ (Stiefel and Scheifele, 1971, p.78). This choice is also related to the time transformation introduced by Heggie (1974) to stabilize the regularized equations in the global three-body formulation. At the present time we therefore recommend the expression (11) together with the condition $\Gamma^*=0$ as the most efficient scheme for practical calculations.

REFERENCES

1. Aarseth, S. J., and K. Zare, 1974, Celes. Mech. 10, 185.
2. Baumgarte, J., 1972, Celes. Mech. 5, 490.
3. Baumgarte, J., and E. Stiefel, 1974, in D. G. Bettis, (ed.) Lecture Notes in Mathematics, Springer-Verlag, Heidelberg, Vol. 362, p. 207.
4. Heggie, D.C., 1974, Celes. Mech. 10, 217.
5. Peters, C.F., 1968, Ph.D. Thesis, Yale University.
6. Stiefel, E.L., and G. Scheifele, 1971, Linear and Regular Celestial Mechanics, Springer-Verlag, Heidelberg.
7. Zare, K., and V. Szebehely, 1975, Celes. Mech. 11, 469.

PART III

THREE AND MANY-BODY PROBLEMS

QUALITATIVE METHODS AND RESULTS IN CELESTIAL MECHANICS

C. Marchal

Office National d'Etudes et de Recherches Aerospatiales (ONERA), 92320 Chatillon, France

ABSTRACT. In a non integrable problem qualitative methods give partial but also rigorously demonstrated properties, they have been essentially developed by Poincaré, Sundman, Chazy, Khilmi, Merman, Sitnikov, Alexeev etc ...

In the n-body problem of Celestial Mechanics very sophisticated results on final evolutions can be obtained qualitatively; there are five main types of final evolution: (A) The triple or multiple collision at a finite instant. These collisions cannot be regularized in general (as can binary collisions) and the solution cannot remain uniquely defined after the collision. (B) The infinite expansion in bounded time. $t \to t_f$ implies $R \to \infty$; R being the diameter of the n-body system. (C) The super-hyperbolic expansion. $t \to \infty$ implies $(R/t) \to \infty$. Types B and C are very particular, they require $n > 4$ and the existence of very strong oscillations; some examples of these types can be built. (D) The hyperbolic expansion. $t \to \infty$ implies $R \sim kt$. The system is subdivided into subsystems escaping linearly and inside which mutual distances are at most of order $t^{2/3}$. (E) The parabolic or sub-parabolic expansion. $t \to \infty$ implies $R = 0\,(t^{2/3})$.

This last type is not completely analysed; there are solutions with parabolic escapes, bounded solutions, oscillatory solutions (for which lim sup $R = +\infty$ and lim inf $R < +\infty$) and even solutions without collision but for which for some mutual distance r_{ij} : lim inf $r_{ij} = 0$ (and thus leading to a collision in real physical cases).

With the measure theory the structure of bounded or oscillatory solutions can be improved, these solutions are almost always of the

V. Szebehely and B. D. Tapley (eds.), Long-Time Predictions in Dynamics, 181–208. *All Rights Reserved. Copyright © 1976 by D. Reidel Publishing Company, Dordrecht-Holland.*

same type for the past and for the future.

1. INTRODUCTION

When a system of differential equations is not analytically integrable there are two possible approaches of the solutions: the quantitative approach and the qualitative approach.

In a quantitative approach solutions are determined by numerical and/or analytical methods and results can be as accurate as desired in an interval of time as large as desired but generally, except for periodic solutions, the accuracy decreases for large times and there is an interval of time out of which the approximate solution has no more interest.

In a qualitative approach the purpose is to obtain some rigorously demonstrated property (generally valid for an infinite period of time) and thus to reduce the uncertainty of solutions.

Qualitative methods can be applied to existence theorems, regularization of singular points, problems of periodicity and stability etc... Integrals of motion are also related to these qualitative methods and their main domain of application is the study of asymptotic properties for infinitely large times, domain where precisely quantitative methods fail to give accurate informations.

2. THE n-BODY PROBLEM OF CELESTIAL MECHANICS NOTATIONS

We will use the usual notations.

The time is the parameter t.

The n masses will be either $m_1, m_2, \ldots m_n$ or sometimes $m_A, m_B, m_C \ldots$, they will be generally all positive and labeled in ascending order from 1 to n:

$$0 < m_1 \leq m_2 \leq m_3 ---- \leq m_{n-1} \leq m_n \tag{1}$$

$$\sum_{i=1}^{n} m_i = M = \text{total mass.} \tag{2}$$

We will generally use the set of axes of the center of mass and hence the radius-vectors $\vec{r}_i$ will be related by:

$$\sum_{i=1}^{n} m_i \vec{r}_i = 0 \tag{3}$$

We will sometimes use the set of axes of the center of mass of a <<subsystem>> or of a <<cluster>>.

Let us define as usual:

$$\vec{r}_j - \vec{r}_i = \overrightarrow{r_{ij}} \quad ; \quad ||\overrightarrow{r_{ij}}|| = r_{ij} \tag{4}$$

$$\vec{V}_i = d\vec{r}_i/dt = i^{th} \text{ velocity vector} \tag{5}$$

We will call G the constant of the law of Newton and we will generally consider the case of point-masses, the equations of motion are then:

$$d^2\vec{r}_i/dt^2 = G \sum_{j=1, j\neq i}^{n} m_j \vec{r}_{ij}/r_{ij}^{3} \tag{6}$$

and the corresponding integrals of motion are:

$$\vec{c} = \text{angular momentum} = \sum_{i=1}^{n} m_i \vec{r}_i \times \vec{V}_i \tag{7}$$

$$h = \text{energy} = \frac{1}{2} \sum_{i=1}^{n} m_i V_i^2 - \sum_{1\leq i<j\leq n} G m_i m_j/r_{ij} \tag{8}$$

The main parameters of the system are:

$$I = \text{half moment of inertia} = \frac{1}{2} \sum_{i=1}^{n} m_i r_i^2 \tag{9}$$

$$U = \text{potential} = \sum_{1\leq i<j\leq n} G m_i m_j/r_{ij} \tag{10}$$

$$T = \text{kinetic energy} = \frac{1}{2} \sum_{i=1}^{n} m_i V_i^2 \tag{11}$$

hence (8) can be written:

$$T = U + h \tag{12}$$

The Lagrange-Jacobi identity (or Viriel identity) is:

$$I'' = d^2I/dt^2 = U + 2h \tag{13}$$

and the Sundman inequality is:

$$4\,I\,T \geq c^2 + I'^2 = c^2 + (dI/dt)^2 \tag{14}$$

R will be the largest mutual distance and r the smallest:

$$R = \sup_{1\leq i<j\leq n} r_{ij} \quad ; \quad r = \inf_{1\leq i<j\leq n} r_{ij} \tag{15}$$

Taking account of (1) and (3) R, I, r and U are related by:

$$m_1 m_2/(m_1+m_2) \leq 2I/R^2 \leq M^*/M \quad ; \quad G\, m_1 m_2 \leq rU \leq GM^* \tag{16}$$

with:

$$M^* = \sum_{1\leq i<j\leq n} m_i\, m_j \tag{17}$$

hence R is of the order of magnitude of $I^{1/2}$ and r of the order of magnitude of U^{-1}.

We will define the length f as the smallest perimeter of triangles of 3 of the n bodies:

$$f = \inf_{1\leq i<j<k\leq n} (r_{ij}+r_{jk}+r_{ki}) \tag{18}$$

it is obvious that:

$$3r \leq f \leq r + 2R \tag{19}$$

We will give the subscript zero to the initial conditions: t_o, r_{io}, V_{io} etc... and we will assume that all r_{ijo} are positive.

The fixed parameters μ, α, p and the variable parameters ρ, ν, λ will be defined in sections (7) and (8.5).

Finally we will use the standard notations of equivalence when the time goes to some limite t_f finite or infinite:

$$A = O(B) \Longleftrightarrow \{A/B \text{ remains bounded when } t \to t_f\} \tag{20}$$

$$A = o(B) \Longleftrightarrow \{t \to t_f \Longrightarrow A/B \to 0\} \tag{21}$$

$$A \sim B \quad \Longleftrightarrow \{t \to t_f \Longrightarrow A/B \to 1\} \tag{22}$$

$$A \approx B \quad \{A/B \text{ and } B/A \text{ remain bounded when } t \to t_f\}. \tag{23}$$

3. SCALE CONSIDERATIONS

From a given solution $\vec{r_i}(t)$, $i = \{1,2,\ldots n\}$ of equations of motion (6) it is easy to deduce other solutions by a rotation, a symmetry or a translation (in space and/or in time). Khilmi has shown in [1] the importance of considerations of dimensions which are essentially related to another transformation of a solution, the following scale transformation:

If $\vec{r_i}(t)$; $i = \{1,2,\ldots n\}$ is a solution of the equations (6) of n-body problem, then $\vec{\rho_i}(t) = k^2\vec{r_i}(t/k^3)$; $i = \{1,2,\ldots n\}$ is another solution of these equations (6).

For this new solution there are: (24)

Velocity: $d\vec{\rho_i}/dt = k^{-1} \,.\, \vec{V_i}(t/k^3)$

Energy integral $= k^{-2}.\, h$

Angular momentum $= k\,\vec{c}$

The product c^2h remains constant in that transformation, it will be an essential parameter in the qualitative analysis of motion. Let us remember that in a two-body case c^2h is directly related to the eccentricity e of motion by:

$$2Mc^2h = G^2m_1{}^3m_2{}^3(e^2-1).$$

4. EXISTENCE AND UNIQUINESS OF SOLUTIONS

Usual existence theorems imply the existence and the uniqueness of the solutions of equations (6) of n-body motion as long as $\vec{r_i}$ vectors remain bounded and r_{ij} distances remain bounded awary from zero. Sundman has shown that binary collisions can be regularized [2] and thus a solution can be extended uniquely after such a collision. On the contrary triple and multiple collisions cannot be regularized in general and infinitely small differences in $\vec{r_i}$ and $\vec{V_i}$ vectors before the collision lead to large differences after the collision. On the other hand a triple or multiple collision corresponds to a zero of the function f(t) defined in (18) and it can be demonstrated [Ref. 3] that if f(t) is bounded away from zero on the finite interval $[t_1,t_2]$ the n $\vec{r_i}$ radius-vectors remain bounded on that interval; thus we obtain the following result:

After regularization of binary collisions the solution of a given n-body problem exists and remains uniquely defined as long as the function f(t) of (18) remains bounded away from zero; the functions $\vec{r_i}(t)$, $r_{ij}(t)$ and f(t) are then absolutely continuous in terms of t (but not the functions $\vec{V_i}(t)$ because of the binary collision).

We will see that f(t) can reach zero at some finite instant t_f by two different ways, either by an ordinary triple or multiple collision or by an <<infinite expansion in finite time>>; the second case is an <<essential singularity>> and the solution cannot remain uniquely defined after that singularity, the first case is also not regularizable for almost all values of the ratios of masses and it even seems that it is never regularizable as soon as we consider the non-collinear problem.

Near triple collisions have been studied numerically in [4-5] and it is easy to show that they represent a major event in a n-body problem and that they can lead to very surprising results.

Let us consider for instance three bodies m_1, m_2 and m_3 in a <<central configuration>> of alined type and in parabolic motion (figure 1), the three masses describe homothetic parabolas and if the two extreme masses are equal the third mass always remain at the center of masses (case of the figure 1); the energy integral h is equal to zero.

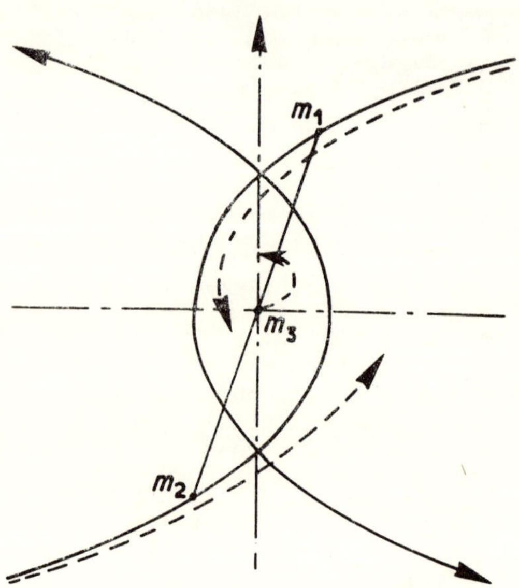

Fig. 1 - Three masses in <<central configuration>> and in parabolic motion.

In the vicinity of this first solution there are solutions of zero energy asymptotic to the parabolic solution for $t \to -\infty$ and leading for $t \to +\infty$ to the formation of a binary and to the hyperbolic escape of the third mass such as the solution in dotted line in figure one where m_2 has a positive final velocitu $\vec{V}_{\infty_2}$.

Let us apply the transformation defined in (24) to these two solutions with the ratio k going to zero; thus the energy remains equal to zero, the angular momentum goes to zero, the parabolic solution goes to a colinear solution in which

$$\vec{r_i} = \vec{A_i}\ |t|^{2/3} \quad ; \quad i = \{1,2,3\}$$

with a triple collision at $t = 0$, the second solution has the same limit for $t \leq 0$ but for $t > 0$ the hyperbolic escape of m_2 has a final velocity $\vec{V}_{\infty_2}/k$ going to infinity ..... Hence in the vicinity of triple collisions there are solutions leading to the formation of a very small binary and to the escape of the third body with a final velocity as large as desired! It is thus not surprising that near triple collisions be difficult to integrate numerically and that triple collisions cannot be regularized in general.....

5. INTEGRALS OF MOTION AND INTEGRAL INVARIANTS

Many people [Ref. 6-16] have tried to look for some new integral of motion independent of the usual integrals (related to the motion of center of masses, the angular momentum and the energy) but for both the restricted and the general problem they have only obtain negative theorems, for instance [Ref 8]: For any n a new integral of n-body problem cannot be algebraic with respect to the velocities (and arbitrary with respect to the positions).

The complexity of numerical studies [Ref. 15, 17-25] lead to think that the only possible integrals are the usual ones.

There are different kinds of integral invariants; we will see in section 9 the properties deduced from the conservative character of the n-body problem which is volume preserving in the phase space of vectors $\vec{r_i}$ and $\vec{V_i}$.

A linear integral invariant is a generalization of integrals to differential forms: it is a <<local integral>> valid to some order in the vicinity of a given orbit.

The following integral invariant is given in [Ref. 10, 24].

$$\sum_{i=1}^{n} m_i[2\ \vec{r_i}\delta\vec{V_i} + \vec{V_i}\delta\vec{r_i}] - 3t\ \delta h \qquad (25)$$

its derivative is of the second order in terms of the differential elements $\delta\vec{r}_i$, $\delta\vec{V}_i$, δh.

With the half moment of inertia I and its derivative $I' = \sum_{i=1}^{n} m_i \vec{r}_i . \vec{V}_i$ it is possible to improve this result, indeed the quantity:

$$\sum_{i=1}^{n} m_i [\vec{r}_i . \delta\vec{V}_i - \vec{V}_i . \delta\vec{r}_i] + 3\delta I' - 6t\, \delta h \qquad (26)$$

has a derivative of the third order in terms of the differential elements $\delta\vec{r}_i$, $\delta\vec{V}_i$, $\delta I'$ and δh.

In the restricted 3-body case there is only the well known Jacobi integral for circular motions of the primaries; on the other hand if the primaries have collinear motions the angular momentum of the third body around the axis of the primaries is again an integral of motion and if furthermore the energy of the primaries is zero the invariant integral (26) becomes:

$$\vec{r}_1\, \delta\vec{V}_1 - \vec{V}_1\, \delta\vec{r}_1 + 3\delta(\vec{r}_1 . \vec{V}_1) - 6t.\delta\left(\frac{V_1^2}{2} - \frac{Gm_2}{r_{12}} - \frac{Gm_3}{r_{13}}\right)$$

the infinitesimal body being m_1 and the time t being zero at the instant of the collision between m_2 and m_3.

6. PARTICULAR SOLUTIONS

The first known solutions of n-body problem are the Lagrange-Euler solutions of 3-body problem in which the configuration is invariable (and called <<central configuration>>) and the 3 bodies have a Keplerian motion of the same eccentricity around the center of masses. These solutions can be generalized to n bodies and we will meet them again for multiple collisions and for parabolic expansions.

The study of periodic solutions essentially began with Poincaré [Ref. 10] with the systematic research and the construction of these solutions; they all present past-future symmetries around proper instants (i.e. mutual distances $r_{ij}(t)$ are even functions of time with respect to these instants) and looking at the <<planetary problem>> (one major mass and two small masses Poincaré first divided the periodic orbits into:

A) Solutions with zero inclinations and small eccentricities.
B) Solutions with zero inclinations and large eccentricities.
C) Solutions with large inclinations and small eccentricities.

There are also solutions with large inclinations and large eccentricities and finally Poincaré conjectured that periodic solutions are dense in the set of bounded solutions [Ref.25].

Modern studies [Ref. 29-93] look for the structure of periodic solutions and for their stability in the linear approximation and when a periodic solution is unstable in that approximation there are solutions asymptotic to the periodic solution of interest.*

We will meet periodic solutions without past-future symmetry at the end of section 12.

7. FINAL EVOLUTIONS

Final evolutions of three-body systems have been classified by Chazy [Ref. 76-78] who pointed out the seven types of final evolution (hyperbolic, hyperbolic-elliptic etc....) and various criteria [Ref.79 - 87] have been built to recognize as early as possible, during a numerical integration, the type of the final evolution.

For instance, with $0 < m_1 \leq m_2$ and for arbitrary m_3, a criterium of final hyperbolic-elliptic evolution is the following [Ref. 84].

At some instant, for a proper length D:

$$r_{12} < D \quad ; \qquad r_3 > m_2 D/M$$

$$\frac{dr_3}{dt} \geq \left\{ \frac{2G(m_1+m_2)^2}{M} \left[\frac{m_1}{Mr_3-m_2D} + \frac{m_2}{Mr_3+m_1D} \right] + \sup\left[0; \quad \frac{2(m_1+m_2)}{Mm_3} \left(h + \frac{Gm_1m_2}{D}\right) \right] \right\}^{1/2} \tag{27}$$

After that instant the difference of the squares of the two members of the last inequation will ever increase and thus m_3 will have an hyperbolic escape. On the other hand the major axis of the binary m_1, m_2 will ever remain smaller than D and thus the final evolution will be of hyperbolic-elliptic type.

For n-body problems much less criteria exist (Ref. 1, 89).

---

*Most studies use the past-future symmetries to built periodic solutions, 2 successive past-future symmetries imply the periodicity.

and their domain of application is very restricted;* however it is still possible to classify the final evolutions with the help of the length λ of [Ref. 3] which is the largest distance between the center of mass of two complementary sets of the n bodies.

For that length λ:

$$\lambda \leq R < 2\lambda$$

and

$$d^2\lambda/dt^2 \geq - G\, M^3/\lambda^2(m_1+m_2)^2$$

These inequalities lead to the five types of final evolution described in the abstract. Consider in particular the variations of the function $(d\ /dt)^2 - 2G\, M^3/\ (m_1+m_2)^2$ and the eventual satisfaction of the condition:

$$d\lambda/dt > [2G\, M^3/\lambda(m_1+m_2)^2]^{1/2} \tag{28}$$

Various properties of the five types of final evolution are described in Ref. 3.

A three-body result of Sundman [Ref. 114] can be generalized to the n-body problem [Ref. 3]: if the angular momentum $\vec{c}$ is not zero the n-body system of interest cannot approach a complete collapse; its moment of inertia 2 I has a positive lower bound.

That result is obvious for the 4 first type of final evolution since I is a continuous function of t as long as the motion is defined, since I = 0 (complete collapse) requires $\vec{c} = 0$ and since I goes to infinity when t goes to its limit in the second, third and fourth types of final evolution.

It remains to study the fifth type, i.e. the parabolic or subparabolic expansion; in that case a positive lower bound of I can be given in terms of G, h, $\vec{c}$ and the n masses m_i.

Let us see it in the simple following way which also leads to various general inequalities and to some limitations on the possible motions.

Let us use again the notation defined in (17):

---

*Most criteria are related to a proper continuous function $G(\vec{x},t)$ of the state $\vec{x}$ and the time t: if along each solution $\vec{x}(t)$ the condition $G > G_0$ imply $dG/dt \geq 0$ then obviously: $G_0 < G[x(t),t]$ imply $\lim_{t \to \infty} > G[\vec{x}(t),t]$.

$$M^* = \sum_{1 \le i < j \le n} m_i m_j \tag{29}$$

and let us defined the following constants:

$$\mu = GM = \text{gravitational constant} \tag{30}$$

$$a = - GM^*/2\,h = \text{<<semi-major axis>>}$$

$$p = Mc^2/GM^{*2} = \text{<<semi-latus rectum>>} \tag{31}$$

the lengths a and p are indeed the semi-major axis and the semi-latus rectum of the relative motion in a two-body problem, in that case $p/a = 1 - e^2 \le 1$; in a n-body case it is possible that $p/a > 1$.

Finally let us define the variable lengths ρ and ν by:

$$M^* \rho^2 = \sum_{1 \le i < j \le n} m_i m_j \, r_{ij}^2 = 2MI \tag{32}$$

ρ is the <<mean quadratic distance>>.

$$M^*/\nu = \sum_{1 \le i < j \le n} m_i m_j / r_{ij} = U/G \tag{33}$$

ν is the <<mean harmonic distance>>.

It is obvious that $\rho \approx R$ and $\nu \approx r$, indeed from (16): $R^2 M\, m_1 m_2/M^*(m_1+m_2) \le \rho^2 \le R^2$ and $r \le \nu \le r\, M^*/m_1 m_2$ on the other hand the quadratic mean of positive numbers is larger than their harmonic mean, hence:

$$0 \le \nu \le \rho \tag{34}$$

the equality $\nu = \rho$ being obtained either for the two-body problem ($\nu = \rho = r_{12}$) or when all r_{ij} are equal ($\nu = \rho =$ all r_{ij}). (All improvement of (34) can be obtain in some particular cases, for instance for a galactic system [Ref. 108]: $0 \le [1, 2]^{3/2}\, \nu \le \rho$).

With these notations the Lagrange-Jacobi identity already written in (13) can also be written:

$$(\rho^2)'' = d^2(\rho^2)/dt^2 = 2\mu(1/\nu - 1/a) \tag{35}$$

and, with $\rho' = d\rho/dt$ the Sundman inequality of (14) becomes:

$$2/\nu \ge 1/a + p/\rho^2 + \rho'^2/\mu \tag{36}$$

The identity (35) and the inequalities (34) and (36) allow to give some limitations on the possible motions, for instance if ν is eliminated we obtain:

$$d^2(\rho^2)/dt^2 \geq \mu \cdot \sup\{2/\rho - 2/a;\ p/\rho^2 + \rho'^2/\mu - 1/a\} \qquad (37)$$

and it leads to the following results: some initial values ρ_o, ρ'_o, t_o are compatible with (37) and with the final values ρ_f, ρ'_f, t_f if and only if exist ρ'_1 and ρ'_2 such that:

I) $\rho'_1 \geq \rho'_o$; $\rho'_2 \leq \rho'_f$

II) the equality:

$$d^2(\rho^2)/dt^2 = \mu \cdot \sup\{2/\rho - 2/a;\ p/\rho^2 + \rho'^2/\mu - 1/a\} \qquad (38)$$

lead from ρ_o, ρ'_1, t_o to ρ_f, ρ'_2, t_f. (39)

Hence, in order to obtain informations on the future of the parameter ρ from the initial conditions ρ_o, ρ'_o, t_o, i.e. informations on the future of the moment of inertia 2 I, we can study the orbits given by the differential equation (39) and starting at ρ_o, ρ'_1, t_o with $\rho'_1 \geq \rho_o$.

Let us define the function j by:

$$j = \rho\, \rho'^2/2\mu + \rho/2a + p/2\rho \qquad (40)$$

j is proportional to the Sundman function $4(I'^2+c^2-4hI)(2I)^{-1/2}$ of Ref. 139].

With (34), (35), (36) it implies:

$$\left.\begin{array}{l}\rho/\nu \geq \sup\{1 \cdot j\}\\[2ex] j' = (\rho'/\rho)\cdot(\rho/\nu - j)\end{array}\right\}\text{hence:}\left\{\begin{array}{l}\rho'\ j' \geq 0\\[2ex] \rho'\ \{\rho(j-1)\}' \geq 0\end{array}\right. \qquad (41)$$

The equation (39) corresponds to $j' = (\rho'/\rho)\cdot\sup\{0, 1-j\}$ and is very easy to integrate, it always leads to a variation of ρ in terms of the time as the distance in a proper two-body problem.

Let $j_1 = \rho_o\rho_1'^2/2\mu + \rho_o/2a + p/2\rho_o$ be the initial value of j at ρ_o, ρ'_1, t_o. If $j_1 \leq 1$ the equation (39) gives $j' = (1-j)\rho'/\rho$ hence $\rho(1-j)$ is constant and equal to $\rho_o(1-j_1)$ and ρ varies as the distance in a two-body motion in which the gravitational constant is μ, the semi-major axis is a and the semi-latus rectum is $p_1 = p + 2\rho_0(1-j_1) \geq p$. If $j_1 \geq 1$ the equation (39) gives $j' = 0$, hence j is constant and equal to j_1 and ρ varies as the distance in a two-body motion in which the gravitational constant is $\mu_1 = \mu j_i$ the semi-major axis is $a_1 = aj_1$ and the semi-latus rectum is $p_1 = p/j_i \leq p$.

In a two-body case j is identical to one and so it is in the three-body Lagrangian motion of equilateral triangle type.

If $a < 0$, i.e. if the energh h is positive, all these orbits are hyperbolic, have a velocity at infinity equal to $\sqrt{-\mu/a}$ and verify for all t: $\rho(t) \geq -\rho_0 + |t - t_0| \sqrt{-\mu/a}$, hence the escape velocity of ρ is at least $\sqrt{-\mu/a}$ that lower limit being obtained if and only if the system has an hyperbolic expansion and if all sub-system have a zero final energy (i.e. they either have only one body or approach a central configuration in parabolic expansion).

If $h = 0$, i.e. if $a = \pm \infty$, the orbits given by (39) are parabolic and ρ has at least a parabolic expansion; in that last case (which corresponds to a n-body system approaching a central configuration in parabolic expansion) it is easy to verify that $\rho \sim At^{2/3}$ and $\nu \sim Bt^{2/3}$ with $A^2B = 9\mu/2$ and $A \geq B$.

If $a > 0$, i.e. if the energy h is negative all the orbits given by (39) are elliptic and their period is $T = 2\pi a^{3/2}\mu^{-1/2}$ if $j_i \leq 1$ and it is $T_1 = Tj_1$ if $j_1 \geq 1$; all this imply:

$$\rho(t_0) + \rho(t_0 + T/2) \geq 2a \tag{42}$$

$$\rho(t_0) \cdot \rho\{t_0 + T(\rho_0/4a + p/4\rho_0)\} \geq ap \tag{43}$$

$$\rho_0\rho_f \leq ap, \text{ and } |t_f - t_0| \geq T\{\rho_0/4a + p/4\rho_0\} \text{ imply: } |t_f - t_0| \geq T\{\rho_f/4a + p/4\rho_f\} \tag{44}$$

The relations (44) are particularly useful if ρ_f is very small.

More information can be obtained if we also use the length λ and the relations (16), (30), (34), (35), (36).

The relations (16) and (30) between I, R and λ can even be improved into:

$$m_1(M-m_1)/M^* \leq \rho^2/\lambda^2 = 2MI/M^*\lambda^2 \leq 2 \sum_{1\leq i<j\leq n} (m_i^2 m_j + m_i m_j^2)/MM^* \tag{45}$$

thus the ration ρ/λ has positive and finite upper and lower bounds always between $\sqrt{2\, m_1/M}$ and $\sqrt{2}$.

In particular if the n-body system of interest has a non-zero angular momentum (and hence $p > 0$) and if it has a parabolic or sub-parabolic expansion the condition(38) is never satisfied and the function $\lambda^{3/2} - \frac{3Mt}{2(m_1+m_2)} \cdot \sqrt{2\,GM}$ is non-increasing, which, taking account of (43) and (45), is impossible if ρ_0 is too small. We obtain thus, when the angular momentum $\vec{c}$ is not zero, a positive lower bound of the mean quadratic distance ρ and the moment of inertia 2 I in terms of $\vec{c}$, h, G and the n masses when the final evolution is a parabolic or sub-parabolic expansion.

8. QUALITATIVE RESULTS AND MEASUREMENT THEORY COMPARISON OF ORIGINAL AND FINAL EVOLUTIONS

This section can be found in Ref. 239.

9. THE THREE-BODY PROBLEM

The three-body problem is of course the most investigated of n-body problems and its qualitative analysis leads to very detailed results [Ref. 16, 94-96].

Original and final evolutions cannot be of the types <<infinite expansion in bounded time>> or <<superhyperbolic expansion>> and it remains the following possibilities:

A) The triple collision at a finite instant.

That case require a zero angular momentum and then a plane motion [Ref. 150], if the system has a collinear motion it of course approaches a central configuration of alined type (sections 8.1.1. and 11) when it goes to the triple collision, if not the only possiblity is the approach of a central configuration of equilateral triangle type.

B) The hyperbolic expansion.

The number of subsystems can be either 2 or 3.

If there are 3 subsystems they have each one body and all mutual distances increase as t, it is the hyperbolic type of Chazy [Ref. 94-96] and the radius vector can be developed in terms of t as in (82):

$$\vec{r_i} = \vec{A_i}t + \vec{B_i}\,\mathrm{Lnt} + \vec{C_i} + \vec{D_i}[\mathrm{Lnt}/t] + \vec{E_i}/t + O(\mathrm{Lnt}/t)^2 \qquad (46)$$

with $\sum_{i=1}^{3} m_i \vec{r_i} = 0$ i.e.: $\sum_{i=1}^{3} m_i \vec{A_i} = 0$ etc... (47)

and: $\vec{r_i}'' = \sum_{j=1;j\neq i}^{3} Gm_j r_{ij}/\vec{r}_{ij}^{\,3}$; that is (with i, j, k = 1, 2, 3):

$$\vec{B_i} = Gm_j(\vec{A_i}-\vec{A_j})\,||\vec{A_i}-\vec{A_j}||^{-3} + Gm_k(\vec{A_i}-\vec{A_k})\,||\vec{A_i}-\vec{A_k}||^{-3} ;$$

$$\vec{D_i} = \ldots ; \vec{E_i} = \ldots \qquad (48)$$

and also: $h = \frac{1}{2}\sum_{i=1}^{3} m_i A_i^2$; $\vec{c} = \sum_{i=1}^{3} m_i \vec{C_i} \times \vec{A_i}$ (49)

If there are only two subsystems one of these subsystems has two bodies and its final energy $h_{s\infty}$ can be either zero or negative.

If $h_{s\infty} = 0$ we obtain the hyperbolic-parabolic type.

If $h_{s\infty} < 0$ we obtain the hyperbolic-elliptic type.

Developments similar to (46) can be done.

C) The parabolic or sub-parabolic expansion.

The energy h is negative or zero.

If h = 0 the system approaches a central configuration either of alined type or of equilateral triangle type and since all these central configurations are regular the 3 radius-vectors are such that:

$$\vec{r_i} = \vec{A_i}\, t^{2/3} + o(t^{2/3}) \qquad (50)$$

It is the tri-parabolic type of Chazy [Ref. 94-96].

If h < 0 the system can sometimes be divided into two cluster in parabolic expansion, $\vec{r_i}$ vectors are then again of the form (50) (two of the three $\vec{A_i}$'s being now equal), R is then equivalent to $[9\,GM/2]^{1/3}\, t^{2/3}$ and in the cluster with two bodies the relative motion goes to an elliptic motion of final energy h. It is the parabolic-elliptic evolution.

Finally if h < 0, if the expansion is parabolic or sub-parabolic and if don't appear a division into clusters in parabolic expansion we obtain: lim sup. $R\,t^{-2/3} \leq [9\ GM/2]^{1/3}$ and: lim inf. $R \leq 3\ c^2/GM\ m_1^2 - 5\ GM^4/48\ m_1^2 h < +\infty$ (a smaller upper bound of lim inf. R can be found in Ref. [105]), the system always come back to small sizes. R can be bounded, and the evolution is then called of elliptic or bounded type, or R can be unbounded and the evolution is called of oscillatory type as in the Sitnikov solutions [Ref. 99,100].

Except for the triple collision and for the hyperbolic-elliptic evolution the sign of the energy is always well defined:

A) hyperbolic and hyperbolic-parabolic evolutions correspond to positive energies.

B) tri-parabolic evolutions correspond to zero energies.

C) parabolic-elliptic, bounded and oscillatory evolutions correspond to negative energies.

An heuristic and more detailed classification of systems of negative energy can be found in [Ref. 108] and Szebehely conjectured in [Ref. 122] from numerical computations that when the ratio of the two small masses is sufficiently near to one and when the product c^2h is non-positive and sufficiently near to zero almost all evolutions are of hyperbolic-elliptic type as it si already the case when the energy h is zero.

Many studies [Ref. 1, 76-79 , 123-134.] deal with the question of the comparison of original and final evolustions in a given solution; Chazy conjuctured in [Ref. 76-78] that these two evolutions were always of the same type with the same escaping body or bodies, but Schmidt [Ref. 123.], Khilmi [1, 79 , 124-126] and Alexeev [126-134] established for particular values of the masses the possibility of exchange, partial capture and escape; for instance the numerical integration of a proper solution leads from positions and velocities satisfying sufficient conditions of hyperbolic original evolution to positions and velocities satisfying sufficient conditions of hyperbolic-elliptic final evolution [Ref. 1,779-87, 123-134].

It is possible to establish these possibilties for almost all values of the ratios of masses in the analytical following way that we will present after the generalization of Hill's regions to the general 3-body problem.

9.1. Hill's regions for the general 3-body problem

Hill's regions are obtained in the circular restricted 3-body problem; they are related to the limitations given by the Jacobi integral.

In the general three-body problem various limitations related to the integrals of motion $\vec{c}$ and h can be given [Ref. 89, 121, 135-137], for instance if $c^2h < -G^2m_2^3n_3^3/2(m_2 + m_3)$ a forbidden zone appear in a real physical space far from the invariable plane and in the vicinity of the axis of the angular momentum, and the allowed latitudes (from the invariable plane) decrease and go to zero when $c^2h \to -\infty$.

An interesting result concern the restrictions on possible triangular configurations of the 3 bodies; in a given 3 body-problem a triangle r_{12}, r_{13}, r_{23} can be compatible with the given integrals of motion $\vec{c}$ and h if and only if it corresponds to a moment of inertia $2I = \sum_{i=1}^{3} m_1 r_i^2$ and a potential $U = G \sum_{1\le i<j\le 3} m_i m_j / r_{ij}$ verifying:

$$c^2 \le 4\,I\,(U + h) \tag{51}$$

With the fixed lengths a and p and the variable lengths ρ and ν defined in (29 - 33) the condition (51) is equivalent to:

$$2\,\rho/\nu \ \ge\ \rho/a + p/\rho \tag{52}$$

The ratio ρ/ν is independent of the scale of the triangle of the three bodies and if we choose the variable distance r_{23} as unit of length we can draw the figure 4 of curves of constant ratio ρ/ν in terms of the position of m_1 with respect to m_2 and m_3. ρ/ν is minimum and equal to one when all r_{ij} are equal, i.e., at the equilateral triangle positions L_4 and L_5; ρ/ν is infinite at m_2, at m_3 and at infinity and finally it has three saddle points at the alined Lagrangian points L_1, L_2, and L_3; thus the figure of curves of constant ratio ρ/ν is a generalization of Hill's curves of the circular restricted three-body problem (which indeed correspond to the limit when m_1 goes to zero).

For a given value of ρ (or a given value of ν) the condition (52) gives the allowed values of the ratio ρ/ν and thus, with figure 4, the allowed configurations of the triangle of the three bodies.

Let us note that if the semi-major axis a is positive (i.e., if the energy h is negative) the minimum of the right member of (52) with respect to ρ is obtained for $\rho = \sqrt{ap}$ and hence:

$$h < 0 \text{ and } (136) \text{ imply: } \rho/\nu \geq \sqrt{p/a} \tag{53}$$

If p/a is larger then one the condition (53) implies that configurations in the vicinity of the equilateral triangle configuration are forbidden for all time and furthermore if $\sqrt{p/a}$ is larger than sup. $\rho/\nu(L_i)$ (i.e. $\sqrt{p/a} > \rho'\nu(L_2)$ if $m_1 < m_2 < m_3$) the allowed configurations are then divided into 3 disconnected parts: around m_2, around m_3 and at a large distance, and the motion remains for all time confined to one of these parts, i.e., there exists a close binary that the third body cannot destroy and motions of exchange type are impossible.

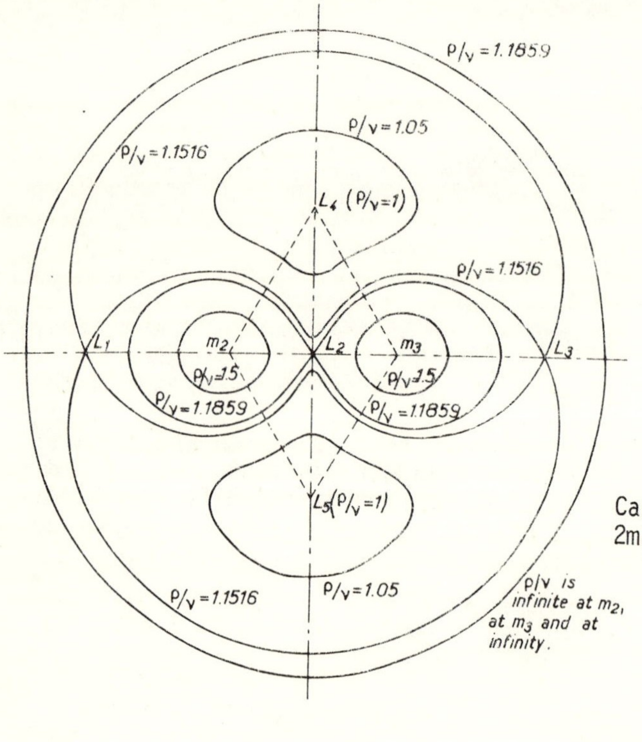

Case of the figure: $2m_1 = m_2 = m_3$.

Figure **2**. Generalization of Hill's curves: curves of constant ratio ρ/ν in terms of the relative position of m_1 with respect to m_2 and m_3

With (115) the condition $p/a > 1$ is equivalent to:

$$c^2h < - G^2M*^3/2M \tag{54}$$

and the condition $\sqrt{p/a} > \rho/\nu\ (L_2)$ is equivalent to:

$$c^2h < -G^2M*^3\ \{\rho/\nu(L_2)\}^2/2M \tag{55}$$

Thus the separation of phase space into 3 disconnected parts and the impossibility of motions of exchange type occur for sufficiently negative produces c^2h (it is easy to see that this is the case of the Solar System if we consider it as a 3-body system with the Sun, Jupiter and Saturn -- 99.98% of the total mass - Jupiter and the Sun being the binary that cannot be destroyed).

Let us note that $0 < m_1 \leq m_2 \leq m_3$ implies:

$$1 = \rho/\nu(L_4)=\rho/\nu(L_5)<\rho/\nu(L_3)\leq\rho/\nu(L_1)<\rho/\nu(L_2)\leq(343/243)^{1/2}=1.1881 \tag{56}$$

the upper limit of $\rho/\nu(L_2)$ is obtained for $m_2 = m_3 = 1.6\ m_1$.

9.2. Examples of motions of exchange type and of partial capture or escape for almost all values of the ratio of masses.

Let us use conditions (118), (119), (124) and (125) for cases of positive or zero energy h (and hence $1/a \leq 0$), (118) and (119) imply:

$$d^2(\rho^2)/dt^2 = 2\mu(1/\nu-1/a) \geq 2\mu/\nu \geq 2\mu/\rho > 0 \tag{57}$$

When $t\to\pm\infty$, ρ goes to infinity at least as $[g\mu/2]^{1/3}\ |t|^{2/3}$ and it has one and only one minimum.

(124) and (125) imply

$$\rho/\nu \geq j = \rho/2a + p/2\rho + \rho\rho'^2/2\mu \qquad \rho'j'\geq 0 \tag{58}$$

Hence if we give the subscript one to the elements obtained at the instant t_1 of the minimum of ρ:

$$\rho(t_1)=\rho_1;\ \ \rho'(t_1)=0;\ j(t_1)=j_1=\rho_1/2a+p/2\rho_1;\ t\neq t_1 \text{ implies}$$

$$\rho>\rho_1 \text{ and } j \geq j_1 \tag{59}$$

and since $\rho/\nu\geq j$, the boundaries $\rho/\nu= j$ obtaines at $t = t_1$ on triangular configurations in Figure 4 are not removing when $t \neq t_1$.

Let us consider especially a case for which $h=0$ and $\vec{c}\neq 0$, the possible original or final evolutions of the system are either of hyperbolic-elliptic type or of tri-parabolic type and in Figure 4 when $t\to\pm\infty$, the point m_1 goes eith to m_2 or to m_3 or to infinity or to one of the Lagrangian points.

if conditions at $t=t_1$, i.e., conditions for which $\rho'=0$, are such that j_1 is sufficiently large, some of the original or final evolutions are impossible because of the boundaries drawn on Figure 4; for instance, since

$0 < m_1 \leq m_2 \leq m_3$ imply $1 < \rho/\nu(L_3) \leq \rho/\nu(L_1) \leq \rho/\nu(L_2)$, if $j_1 > \rho/\nu(L_2)$ the possible positions of m_1 in Figure 4 are divided into three disconnected parts and the motion is necessarily of hyperbolic elliptic type for both $t\to-\infty$ and $t\to+\infty$ without exchenge; on the other hand if $\rho/\nu(L_1)<j_1<\rho/\nu(L_2)$ e.g., $j_1 = 1.17$ in the case of Figure 4), the possible positions of m_1 are divided into only two disconnected parts: an outer part further than the Lagrangian points and an inner part containing m_2. L_2 and m_3 are leading necessarily for both $t\to\pm\infty$ either to the hyperbolic-elliptic escape of m_3 or to the hyperbolic elliptic escape of m_2 or to an evolution or tri-parabolic type, m_1 being between m_2 and m_3 and going to L_2.

Let us now compare the original and the final evolutions of given solutions of that last type of motion of zero energy; we know already that:

A) Motions of tri-parabolic type represent only sub-manifolds of motions of zero energy.

B) Three-body motions are reversible.

Hence there are only two possibilities:

Either exist motions of the two exchange types leading from an hyperbolic-elliptic escape of m_3 or conversely.

Or no such motions exist in the class of motion of interest and all motions have the same type of past and for future, the tri-parabolic motions lying at the boundary between the two types of hyperbolic-elliptic motions (either m_2 or m_3 being the escaping body for both past and future).

The essential point is thus the existence of exchange motions; it can be verified easily when $m_2=m_3$ by proper numerical tests, indeed in that case exist space-time symmetrical solutions around the instant t_1 and around some proper axis or plane of symmetry, the mass m_2 being exchanged with m_3 when t is exchanged with $2t_1-t$; these solutions have then symmetrical positions and

antisymmetrical velocities at $t = t_1$ (which imply $\rho'(t_1) = 0$), hence then: $r_{12} = r_{13} < r_{23}$; $h = 0$ and $j_1 = p/2\rho_1 > \rho/\nu\ (L_1)$.
Because of the symmetry these solutions are either of tri-parabolic type for both past and future (it is the case of the Lagrangian solution) or they present an exchange, m_1 being with m_2 on one side and with m_3 on the other side.

An example of these symmetrical solutions of exchange type has been computed by Szebehely [Ref. 173], the motion is plane and the conditions $t_1=0$ are the following:

Constant of gravitation: $G = 1$

Masses: $m_1 = 1$; $m_2 = m_3 = 2$ (this corresponds to Figure 4).

Positions at $t = t_1 = 0$:

$$x_1 = y_1 = y_2 = y_3 = 0;\quad x_2 = -3;\quad x_3 = +3 \tag{60}$$

Velocities at $t = 0$:

$$x_1' = -28/123;\quad y_1' = 0;\quad x_2' = x_3' = 7/123\ ;\quad y_2' = -y_3' = -122/123 \tag{61}$$

The space-time symmetry is:

t	x_1	y_1	x_2	y_2	x_3	y_3
$-t$	$-x_1$	y_1	$-x_3$	y_3	$-x_2$	y_2

(62)

and we obtain: $h=0$; $j_1 = p/2\ \rho_1 = 372\sqrt{10}/10086 = 1.171\ \ldots$
$> \rho/\nu\ (L_1) = 1.1516\ \ldots$

Figure 5 presents the results of the numerical integration, m_1 forms a binary with m_3 for $t < 0$ and with m_2 for $t > 0$.

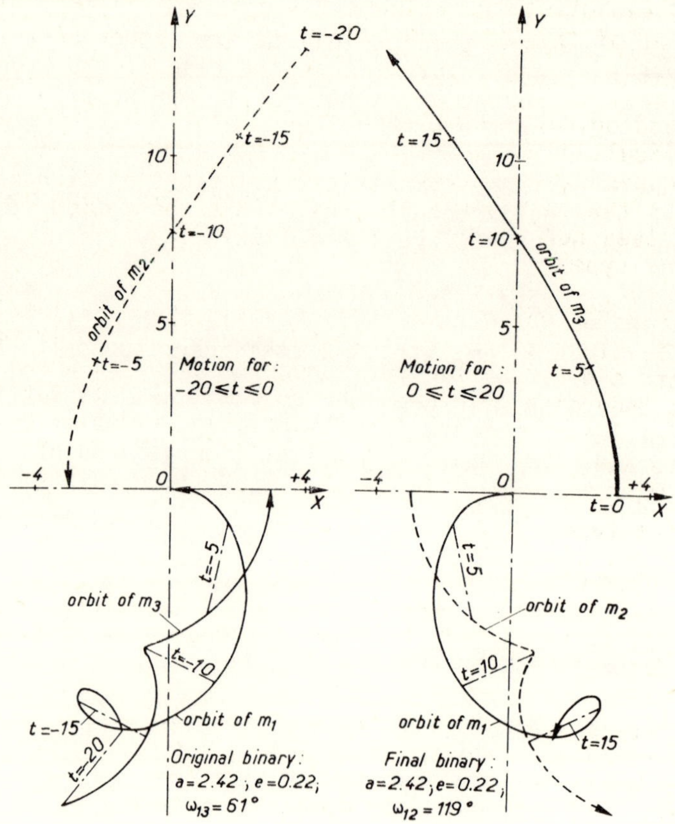

Figure **3**. Example of symmetrical exchange motion $G = m_1 = 1$; $m_2 - m_3 = 2$.

Conditions at $t = 0$: $x'_1 = y_1 = y_3 = 0$; $x_3 = -x_2 = 3$; $x'_1 = \frac{-28}{123}$; $y'_1 = 0$; $x'_2 = x'_3 = \frac{7}{123}$; $y'_3 = y'_2 = \frac{122}{123}$.

9.2.1 Qualitative analysis for cases of negative energy

If the energy h is negative and if the angular momentum $\vec{c}$ is not zero we already know [see the Ref. 76-78 or the section 10] that there are eight original or final evolutions: the three hyperbolic-elliptic evolutions HE_i the three corresponding parabolic=elliptic evolutions PE_i, the bounded evolutions (or "elliptic" evolution) B and the oscillatory evolution) B and the oscillatory evolution 0 either of the type "unbounded mutual distances" or of the type "unbounded velocities" (i.e.,: lim inf. $r_{ij}=0$) or both.

The topology of these types of evolution in phase space is a very difficult problem, we agian obtain of course that HE_i types correspond to open sets and that PE_i types belong to submanifolds at the boundaries of HE_i types, but these boundaries have themselves not necessarily a measure zero and they contain solutions of types 0 and B.

Solutions of the six exchange types $(HE_i \to HE_j)$ exist for almost all values of the ratios of masses, at least when the product c^2h is sufficiently near to zero (as shown is section 10.2) but they disappear when the product c^2h is sufficiently negative (at least when the condition (55) is satisfied). It even possible to add: taking account of the figure 4 and of: "$0 < m_1 \leq m_2 \leq m_3$ imply: $1 < \rho/\nu(L_3) \leq \rho/\nu(L_1) < \rho/\nu(L_2)$" the exchanges of types $HE_2 \to HE_3$ and $HE_3 \to HE_2$ are impossible when $c^2h \leq -G^2M^{*3}(\rho/\nu(L_2))^2/2M$ and the four other types of exchange are impossible earlier, when: $c^2h \leq -G^2M^{*3}(\rho/\nu(L_1))^2/2M$. The exact limits of the possibility of exchange seem difficult to obtain.

Sitnikov solutions [Ref. 122-124 or section 12] shows that, at least for particular values of masses, all possible original-final evolutions exist but let us remember that the "complete capture" (i.e.,: $HE_i \to B$) correspond to a set of measure zero as well as evolutions of types $B \to HE_i$; $HE_i \to 0$; $0 \to HE_i$; $B \to 0$ and $0 \to B$; on the other hand solutions of types $B \to B$ and $0 \to 0$ have almost always the strong properties of stability described in section 9 and in Ref. 144-149.

Let us remember the Szebehely conjecture: "If the ratio of the two small masses is sufficiently near to one and if the product c^2h is non-positive and sufficiently near to zero almost all solutions are of hyperpolic-elliptic type for both $t \to +\infty$ and $t \to -\infty$". But when c^2h is sufficiently negative solutions of type $B \to B$ fill a set positive measure [Ref. 133-138].

Let us also remember the conjectures on oscillatory motions: oscillatory solutions of the type "unbounded mutual distances" seem to fill a set of measure zero, but oscillatory solutions of the type "unbounded velocities" seem to fill a set of positive measure.

REFERENCES

1. Khilmi, G. F., Qualitative methods in a many-body problem, Russian Tracts on Advanced Math. and Physics, Vol. 3, Gordon and Breach Science Publishers, New York (1961).
2. Sundman, K.F., Acta Mathematica 36, 105, 1912.
3. Marchal, C., and D. G. Saari, J. Diff. Equations, January 1976.
4. Szebehely, V., Astr. J. 79, 981, 1974.
5. Szebehely, V., Astr. J. 79, 1449, 1974.
6. Bruns, H., Berichte der Kon Sachs. Gesellschaft der Wiss. , 1-55, 1887.
7. Bruns H., Acta Mathematica, 11, 25, (1887-88).
8. Painlevé P., Bulletin Astronomique, 15, 81, 1898.
9. Poincaré H., Acta Mathematica, 13, 259, (1890-91).
10. Poincaré H., Les methodes nouvelles de la mecanique celeste, Bauthier-Villars, Paris, 1892-1893-1899).
11. Painlevé P., C. R., 80, 1699, 1900.
12. Whittaker, E. T., A treatise on the Analytcal Dynamics of particles and rigid bodies with an introduction to the problem of 3-bodies, Cambridge at the University Press (1927).
13. Siegel, C. L., T. Am. Math. Soc. 39, 225, 1936.
14. Contopoulos, G., Bul. Astr. Serie 3, 2, 223, 1967.
15. Henon, M., Bul. Astron. 1, fasc. 1, 57,1966.
16. Henon, M., Bul. Astron. 1, fasc. 2, 49, 1966.
17. Henon M., Quar. Appl. Math., Vol. 27, 291 (1969).
18. Henon M., Astron. Astrophy., 9, 24 (1970).
19. Froeschle, C., Astron. Astrophy.9, 15. 1970.
20. Szebehely V., Celestial Mechanics, 4, 116, 1971.
21. Brahic A., Astron. Astrophy. 12, 98, 1971.
22. Jenkins B. Z., J. H. Bartlett, Celestial Mechanics, 5, 407,1972.
23. Rannou, F., Astro. Astrophy. 31, 289, 1974.
24. Cartan, E., Lecons sur les invariants integraux., Hermann, Paris, 1971.
25. Pars, L. A., A treatise on analytical dynamics, Heinemann, London, 435 (1965).
26. Darwin, G. H., Periodic orbits. Acta Mathematica, 21, 99, 1897.
27. Jackson, J., M. N. R. S., 74, 62, 1913.
28. Moulton F. R., Periodic orbits., Carnegie Foundation, Washington, 1920.
29. Charlier C. L., Die Mechanik des Himmels, 2. Auflage - vol. 11 - Walter de Gruyter, Berlin, 1927.
30. Matukuma,T., Proceedings of the Imperial Academy of Japan, 6, 6, 1930.
31. Matukuma, T., Proceedings of the Imperial Academy of Japan, 8, 147, 1932.
32. Matukuma,T., Proceedings of the Imperial Academy of Japan, 9, 364, 1933.
33. Stromgren, E., Bull. Astron. 2, 87, 1933.
34. Stromgren, E., Publication of the Copenhagen Observatory, No. 100, 1935.

35. Uno, T., Japan J. Astron. Astrophy., 15, 149, 1937.
36. Moser, J. D., Mathematical Analysis, 126, 325, 1953.
37. Arenstorf, R. F., Am. J. Math., 85, 27, 1963.
38. Arenstorf, R. F., AIAA J., 1, 238, 1963.
39. Henon, M., Annales d'Astrophysique, 28, 499, 1965.
40. Henon, M., Annales d'Astrophysique, 28, 992, 1965.
41. Bartlet, J. H., C. A. Wagner, Copenhagen Observatory Publication No. 183, 1965.
42. Barrar, R. B., Atron. J., 70, 3, 1965.
43. Deprit A., J. Henrard, Astron. J. 70, 271, 1965.
44. Szebehely V., C. F. Peters, Atron. J., 72, 876, 1967.
45. Szebehely, V., C, F., Peters, Astron. J., 72, 1187, 1967.
46. Giacaglia, G. E. O., Astron. J., 72, 386, 1967.
47. Deprit.,A., J. Henrard, Astron. J., 74, 308, 1968.
48. Deprit., A, E. Rabe, Astron. J., 74, 317, 1968.
49. Henon, M., Bull. Astron., 3, 377, 1968.
50. Guillaume, P., Atron. Astrophy., 3, 57, 1969.
51. Roels, J., Astron. Astrophy. 2, 52, 1969.
52. Henon, M., Astron. Astrophy. 1, 223, 1969.
53. Broucke, R., AIAA, 7, 1003, 1969.
54. Giacaglia, G.E.O., Periodic orbits, stability and resonances, Reidel Co., Dordrecht,Holland, 1970.
55. Sinclair, A. T., Celestial Mechanics, 2, 350, 1970.
56. Bozis, G., Atron. Astrophy., 11, 320, 1971.
57. Broucke, R., Celestial Mechanics, 3, 461, 1971.
58. Siegel C. L., J. K., Moser, Lectures on Celestial Mechanics, Springer-Verlag, Berlin, 1971.
59. Shelus, P. J., Celestial Mechanics, 5, 483, 1972.
60. Kozai, Y., K. Hiroski K., Celestial Mechanics, 7, 156, 1973.
61. Henon, M., Astron. Astrophy., 28, 415, 1973.
62. Henon, M., Astron. Astrophy., 30, 317, (1974).
63. Henon, M., Celestial Mechanics, 10, 375, 1974.
64. Guillaume, P., Astron. Astrophy. 37, 209, 1974.
65. Guillaume,P., Celestial Mechanics, 10, 475, 1974.
66. Markellos, V.V., W. Black, P. E., Moran, Celestial Mechanics, 9, 507, 1974.
67. Markellos, V.V., Celestial Mechanics, 9, 365, 1974.
68. Perko, L.M., SIAM J. Appl. Math., 27, 200, 1974.
69. Broucke, R., D. Boggs, Celestial Mechanics, 11, 13, 1975.
70. Sergysels-Lamy, A., Celestial Mechanics, 11, 43, 1975.
71. Brouwer D., G. M. Clemence., Methods of Celestial Mechanics, Academic Press, New York, 1961.
72. Danby, J.M.A., Astron. J., 69, 165, 1964.
73. Bennett, A., Icarust., 4, 177, 1965.
74. Szebehely V., Theory of Orbits, The restricted problem of 3 bodies, Academic Press, New York (1967).
75. Vinh, N. X., Celestial Mechanics, 6, 305, 1972.
76. Chazy, J., Ann. l'Ecole Norm. Sup., 3e serie, 39, 29 (1922).
77. Chazy, J., J. Math. pure et Appl., 8, 353, 1929.

78. Chazy, J., Bull. Astronomi. 8, 403, (1932).
79. Khilmi, G. F., Doklady Akad. Nauk SSSR, 78, 653, 1951.
80. Merman, G. A., Byull. Inst. Teor. Astron. 5, 594, 1954.
81. Tevzadze, G. A., Izv. Akad. Nauk Armyan. SSR 15 - No. 5, 67, 1962.
82. Standish, E. M., Jr., Celestial Mechanics, 4, 44, 1971.
83. Standish, E. M., Jr., Celestial Mechanics, 6, 352, 1972.
84. Yoshida, J., Publ. of Astron. Soc. of Japan, 24, 391, 1972.
85. Yoshida, J., Publ. Astron. Soc. of Japan, 26, 367, 1974.
86. Marchal, C., Celestial Mechanics, 9, 381, 1974.
87. Merman, G. A., Byull. Inst. Teor. Astron., 6, 687, 1958.
88. Marchal, C., AStron. and Astrophy., 10, 278, 1971.
89. Saari, D.G., Trans. Amer. Math. Soc., 162, 267, 1971.
90. Saari, D. G., Trans. Amer. Math. Soc., 168, 1972.
91. Saari, D. G., Trans. Amer. Math. Soc., 181, 351, 1973.
92. Painleve, P., Lecons sur la theorie analytique des equations differentielles. Paris, A. Hermann, 1897.
93. Von Zeipel, H., Sur les singularities du probleme des n-corps. Ark. Mat. Astr. Fys., 4, No. 32, 1908.
94. Pollard, H., D. G. Saari, Arch. Rat. Mech. Anal. 30, 263, 1968.
95. Sperling, H., J. Reine Angew. Math, 245, 15, 1970.
96. Saari, D. G., Arch. Rat. Mech. Anal., 49, 311, 1973.
97. Wintner, A., The analytical foundations of Celestial Mechanics, Princeton University Press, Princeton, New Jersey, 1941.
98. Hagihara, Y., Celestial Mechanics., Vol. 1 - Dynamical principles, Chapter 3, 228-284, MIT Press, 1970.
99. Sitnikov, K.A., Kokladij Akademii Nauk SSSR, 133, 303, (1960) Soviet Physics Dokl. 5, 647, 1961.
100. Moser, J., Stable and Random Motion in Dynamical Systems, Princeton University Press, Princeton, 1973.
101. Pollard H., J. Math. Mech., 17, 601, 1967.
102. Saari, D. G., Tran. Am. Math. Soc., 156, 1, 1971.
103. Boas, R. P., Amer. J. of Math., 61, 161, 1939.
104. Pollard, H., Proc. Amer. Math. Soc., 18, 399, 1967.
105. Pollard H., D. G. Saari, Proc. Amer. Math. Soc., 24, 593, 1970.
106. Saari, D. G., Proc. Amer. Math. Soc., 21, 459, 1969.
107. Saari, D. G., SIAM Math. Anal., 5, 649, 1974.
108. Szebehely, V., Celestial Mechanics, 4, 116, 1971.
109. Arnold,V.I., Usp. Math. Nauk SSSR, 18, 13, 1963.
110. Arnold, V. I., Russian Mathematical Surveys, 18, 9, 1963.
111. Arnold, V. I., Soviet Math. Doklady, 6, 331, 1965.
112. Arnold, V. I., A, Avez, Problemes ergodiques de la mecanique classique. Gauthier-Villars, Paris, 1967.
113. Arnold, V. I., A. Avez, Ergodic problems of Classical Mechanics, Benjamin, 1968.
114. Sundman, K., Acta Mathematica, 36, 14, 1912.
115. Kolmogorov, A. N., Theorie generale des systemes dynamiques et mecaniques classiques. Proc. Int. Congress of Math. Amsterdam, 1957.

116. Moser, J., Nach. Akad. Wiss. Gottingen, Math. Phys. K.1 1, 1962.
117. Moser, J., Bull. Astr., 3, 53, 1968.
118. Halmos, P. R., Lectures on ergodic theory, Chealsea publishing company, New York, 1958.
119. Oxtoby, J.C., Measure and Category, Graduate texts in Mathematics. Springer Verlag, New York, Heidelberg, Berlin, 1970.
120. Marchal, C., Celestial Mechanics, 4, 406, 1971.
121. Dziobeck, O., Die Mathematischen Theorien der Planeten, Bewegungen, Leipzig, Johann Ambrosius Barth, 68, 1888.
122. Szebehely, V., Celestial Mechanics, 9, 359, 1974.
123. Schmidt, O. J., Dokl. Akademii Nauk SSSR, 58, 213, 1947.
124. Khilmi, G. F., Doklady Akademii Nauk SSSR, 62, 39, 1948.
125. Schmidt, O. J., G.F. Khilmi., Uspekhi mat. Nauk 3,4,(26), 157, 1948.
126. Alexeev, V. M., Doklady Akademii Nauk SSSR, 108, 599, 1956.
127. Alexeev, V. M., Astron. Zh., 38, 325, 1961.
128. Alexeev, V. M., Soviet Astron. AJ. 5, 242, 1961.
129. Alexeev, V. M., Astron. Zh., 38, 726, 1961.
130. Alexxev, V. M., Astron. Zh., 38, 1099, 1961.
131. Alexxev, V. M., Vestn MGU. ser. fiz. i. astr. 1, 67, 1961.
132. Alexeev, V. M., Soviet Astr. AJ., 5, 550, 1962.
133. Alexeev, V. M., Soviet Astr. AJ., 5, 841, 1962.
134. Alexeev, V. M., Vestn. MGU ser. matem. 4, 17, 1962.
135. Easton R., J. Diff. Eq., 10, 92, 1971.
136. Saari, D. G., SIAM J. Appl. Math., 26, 806, 1974.
137. Saari, D. G., Celestial Mechanics, 9, 175, 1974.
138. Szebehely, V., A three-body symmetrical solution of exchange type, private correspondnece, 1975.
139. Chazy, J., Bull. Soc. Math. France, 55, 22, 1927.
140. Nahon, F., Celestial Mechanics, 8, 169, 173.
141. Irigoyen, M., Celestial Mechanics, 9, 491, 1974.
142. Sperling, H. J., Celestial Mechanics, 5, 396, 1972.
143. Saari, D. G., Celestial Mechanics, 1, 243, (1970).
144. Saari, D. G., J. Diff. Eq., 14, 275, (1973).
145. Whittaker, E. T., A treatise on the Analytical Dynamics of Particles and Rigid Bodies. Cambridge University Press, London and New York, 1904.
146. Lagrange, J. L., Theorie de la libration de la Lune, Oeuvres V.
147. Duboshin,G. N., Astr. J., USSR, 35, No. 2, 1958.
148. Duboshin,G. N., Soviet Astron., 35, 239 (1958).
149. Duboshin,G. N., Soviet Astron., 36, 154 (1959).
150. Duboshin,G. N., Soviet Astron., 3, 702 (1959).
151. Duboshin,G. N., Theora of Attraction. Russ. Fismatgis, 1961.
152. Duboshin,G. N., Celestial Mechanics, 4, 423, 1971.
153. Duboshin,G. N., Celestial Mechanics, 6, 27, 1972.
154. Anchev, A.A., Flywheel stabilization of relative equilibrium of a satellite. Kosmicheskie Issledovania, 4, 192, 1966.

155. Barlier, F., J. Kovalevsky, Space Science Reviews, 7, 63, 1967.
156. Cetaev, N. G., The stability of motion, Pergamon Press, New York, 1961.
157. Kinoshita, H., Publ. Ast. Soc. Jap., 22, 3, 1970.
158. Kondurar, V. T., Soviet Astr. 36, 863, 1959.
159. Kondurar, V. T., T. K. Shinkarik, Bull. l'Inst. Astron., 13, 102, 1971.
160. Likins, P. W., J. Astron. Sci., 12, 18 (1965).
161. Lilov, L. K., P.M.M., 36, 977, 1972.
162. Marchal, C., Bull. Astron. serie 3, 1, 189 (1966).
163. Modi, V. J., J. E. Neilson, Celestial Mechanics, 5, 126, 1972.
164. Modi, V. J., R. C. Brereton, AIAA J., 7, 1217, 1969.
165. Modi, V. J., R. C. Brereton, AIAA J., 7, 1465. 1969.
166. Baker, RML, Jr., ARS J. 30, 124, 1960.
167. Mohan, S. N., J. V. Breakwell, B. O. Lange, Celestial Mechanics, 5, 157, 1972.
168. Pascal, M., C. R. Acad. Sc., Paris, 272, 1971.
169. Pascal, M., C. R. Acad. Sc., Paris, 272, 1971.
170. Pascal, M., J. Mec., 1, 147, 1972.
171. Pascal, M., C. R. Acad. Sc., Paris, 274, 1972.
172. Roberson, R. E., J. of Astronautical Sciences, 15, 1968.
173. Stepanov, S. J., P.M.M., 33, 121, 1969.
174. Sulikashvili, R. S., P.M.M., 33, 1119, 1969.
175. Smale, S., Inventiones Math., 11, 45 1970.
176. Stiefel, E., G. Scheifele, Linear and regular Celestial Mechanics, Springer-Verlag, Berlin, 1971.

ON THE CHARACTERISTIC EXPONENTS OF THE GENERAL THREE-BODY PROBLEM

R. Broucke

Jet Propulsion Laboratory, Pasadena, California

1. INTRODUCTION

In this article we describe some properties of the characteristic exponents of the general three-body problem. Our analysis is based on the variational equations written in Lagrangian form (section 3) rather than the more usual Hamiltonian form. It is shown that the basic property of the matrix of the linearized equations of motion is the skew-symplecticity exhibited in equation (18). This property generates the symplectic property (equation 21, section 4) of the fundamental solution matrix R and of the monodromy matrix. These properties are not unique to the general three-body problem, but they hold for all dynamical systems with a Lagrangian of the form (1).

In section 5 we describe the properties of the characteristic equation and of the roots. We have an eight-degree equation of the reciprocal form, with four unit roots and two other pairs of roots, of the form λ_1, $1/\lambda_1$, λ_2, $1/\lambda_2$. These properties of the roots are similar to those of the elliptic restricted three-body problem (Broucke,1969). In particular, there are seven types of root configurations in the complex plane and twelve possible transitions between these regions of stability and instability.

In section 6 we develop in detail the three trivial solutions of the linearized equations of motion. It is known that these solutions are intimately related to the theory of Integral Invariants (Cartan, 1971) and to the so-called eleventh integral of the motion (Losco, 1974).

V. Szebehely and B.D. Tapley (eds.), Long-Time Predictions in Dynamics, 209–222. *All Rights Reserved. Copyright © 1976 by D. Reidel Publishing Company, Dordrecht-Holland.*

Finally in the last two sections we show several properties that hold for symmetric periodic solutions. These properties allow the stability calculation of a periodic solution by integrating the variational equations over a half revolution only.

Our results are presented on the basis of a non-rotating frame of reference. However they can be applied to relative periodic solutions as well, by taking a uniformly rotating coordinate system. A constant rotation matrix will then transform the monodromy matrix of the non-rotating system to the relative system of reference. Other methods have also been developed recently for computing the stability of relative periodic orbits (Hadjidimetriou, 1975).

2. EQUATIONS OF MOTION

Let us designate the three strictly positive point-masses by m_0, m_1 and m_2. We will restrict ourselves to planar motions only. It is known that we can associate with the problem a conservative Lagrangian in four variables q^i $(i = 1,2,3,4)$:

$$L = \frac{1}{2} g_{ij} \dot{q}^i \dot{q}^j + V(q^k) , \tag{1}$$

where the metric tensor g_{ij} is constant and depends on the values of the three masses only. The potential function depends only on the three masses and their mutual distances r_{01}, m_{02}, m_{12}:

$$V = \frac{m_0 m_1}{r_{01}} + \frac{m_0 m_2}{r_{02}} + \frac{m_1 m_2}{r_{12}} . \tag{2}$$

The above Lagrangian (1) is of a classical type that is well known in dynamics: it is quadratic and homogeneous in the velocity components, while the potential function is homogeneous and of degree -1 in the coordinates. We also note the absence of terms that are linear in the velocity components (Coriolis terms) and the problem may thus be considered reversible.

There exist several coordinate systems which allow a formulation of the Lagrangian of the planar three-body problem in the form (1), with a constant metric tensor g_{ij} with four degrees of freedom. One of the best known systems is the so-called heliocentric rectangular coordinates, where m_0 is assumed to be the origin of axes and where m_1 and m_2 have the coordinates $q^i = (x_1, y_1, x_2, y_2)$ which are the components of the vectors $m_0 m_1$ and $m_0 m_2$. The covariant components of the metric tensor are then given by the following non-singular constant 4 x 4 matrix, (where m is the sum of the three masses):

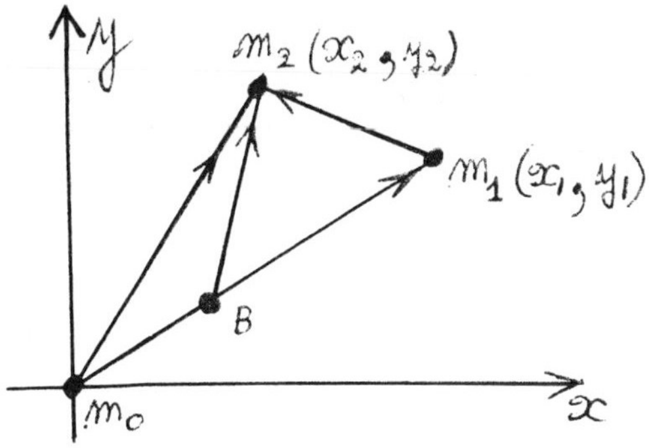

$$g_{ij} = \begin{bmatrix} m_1(1-\frac{m_1}{m}) & 0 & \frac{-m_1m_2}{m} & 0 \\ 0 & m_1(1-\frac{m_1}{m}) & 0 & \frac{-m_1m_2}{m} \\ \frac{-m_1m_2}{m} & 0 & m_2(1-\frac{m_2}{m}) & \\ 0 & \frac{-m_1m_2}{m} & 0 & m_2(1-\frac{m_2}{m}) \end{bmatrix} \qquad (3)$$

Unfortunately the above matrix is not diagonal. However, it is possible to use a slightly different coordinate system which will result in a diagonal metric tensor. This is achieved by the well-known Jacobian Coordinates. They are the components of the vectors $m_0 m_1$ and Bm_2, where B is the barycenter of m_0 and m_1. It is then found that the Lagrangian is still of the form (1) with a constant diagonal g_{ij}-matrix:

$$g_{ij} = \begin{bmatrix} \frac{m_0m_1}{m_0+m_1} I_2 & 0_2 \\ 0_2 & \frac{m_2(m_0+m_1)}{m} I_2 \end{bmatrix}, \qquad (4)$$

where I_2 and 0_2 are respectively the unit and null matrices of order 2.

Using standard tensor notations, the four Lagrangian equations of motion can now be written in contra-variant form

$$g_{ij}\, \ddot{q}^j = V_i = \frac{\partial V}{\partial q^i}\,, \qquad (5)$$

or in an equivalent co-variant form

$$\ddot{q}^i = V^i = g^{ij} V_j\,. \qquad (6)$$

The conjugate momenta, defined by $p_i = g_{ij}\dot{q}^j$, allow us to write the Hamiltonian

$$H = \frac{1}{2} g^{ij} p_i p_j - V \tag{7}$$

and the corresponding eight canonical equations of motion:

$$\dot{q}^i = g^{ij} p_j \quad , \quad \dot{p}_i = V_i \ . \tag{8}$$

In the above equations, the standard summation conventions are freely used and the matrix g^{ij} is the inverse of the matrix g_{ij}.

3. THE VARIATIONAL EQUATIONS

The variational equations are obtained by linearizing the above Lagrangian equations of motion in the neighborhood of a given known solution. They are the following four second-order linear differential equations with time-dependent coefficients:

$$\delta\ddot{q}^i = g^{ij} V_{jk} \, \delta q^k \ . \tag{9}$$

The matrix A' of coefficients of the above linear system is thus

$$A' = g^{ij} V_{jk} \ , \tag{10}$$

where V_{jk} represents the sixteen second partial derivatives of the potential function.

The variational equations are more conveniently written in the form of a system of first order differential equations, in the present case in eight variables δq^i, $\delta\dot{q}^i$. In matrix form, this system of eight equations is:

$$\begin{bmatrix} \delta\dot{q} \\ \delta\ddot{q} \end{bmatrix} = A \begin{bmatrix} \delta q \\ \delta\dot{q} \end{bmatrix} = \left[\begin{array}{c|c} 0_4 & I_4 \\ \hline A' & 0_4 \end{array}\right] \cdot \begin{bmatrix} \delta q \\ \delta\dot{q} \end{bmatrix} \ , \tag{11}$$

where the (4 x 4)-matrix A' has been defined in (10).

We will now prove an important property of the A-matrix of the Lagrangian variational equations. Let us first define the following matrices:

$$T = \left[\begin{array}{c|c} I_4 & 0_4 \\ \hline 0_4 & g_{ij} \end{array}\right] \tag{12}$$

and

$$S = T^{-1} J T^{-T} , \tag{13}$$

where

$$J = \left[\begin{array}{c|c} 0_4 & I_4 \\ \hline -I_4 & 0_4 \end{array}\right] , \tag{14}$$

and where the superscript -T indicates the inverse of the transpose of the matrix. The inverse of T is given by

$$T^{-1} = \left[\begin{array}{c|c} I_4 & 0_4 \\ \hline 0_4 & g^{ij} \end{array}\right] . \tag{15}$$

The explicit form of the matrix S and it's inverse are then

$$S = \left[\begin{array}{c|c} 0_4 & g^{ij} \\ \hline -g^{ij} & 0_4 \end{array}\right] , \quad (16) \qquad S^{-1} = \left[\begin{array}{c|c} 0_4 & -g_{ij} \\ \hline g_{ij} & 0_4 \end{array}\right] . \tag{17}$$

With the aid of the matrices which have just been defined, it is now an easy matter to verify the most important identity satisfied by the matrix A:

$$S\, A^T A^{-1} = -A . \qquad *18)$$

Because of this identity, we will say that the matrix A is skew-symplectic. The skew-symplectic property will be useful in the following section for proving the symplectic properties of the fundamental solution matrix. Let us also note here that the basic property (18) holds not only for both the heliocentric and the Jacobian forms of the three-body problem but also for all dynamical systems with a Lagrangian of the form (1) with a constant metric tensor and an arbitrary potential function $V(q^k)$.

4. THE FUNDAMENTAL MATRIX OF SOLUTIONS

We will designate by R any (8 x 8)-matrix whose columns are independent solutions of the variational equations. It is known (and seen by direct substitution in the variational equations) that, when R is a fundamental matrix, then RC is also a fundamental matrix, for any constant non-singular matrix C. The matrix C could be chosen in such a way as to form what we call the principal fundamental matrix R(t) which is such that R(0)=1. In other words the initial values of the principal fundamental matrix form a unit matrix. In what follows we will only use the principal fundamental matrix.

By definition, any fundamental matrix R verifies the corresponding variational equations:

$$\dot{R} = AR , \tag{19}$$

and R^{-1} is solution of the adjoint equation:

$$\frac{dR^{-1}}{dt} = -R^{-1}A . \tag{20}$$

This solution is seen to be

$$R^{-1} = SR^TS^{-1} , \tag{21}$$

by direct substitution, and by using the skew-symplectic property (18) of A. The last equation is a remarkably simple expression for the inverse of R, directly in terms of the elements of R. This equation also expresses that R is symplectic. It contains one of the most important properties of the solution of the variational equations of the three-body problem. A consequence of the above relation is that R, R^T and R^{-1} have the same eigenvalues. Thus the eight eigenvalues of R form four reciprocal pairs. Zero eigenvalues are excluded here, because R is nonsingular. The reciprocal pair property is true for all values of time t, and for all solutions, periodic and non-periodic.

5. THE CHARACTERISTIC EQUATION

There is another property of the eigenvalues of R which is valid only for the variational equations of periodic solutions, and only for those values of t which are an integer multiple of the period T: for every independent isolating first-integral, there is a pair of unit eigenvalues for the fundamental matrix, Moulton (1920).

In the planar heliocentric three-body problem, there are only two first integrals, (Energy and Angular Momentum), so that there will be four unit eigenvalues. The eigenvalues of R are always of the following form:

$$\lambda_1, \frac{1}{\lambda_1}, \lambda_2, \frac{1}{\lambda_2}, 1, 1, 1, 1,$$

where of course λ_1 and λ_2 could be complex. These properties of the roots result in a special form of the characteristic equation. First of all the quadruple root +1 is solution of

$$(s - 1)^4 = s^4 - 4s^3 + 6s^2 - 4s + 1 = 0. \tag{22}$$

Next, the four roots $\lambda_1, \frac{1}{\lambda_1}, \lambda_2, \frac{1}{\lambda_2}$ are solutions of the equation

$$s^4 + a_1 s^3 + a_2 s^2 + a_1 s + 1 = 0, \tag{23}$$

where the two real stability coefficients a_1 and a_2 are defined in terms of the two stability indices k_1 and k_2:

$$\begin{aligned} a_1 &= -(k_1 + k_2) \\ a_2 &= 2 + k_1 k_2 , \\ k_1 &= \lambda_1 + \frac{1}{\lambda_1}, \\ k_2 &= \lambda_2 + \frac{1}{\lambda_2}. \end{aligned} \tag{24}$$

Multiplying the two fourth-degree equations side by side gives us the eight-degree characteristic equation (which is of course a reciprocal equation):

$$\begin{aligned} &(s^8 + 1) + (a_1 - 4)(s^7 + s) + (a_2 - 4a_1 + 6)(s^6 + s^2) \\ &\quad -(4a_2 - 7a_1 + 4)(s^5 + s^3) + (6a_2 - 8a_1 + 2)s^4 = 0. \end{aligned} \tag{25}$$

In general the principal fundamental matrix R of a periodic solution will be obtained by solving the variational equations eight times with different initial conditions. The coefficients of the characteristic equation can then all be found in terms of the determinant of R and several subdeterminants. When these coefficients have been obtained, we can determine the two stability coefficients a_1 and a_2.

The discussion of the stability of a periodic orbit will then proceed in the same way as for the elliptic restricted three-body problem (Broucke, 1969). All the conclusions which have been given by Broucke for the elliptic restricted problem apply to the planar general three-body problem. In particular, the stability can be discussed most conveniently in the (a_1, a_2) - plane. As has been found previously there is one region of the (a_1, a_2) - plane corresponding to stable roots and six different unstable regions or types of unstable orbits. There are also nine general cases of transitions between regions in addition to three particular cases.

6. THE THREE TRIVIAL SOLUTIONS OF THE VARIATIONAL EQUATIONS OF THE THREE-BODY PROBLEM

It is well known that if a dynamical system has a solution q(t,c) depending on an arbitrary constant c not contained in the original equations of motion, then the corresponding variational equations have a known solution of the form:

$$\delta q(t) = \lim_{\varepsilon \to 0} \frac{q(t,c+\varepsilon) - q(t,c)}{\varepsilon} = \frac{\partial q(t,c)}{\partial} . \qquad (26)$$

We now intend to show that the use of this principle provides us with three immediate solutions of the variational equations of the three-body problem. This result has its practical importance because it allows us to construct a monodromy matrix R(T) (of dimension 8 x 8) of a periodic solution by numerically integrating only five independent solutions of the variational equations instead of eight. The arbitrary constants available in the three-body problem correspond to a change of the time origin of a solution, a rotation of this solution or finally a change by a scale factor.

The change of the origin of time is possible because the original equations of motion do not contain the time explicitly and a solution q(t) may also be written in the form $q(t - t_0)$. Consequently if the eight components of a solution are $[q(t), \dot{q}(t)]$, then the components of an immediate solution w(t) of the variational equations are obtained by taking time-derivatives:

$$\omega(t) = [\dot{q}(t), \ddot{q}(t)] = [\dot{x}_1,\dot{y}_1,\dot{x}_2,\dot{y}_2,\ddot{x}_1,\ddot{y}_1,\ddot{x}_2,\ddot{y}_2]. \qquad (27)$$

Another trivial solution is obtained by applying a constant rotation of angle θ to a known solution:

$$x_1' = x_1 \cos\theta + y_1 \sin\theta ,$$

$$y_1' = -x_1 \sin\theta + y_1 \cos\theta . \qquad (28)$$

Taking partial derivatives with respect to the constant θ and next setting θ equal to zero gives the following solution of the variational equations (the same rotation also applies to x_2, y_2 and their derivatives):

$$z(t) = [y_1, -x_1, y_2, -x_2, \dot{y}_1, -\dot{x}_1, \dot{y}_2, -\dot{x}_2] . \qquad (29)$$

The third explicit solution of the variational equations is derived by considering the arbitrary scale factor α which may be applied to any solution of the three-body problem. More precisely, when a solution is $\psi(t) = [q(t), \dot{q}(t)]$, a similar solution is obtained by application of the scale factors α, $\alpha^{-\frac{1}{2}}$ and $\alpha^{-3/2}$ respectively to the coordinates q(t), the velocities $\dot{q}(t)$ and the time t. A new solution is thus

$$\psi'(t) = [\alpha q(\alpha^{-3/2}t), \alpha^{-\frac{1}{2}} \dot{q}(\alpha^{-3/2}t)] . \qquad (30)$$

Taking the partial derivative with respect to α and then setting $\alpha=1$ gives the anticipated new solution of the variational equations:

$$s(t) = \frac{\partial\psi'(t,\alpha)}{\partial\alpha}\Big]_{\alpha=1} = [q(t) - \tfrac{3}{2}t\dot{q}(t)\ ,\ \tfrac{-1}{2}\dot{q}(t) - \tfrac{3}{2}t\ddot{q}(t)]\ . \quad (31)$$

We see the presence of a linear term in t and this shows that we have a secular solution. We have now obtained three particular solutions of the variational equations of the three-body problem, w(t), z(t) and s(t). Comparison of w(t) and s(t) shows the following relationship:

$$s(t) = [q(t) - \tfrac{3}{2}tw(t)\ ,\ - \tfrac{1}{2}\dot{q}(t) - \tfrac{3}{2}tw(t)]\ . \quad (32)$$

Let us also show how these three solutions are related to the eigenvectors of the monodromy matrix M = R(T). First of all, any solution u(t) of the variational equations can be written as a linear combination of columns of the fundamental matrix:

$$u(t) = R(t)\ u_0. \quad (33)$$

Applying this to the periodic solution w(t) gives:

$$w(T) = R(T)w_0 = R(T)w(T), \quad (34)$$

showing that w(T) is an eigenvector corresponding to a unit eigenvalue of the monodromy matrix M = R(T). The same proof shows that z(T) is another such eigenvector of R(T):

$$R(T)z(T) = z(T). \quad (35)$$

Let us finally show the corresponding property of the solution s(t) associated with the change of scale α. We have again:

$$s(t) = R(t)s_0 \quad (36)$$

and consequently:

$$s(T) = R(T)s_0 = s_0 - \tfrac{3}{2}Tw_0 \quad (37)$$

or

$$(R(T) - I)s_0 = - \tfrac{3}{2}Tw_0\ . \quad (38)$$

This equation shows that s_0 is not an eigenvector of the monodromy matrix M = R(T), (corresponding to a unit eigenvalue). However, if we multiply both sides of the previous equation by R(T)-I, we obtain

$$(R(T) - I)^2 s_0 = 0, \quad (39)$$

which shows that s_0 is a principal vector of degree two of the monodromy matrix.

In summary, we have obtained three important solutions of the variational equations of a periodic solution of the three-body problem, w(t), z(t), s(t), the first two being pure periodic and the last one containing a secular term. The initial values of these solutions are intimately related to the eigenvectors of the monodromy matrix M; more precisely we have the following four identities:

$$\begin{aligned}(M - I)w_0 &= 0,\\ (M - I)z_0 &= 0,\\ (M - I)^2 s_0 &= 0,\\ (M - I)s_0 &= -\frac{3}{2}Tw_0.\end{aligned} \tag{40}$$

These properties have been successfully used by D. Boggs (1975) to facilitate the computation of eigenvalues of M. They make it possible to reduce the 8 x 8 eigenvalue problem to a 5 x 5 eigenvalue problem.

7. SYMMETRIC PERIODIC ORBITS

We intend to show here the special properties of the variational equations for symmetric periodic orbits. These properties will allow us to determine the stability and characteristic exponents of a periodic solution by integrating only a half rather than a complete revolution. We will use heliocentric coordinates here.

If the solution under consideration is symmetric with respect to the x-axis, we have the following properties for the above eight functions:

$$\begin{aligned}x_1(-t) &= + x_1(t) ,\\ y_1(-t) &= - y_1(t) ,\\ x_2(-t) &= + x_2(t) ,\\ y_2(-t) &= - y_2(t) ,\\ \dot{x}_1(-t) &= - \dot{x}_1(t) ,\\ \dot{y}_1(-t) &= + \dot{y}_1(t) ,\end{aligned} \tag{41}$$

$$\dot{x}_2(-t) = -\dot{x}_2(t) ,$$
$$\dot{y}_2(-t) = +\dot{y}_2(t) . \tag{41}$$

Accordingly, the transition from the orbit to its symmetric image (w.r. to the x-axis) may be considered as a linear transformation with constant coefficients, represented by the following (8 x 8) diagonal matrix:

$$\text{diag } B = [+1, -1, +1, -1, -1, +1, -1, +1] . \tag{42}$$

The above simple matrix B has the trivial properties:

$$B^2 = +1 ; \quad B^{-1} = B . \tag{43}$$

We will call this matrix B the "symmetric matrix." We also see that, by defining the following (4 x 4)-matrix:

$$B' = \text{diag}[+1, -1, +1, -1] , \tag{44}$$

we can write:

$$B = \left[\begin{array}{c|c} B' & 0_4 \\ \hline 0_4 & -B' \end{array}\right] . \tag{45}$$

It can now be seen, by inspection of each one of the sixteen terms, that the 4 x 4-matrix A' of the variational equations given in (10) has the following symmetry properties

$$a_{ij}(-t) = \begin{bmatrix} +a_{11}(t) & -a_{12}(t) & +a_{13}(t) & -a_{14}(t) \\ -a_{21}(t) & +a_{22}(t) & -a_{23}(t) & +a_{24}(t) \\ +a_{31}(t) & -a_{32}(t) & +a_{33}(t) & -a_{34}(t) \\ -a_{41}(t) & +a_{42}(t) & -a_{43}(t) & +a_{44}(t) \end{bmatrix} . \tag{46}$$

We can easily verify now, by using (44), that the relations between $a_{ij}(t)$ and $a_{ij}(-t)$ can be written in matrix form:

$$[a_{ij}(-t)] = B'[a_{ij}(t)]\, B' . \tag{47}$$

Using this result also allows us to write the corresponding property for the (8 x 8)-matrix A of the variational equations:

$$A(-t) = -B\, A(t)\, B \tag{48}$$

8. THE FUNDAMENTAL MATRIX OF A SYMMETRIC SOLUTION

The variational equations (19) may be written in the form

$$\frac{d}{dt} R(t) = A(t)\, R(t) \tag{49}$$

Let us now see what is the effect of a change of t into -t, (and dt into -dt):

$$\frac{-d}{dt} R(-t) = A(-t)\, R(-t)\ , \tag{50}$$

or

$$\frac{d}{dt} R(-t) = B\, A(t)\, B\, R(-t). \tag{51}$$

Multiplying both sides of the above equation by B (on the left) gives

$$\frac{d}{dt} [B\, R(-t)] = A(t)[B\, R(-t)]\ . \tag{52}$$

This result shows that B R(-t) is a solution of the variational equations. Consequently, B R(-t) must be equal to R(t) multiplied on the right-side by the appropriate constant initial condition matrix, which turns out to be B in the present case:

$$B\, R(-t) = R(t)\, B. \tag{53}$$

Multiplying now both sides of the previous identity on the left by B gives the important property

$$R(-t) = B\, R(t)\, B \tag{54}$$

9. HALF-PERIOD PROPERTIES OF SYMMETRIC PERIODIC ORBITS

We will now use the above results to show that for symmetric solutions, there is a remarkable relationship between the monodromy matrix R(T) and the matrix $R(\frac{T}{2})$. This relationship is important because it allows us to construct the monodromy matrix R(T) of a periodic solution of period T by only integrating the orbits and their associated variational equations over a half-period T/2.

We first have to show an important general property of the monodromy matrix R(T) of a periodic solution. We know that for such a solution

$$A(t + T) = A(t)\ , \tag{55}$$

but what is the corresponding property for the matrix R? We can directly write

$$\frac{d}{dt} R(t + T) = A(t + T)\ R(t + T) = A(t)\ R(t + T). \qquad (56)$$

This shows thus that R(t + T) is a solution of the variational equations. The columns of R(t + T) must thus be linear combinations (with constant coefficients) of the columns of R(t). In other words R(t + T) must be R(t) multiplied at the right side by a constant non-singular matrix, which by considering the initial conditions is found to be R(T):

$$R(t + T) = R(t)\ R(T)\ . \qquad (57)$$

This is the relation between R(t) and R(t + T), which is true for all periodic solutions, symmetric or not. Replacing in the above identity, t by -T/2 gives the relation:

$$R(\tfrac{T}{2}) = R(-\tfrac{T}{2})\ R(T)\ , \qquad (58)$$

which can also be written

$$R(T) = R(-\tfrac{T}{2})^{-1}\ R(\tfrac{T}{2})\ . \qquad (59)$$

In the last formula, $R(-\frac{T}{2})$ can be expressed in terms of $R(+\frac{T}{2})$ by using the result of the previous section:

$$R(-\tfrac{T}{2}) = B\ R(\tfrac{T}{2})\ B\ . \qquad (60)$$

We have thus:

$$R(T) = [B\ R(\tfrac{T}{2})\ B]^{-1}\ R(\tfrac{T}{2}) = B\ R(\tfrac{T}{2})^{-1}\ B\ R(\tfrac{T}{2})\ . \qquad (61)$$

Finally the inverse matrix can be expressed in terms of the matrix itself by using the symplectic property of R:

$$R(\tfrac{T}{2})^{-1} = S\ R(\tfrac{T}{2})^{T}\ S^{-1}\ , \qquad (62)$$

where S is a known constant matrix depending on the g_{ij}'s only (in the present problem, S is a function of the three masses m_0, m_1, m_2 only). The expression for R(T) takes on the final form:

$$R(T) = B\ S\ R\ (\tfrac{T}{2})^{T}\ S^{-1}\ B\ R\ (\tfrac{T}{2})\ , \qquad (63)$$

or

$$R(T) = (B\ S)\ R\ (\tfrac{T}{2})^{T}\ (B\ S)^{-1}\ R(\tfrac{T}{2})\ . \qquad (64)$$

In the above formula, the superscript T indicates the transpose of the matrix. Keeping in mind that the matrix B S is constant, we see that R(T) is related to the half-period fundamental matrix $R(\frac{T}{2})$ by a very simple and elegant formula. We find the following explicit formula for the matrix B S:

$$B\,S = \left[\begin{array}{c|c} 0 & B'\,g^{ij} \\ \hline B'\,g^{ij} & 0 \end{array}\right] \qquad (65)$$

where

$$B'\,g^{ij} = \begin{bmatrix} g^{11} & g^{12} & g^{13} & g^{14} \\ -g^{21} & -g^{22} & -g^{23} & -g^{24} \\ g^{31} & g^{32} & g^{33} & g^{34} \\ -g^{41} & -g^{42} & -g^{43} & -g^{44} \end{bmatrix} . \qquad (66)$$

We see thus that the matrix B S has a remarkably symmetric form. A similar result exists for the matrix S^{-1} B.

REFERENCES

1. Boggs, D., "Periodic Orbits in the General-Three-Body Problem," Ph.D. Dissertation, UCLA, California, 1975.
2. Broucke, R., "Stability of Periodic Orbits in the Elliptic Restricted Three-Body Problem," AIAA Journal, Vol. 7, 1969, pages 1003-1009.
3. Cartan, E., "Legons sur la theory des Invariants Integraue," Hermann, Paris, 1971.
4. Loseo, L., "Sur un invariant integral du probleme des n corps," IAU Symposium No. 62, Warsaw, September 5-8, 1973; Proceedings, pages 249-255 (1974).
5. Moulton, R. R., "Periodic Orbits," 1920.

FAMILIES OF PERIODIC ORBITS IN THE GENERAL N-BODY PROBLEM

J. D. Hadjidemetriou

University of Thessaloniki, Thessaloniki, GREECE

INTRODUCTION. It is well known that a detailed study has been made since long ago of the families of periodic orbits of the restricted circular 3-body problem. An extensive review is contained in the "Theory of Orbits" by Szebehely (1967). Also, families of periodic orbits of the elliptic restricted 3-body problem have been studied by Broucke (1969). However, no similar study has been made for the periodic orbits of the general N-body problem with the exception of a few cases of the general 3-body problem studied recently by Szebehely (1969, 1971) and Standish (1969). These were periodic orbits in a fixed system and involved collisions of two bodies. Recently Broucke and Boggs (1975) have made a more detailed study of periodic orbits of the general 3-body problem in a fixed system.

In the following we shall show that families of periodic orbits of the general N-body problem ($N \geq 3$) exist in a rotating frame of reference and we shall develop a systematic way to obtain them. These results are based on research work carried out at the University of Thessaloniki. In the 3-body problem similar work has been done independently by Broucke (1975) and Henon (1974, 1975).

In what follows, we shall start first with the 3-body problem and we shall extend these results to the N-body problem for any value of N.

V. Szebehely and B.D. Tapley (eds.), Long-Time Predictions in Dynamics, 223–240. *All Rights Reserved. Copyright* ©1976 *by D. Reidel Publishing Company, Dordrecht-Holland.*

1. THREE-BODY PROBLEM

1.1 Reduction to Three Degrees of Freedom

We consider three bodies P_1, P_2, P_3 with nonzero masses moving in the plane under their mutual gravitational attraction and define a rotating frame of reference XOY such that its origin coincides with the center of mass of P_1 and P_2 (hereafter called primaries) and its x axis contains always these primaries (the positive direction being from P_2 to P_1). For $m_3 = 0$ the above system XOY reduces to the well known rotating system of the restricted problem.

Without loss of generality we can consider the center of mass G of the whole system fixed with respect to an inertial frame. In this case we have four degrees of freedom and we can choose as generalized coordinates the cartesian coordinates x_1 of P_1 and x_3, y_3 of P_3 and the angle θ between the x axis and a fixed direction in the inertial frame. The lagrangian of the system is

$$L = \frac{1}{2}(m_1 + m_2)\{q(\dot{x}_1^2 + x_1^2\dot{\theta}^2) + \frac{m_3}{m}[\dot{x}_3^2 + \dot{y}_3^2 + \dot{\theta}^2(x_3^2 + y_3^2) + + 2\dot{\theta}(x_3\dot{y}_3 - \dot{x}_3y_3)]\} - V \tag{1}$$

where

$$V = -\frac{k^2m_1m_3}{r_{13}} - \frac{k^2m_2m_3}{r_{23}} - \frac{k^2m_1m_2}{r_{12}}, \tag{2}$$

$$m = m_1 + m_2 + m_3, \qquad q = m_1/m_2 \tag{3}$$

r_{ij} (i, j = 1, 2, 3) is the distance between P_i and P_j and k^2 is the gravitational constant. From (1) we can see that θ is an ignorable coordinate and consequently we have, apart from the energy (or Jacobi) integral, the angular momentum integral, which takes the form

$$\dot{\theta} = \frac{\dfrac{p}{m_1 + m_2} - \dfrac{m_3}{m}(x_3\dot{y}_3 - \dot{x}_3y_3)}{qx_1^2 + \dfrac{m_3}{m}(x_3^2 + y_3^2)}, \tag{4}$$

where p is the angular momentum constant. This means that the angle θ can be eliminated completely by properly choosing the Lagrangian function (i.e., the Routhian) and in this way the problem reduces to three degrees of freedom. In practice however, it was more convenient to use in the numerical computations the system derived from the original Lagrangian (1). This system is, for $k^2 = 1$,

$$\begin{aligned}
\ddot{r} - r\dot{\theta}^2 &= -\frac{m_1 + m_2}{r^2} - m_3 A , \\
\ddot{x}_3 - 2\dot{\theta}\dot{y}_3 - \dot{\theta}^2 x_3 - \ddot{\theta} y_3 &= mB , \\
\ddot{y}_3 + 2\dot{\theta}\dot{x}_3 - \dot{\theta}^2 y_3 + \ddot{\theta} x_3 &= mC , \\
\ddot{\theta} + \frac{2\dot{r}\dot{\theta}}{r} &= -\frac{m_3}{\mu(1-\mu)r^2} [Cx_3 - By_3] ,
\end{aligned} \tag{5}$$

where r is the distance between the primaries, related to x_1 by

$$x_1 = \mu r , \tag{6}$$

and

$$\mu = m_2/(m_1 + m_2) \tag{7}$$

$$\begin{aligned}
A &= -\frac{x_3 - \mu r}{r_{13}^3} + \frac{x_3 + (1-\mu)r}{r_{23}^3} , \\
B &= -(1-\mu)\frac{x_3 - \mu r}{r_{13}^3} - \mu \frac{x_3 + (1-\mu)r}{r_{23}^3} , \\
C &= -(1-\mu)\frac{y_3}{r_{13}^3} - \mu \frac{y_3}{r_{23}^3} .
\end{aligned} \tag{8}$$

The elimination of the angle θ from the first three equations of (5) can be made at this stage by making use of (4), but this results in a more complicated system of differential equations.

1.2 Analytic Continuation

For $m_3 = 0$ the first equation of (5) becomes independent of the rest system and gives, in fact, the motion of the two primaries. For circular motion (r = const.) and $m_3 = 0$, the last equation (5) gives $\dot{\theta}$ = const. and the second and third equations (5) reduce in this case to the well known equations of the restricted circular three-body problem.

Let us assume now that for the case $m_1, m_2 \neq 0$ and $m_3 = 0$ the system (5) admits the following solution:

$$r = 1 , \qquad \dot{\theta} = 1 \tag{9}$$

and

$$x_3 = x_3^*(t), \qquad y_3 = y_3^*(t), \tag{10}$$

where $x_3^*(t)$ and $y_3^*(t)$ are periodic functions with period T. It can be easily seen that this latter part (10) of the solution represents a periodic orbit of the circular restricted three-body problem. Further, we assume that this is a symmetric periodic orbit,

$$x_3^*(t) = x_3^*(-t), \quad y_3^*(t) = -y_3^*(t). \tag{11}$$

We allow now m_3 to increase slightly, $m_3 \ll 1$, while m_1 and m_2 are kept fixed, and express the solution of (5) in the form

$$r = 1+\xi_1, \quad x_3 = x_3^*(t)+\xi_2, \quad y_3 = y_3^*(t)+\xi_3, \quad \dot{\theta} = 1+\xi_4 \tag{12}$$

where $\xi_1(0),\ldots\xi_4(0)$ are considered small, of the order of m_3. It can then be shown to a linear approximation (Hadjidemetriou 1975a) that one can choose the initial conditions $\xi_1(0)\ldots\xi_4(0)$, $\dot{\xi}_1(0),\ldots\dot{\xi}_4(0)$ in such a way that the new solution, corresponding to $m_3>0$, is of the form

$$\begin{aligned} r &= 1 + m_3\Phi_1(t), \\ x_3 &= x_3^*(t) + m_3\Phi_2(t), \\ y_3 &= y_3^*(t) + m_3\Phi_3(t), \end{aligned} \tag{13}$$

where the functions $\Phi_1(t)$, $\Phi_2(t)$ and $\Phi_3(t)$ are periodic with the same period T as the solution (10). It can be also shown that $\Phi_1(t)$ and $\Phi_2(t)$ are even functions and $\Phi_3(t)$ is an odd function of time. Thus we come to the conclusion that an arbitrary symmetric periodic orbit (10) of the circular restricted three-body problem is continued analytically, when $m_3>0$, to a symmetric periodic orbit of the general three-body problem, with respect to the rotating frome XOY defined above.

The above reasoning holds for an arbitrary orbit of the restricted problem. Thus, one can readily extend the above to state that a whole (monoparametric) family of symmetric periodic orbits of the circular restricted three-body problem can be extended to a monoparametric family of periodic orbits of the general three-body problem, for m_1, m_2, m_3 fixed ($m_3<<1$), with respect to the rotating frame XOY.

The proof of the above is based on a linear theory. The same statement can be proved by making use of the continuation theorem. In this latter case the symmetry of the orbit (10) is no longer needed. A full exposition of the method is contained in Hadjidemetriou (1975a). The continuation theorem is not applicable when the period of (10) is a multiple of 2π.

1.3 Numerical Continuation

The proof of the analytic continuation of a periodic orbit of the restricted problem to the general problem holds for sufficiently small masses of the third body P_3. The numerical computations however have revealed that this continuation continuation can be carried out for large values of m_3 as well, (Hadjidemetriou and Christides, 1975).

Before we proceed to the numerical computations we must normalize the units of length, mass and time. This can be done in many ways and we shall mention three normalization schemes which we think are more convenient. We need three conditions for the normalization and two of them can be taken to be, for all cases,

$$m = m_1 + m_2 + m_3 = 1, \quad k^2 = 1. \tag{14}$$

As for the third condition we can normalize the energy (Henon 1974) to a fixed value, say

$$E = \frac{1}{2}(m_1m_2 + m_1m_3 + m_2m_3), \tag{15}$$

or we can normalize the angular momentum p to a fixed value, say

$$p = m_1m_2 + m_1m_3 + m_2m_3, \tag{16}$$

Another possibility would be to take $\dot{\theta}$ at $t=0$ equal to a fixed value, e.g.,

$$\dot{\theta}_0 = 1. \tag{17}$$

Conditions (14), (15) or (14), (16), or (14), (17) are normalization schemes such that for $m_3 = 0$ we obtain the usual normalized values of the restricted three-body problem.

In the numerical computations we shall restrict ourselves to symmetric periodic orbits only, and we shall assume that one of the above normalization schemes, say (14), (17) has been applied. We shall use from now on x_1 instead of r. These variables are related by Eq. (6).

We must note at this point that the system (5) is invariant under the transformation

$$x_1 \to x_1, \quad x_3 \to x_3, \quad y_3 \to -y_3, \quad t \to -t. \tag{18}$$

This means that if at $t = 0$ we have

$$x_1(0)\neq 0,\ x_3(0)\neq 0,\ y_3(0)=0,\ \dot{x}_1(0)=0,\ \dot{x}_3(0)=0,\ \dot{y}_3(0)\neq 0 \tag{19}$$

and at $t=T/2$ we have, at the first crossing of P_3 with the x axis,

$$x_1(T/2)\neq 0,\ x_3(T/2)\neq 0,\ y_3(T/2)=0,\ \dot{x}_1(T/2)=0,\ \dot{x}_3(T/2)=0,$$
$$\dot{y}_3(T/2)\neq 0, \tag{20}$$

the orbit is periodic with respect to the rotating frame XOY with period T. Consequently, for fixed masses of the three bodies, a symmetric periodic orbit with respect to the rotating frame XOY is determined by the initial conditions

$$x_1 = x_1(0), \quad x_3 = x_3(0), \quad \dot{y}_3 = \dot{y}_3(0). \tag{21}$$

Clearly, $\theta(0)$ can be arbitrary and $\dot{\theta}(0)$ is uniquely determined by the normalization scheme used. Also, because of the normalization condition m-1, we have two parameters for the masses, and we can choose as such the ratio m_1/m_2 and the mass m_3.

We can write now the solution of the system (5) corresponding to the initial conditions (19) as

$$x_1 = x_1(x_1(0),\ x_3(0),\ \dot{y}_3(0),\ m_1/m_2,\ m_c;\ t)\ , \tag{22}$$

etc.

Consequently, the periodicity conditions are of the form

$$y_3(x_1(0),\ x_3(0),\ \dot{y}_3(0),\ m_1/m_2,\ m_3;\ T/2) = 0\ ,$$
$$\dot{x}_1(x_1(0),\ x_3(0),\ \dot{y}_3(0),\ m_1/m_2,\ m_3;\ T/2) = 0\ , \tag{23}$$
$$\dot{x}_3(x_1(0),\ x_3(0),\ \dot{y}_3(0),\ m_1/m_2,\ m_3;\ T/2) = 0\ .$$

This is a system of three equations in the unknowns $x_1(0)$, $x_3(0)$, $\dot{y}_3(0)$. Its solution, if it exists, provides the initial conditions of a periodic orbit of the 3-body problem, with period T and masses obtained from m_1/m_2, m_3, $(m_1+m_2+m_3=1)$. Consequently, this solution can be expressed as

$$x_1(0) = f_1(m_1/m_2,\ m_3,\ T)\ ,$$
$$x_3(0) = f_2(m_1/m_2,\ m_3,\ T)\ , \tag{24}$$
$$\dot{y}_3(0) = f_3(m_1/m_2,\ m_3,\ T)\ .$$

Clearly, for fixed values of the masses, the initial conditions (24) form a monoparametric family of symmetric periodic orbits of the general 3-body problem with the period T as a parameter. It should be emphasised however that the above argument does not constitute a rigorous proof, since the existence of the solution (24) of the system (23) has not been proved. The numerical computations however, have revealed that this solution does exist in general.

From the above reasoning we can see that a symmetric periodic orbit can be represented by its initial conditions by a point in the space x_{10}, x_{30}, $\dot{y}_{30}$. Thus, for fixed values of the three masses, a monoparametric family can be represented by a space curve in the above mentioned space of initial conditions.

In order to obtain a direct comparison with the restricted problem, we shall assume that the ratio m_1/m_2 is fixed and m_3 varies (m=1 always), since the families of the restricted problem have been studied for a fixed value of the ratio of the masses of the primaries. In this case the family of periodic orbits is two-parametric, with parameters m_3 and T, and can be represented in space x_{10}, x_{30}, $\dot{y}_{30}$ by a surface, which we shall call a <u>characteristic surface</u>. One can assume on this surface coordinate lines m_3=const. and T=const. The former represent monoparametric families for fixed masses and the latter monoparametric families with fixed period. In particular, the coordinate lines $m_3=0$ represent families of period orbits of the restricted problem, circular or elliptic. A description of the characteristic surface is given by Bozis and Hadjidemetriou (1975).

Also, a systematic study of periodic orbits of the 3-body problem of the planetary type (i.e., one large mass and two much smaller but not negligible masses) has been made by Hadjidemetriou (1975d) and Delibaltes (1975).

1.4 Stability

We shall develop now a linear method to study the stability of a periodic orbit of the 3-body problem based on the notion of surface of section. We shall restrict ourselves to symmetric periodic orbits only but the method can be extended to nonsymmetric orbits as well. We shall outline the main points of the method. A full exposition is given by Hadjidemetriou (1975b).

We assume that the problem has been reduced to three degrees of freedom by making use of the angular momentum integral to eliminate the angle θ. Thus the angular momentum constant p appears as a fixed parameter in the reduced system of differential equations. The phase space is now the 6th dimensional space $x_1x_3y_3\dot{x}_1\dot{x}_3\dot{y}_3$ and because of the energy integral, the motion takes place in the hypersurface defined by the energy integral in the above space. Let us consider now a symmetric periodic orbit with period T and take $t=0$ when $y_3=0$ The initial conditions of this periodic orbit are

$$x_1 = x_{100}, \quad x_3 = x_{300}, \quad \dot{y}_3 = \dot{y}_{300}. \tag{25}$$

At the time $t = T$ the variables $x_1, x_3, \dot{y}_3$ assume again their initial values (25). We take now a neighboring orbit with initial conditions

$$x_{10}=x_{100}+\xi_{10}, \quad x_{30}=x_{300}+\xi_{20}, \quad y_{30}=0, \quad \dot{x}_{10}=\xi_{30}, \quad \dot{x}_{30}=\xi_{40} \tag{26}$$

and $\dot{y}_{30}$ such that the energy constant is the same as that of the periodic orbit. When the variable y_3 becomes again equal to zero and the orbit crosses the x axis in the same direction as at t=0, we have

$$x_1=x_{100}+\xi_1, \quad x_3=x_{300}+\xi_2, \quad \dot{x}_1=\xi_3, \quad \dot{x}_3=\xi_4 \ . \tag{27}$$

In this way a mapping is established in the four dimensional space $x_1x_3\dot{x}_1\dot{x}_3$ between the initial conditions (26) of the perturbed orbit and the final conditions (27) at the next crossing in the same direction with the x axis, of the form

$$x_1 = g_1(x_{10}, x_{30}, \dot{x}_{10}, \dot{x}_{30}), \quad \text{etc.} \tag{28}$$

The corresponding mapping between the pertubations ξ_{io} and ξ_i in linearized form is

$$\xi = A\xi_o, \tag{29}$$

where ξ, ξ_o are column vectors with elements ξ_i, ξ_{io} respectively and A is a matrix whose elements are given by $\partial g_i/\partial x_{10}$, etc.

Evidently, the fixed points of the mapping (28) represent periodic orbits of the general problem. The linear stability of these orbits can be studied from the linearized mapping (29) by studying the eigenvalues of the matrix A, corresponding to a fixed point. Several properties of the matrix A can be proved for symmetric periodic orbits (Hadjidemetriou 1975b). First of all we have

$$\det(A) = 1 \tag{30}$$

which means that the mapping (29) is volume preserving. Another useful property is that the matrix A can be obtained by computing perturbed orbits for half a period only. If B is the corresponding matrix, analogous to A, for the case the final conditions (27) are considered at the next intersection with the x axis (with opposite direction), the matrix A is obtained from the relation

$$A = JB^{-1}JB \tag{31}$$

where

$$J = \begin{pmatrix} 1 & 0 & 0 & 0 \\ 0 & 1 & 0 & 0 \\ 0 & 0 & -1 & 0 \\ 0 & 0 & 0 & -1 \end{pmatrix} \tag{32}$$

Relation (31) is very useful because we can obtain the matrix A by performing the numerical computations for half the period only. This has also the advantage that a better accuracy is obtained.

The characteristic equation of the matrix A, corresponding to a fixed point of the mapping (29) is a reciprocal equation and the conditions for stability or instability are given in Hadjidemetriou (1975b). In this paper a family of periodic orbits with equal masses of the three bodies has been computed. All orbits of the family consist of a hierarchy of binaries such that the third body revolves around the inner binary in the same direction. It was found that a large part of the family represents stable motion. Instability starts when the third body comes close to the inner binary.

In order for the stability analysis of a planar periodic motion of the three-body system to be complete, the vertical stability must be also studied. A method to do that has been developed by Delibaltas (1975) in his study of families of periodic orbits of the planetary type for the Sun-Jupiter-Saturn mass ratio.

2. N-BODY PROBLEM

2.1 Differential Equations of Notion

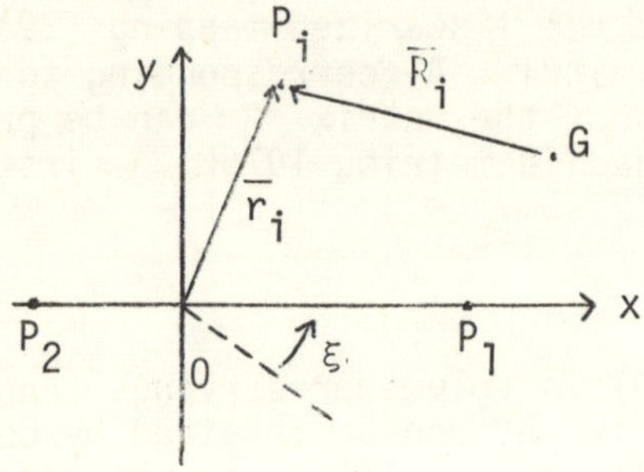

Fig. 1: The rotating frame XOY

We consider N bodies (N>3) moving in the plane. The center of mass G of the system is considered to be at rest with respect to an inertial frame. We define now a rotating coordinate system XOY such that its origin coincides with the center of mass 0 of the bodies P_1 and P_2 and its x axis contains always these bodies (Fig. 1). We also define the vectors $\overline{r}_i$ and $\overline{R}_i$ as

$$\overline{r}_i = O\overline{P}_i = x_i\overline{i} + y_i\overline{j} \quad , \quad \overline{R}_i = G\overline{P}_i = \xi_i\overline{i} + \eta_i\overline{j}. \tag{33}$$

We have 2(N-1) degrees of freedom and we choose as generalized coordinates the coordinate x_1 of P_1, ξ_i, η_i (i=3,...N) defined in (33) for the bodies $P_3,..P_N$ and also the angle θ between the x axis and a fixed direction in the inertial system.

The kinetic energy of the system can be expressed in terms of the above variables (Hadjidemetriou 1975c) as

$$T = \frac{1}{2} m_1(1+q)(\dot{x}_1^2+\dot{\theta}^2 x_1^2)+\frac{1}{m_1+m_2}\left\{\left[\sum_{i=3}^{N} m_i(\dot{\xi}_i-\dot{\theta}\eta_i)\right]^2 + \left[\sum_{i=3}^{N} m_i(\dot{\eta}_i+\dot{\theta}\xi_i)\right]^2\right\} + \frac{1}{2}\sum_{i=3}^{N} m_i\left[(\dot{\xi}_i-\dot{\theta}\eta_i)^2+(\dot{\eta}_i+\dot{\theta}\xi_i)^2\right] \tag{34}$$

where q is given by (3). The potential energy is

$$V = -\sum_{i<j}^{1,\ldots N} \frac{m_i m_j}{r_{ij}} , \tag{35}$$

where r_{ij} is the distance between the bodies P_i and P_j and the gravitational constant is taken equal to unity. Using (34) and (35) we obtain the Lagrangian equations of motion, after some algebra, in the form

$$\ddot{x}_1 = x_1\dot{\theta}^2 = -\frac{1}{m_1(1+q)}\frac{\partial V}{\partial x_1} , \tag{36}$$

$$\ddot{x}_\kappa - \ddot{\theta}y_\kappa - 2\dot{\theta}\dot{y}_\kappa - \dot{\theta}^2 x_\kappa = -\frac{1}{m_\kappa}\frac{\partial V}{\partial \xi_\kappa} \tag{37}$$

$$\ddot{y}_\kappa + \ddot{\theta}x_\kappa + 2\dot{\theta}\dot{x}_\kappa - \theta^2 y_\kappa = -\frac{1}{m_\kappa}\frac{\partial V}{\partial \eta_\kappa} \qquad (\kappa=3\ldots N).$$

We note from (34), (35) that the angle θ is ignorable. Consequently the angular momentum integral exists, in the form

$$p = \sum_{i=3}^{N} m_i(x_i\dot{\eta}_i - y_i\dot{\xi}_i) + \dot{\theta}\left[m_1(1+q)x_1^2 + \sum_{i=3}^{N} m_i(\xi_i x_i + y_i\eta_i)\right], \tag{38}$$

where p is the angular momentum constant.

The equations (36), (37), (38) are the differential equations for the motion of the N-body system. We may consider as new generalized coordinates the variables

$$x_1, x_3, \ldots x_N, y_3, \ldots y_N, \theta. \tag{39}$$

where the x_i, y_i $(i \geq 3)$ have been defined by (33). The ξ_i, η_i appearing in (37), (38), (39), are expressed in terms of (39) by

$$\xi_i = \sum_{\kappa=3}^{N} (\delta_{i\kappa} - \frac{m_\kappa}{m})\, x_\kappa \quad , \quad \eta_i = \sum_{\kappa=3}^{N} (\delta_{i\kappa} - \frac{m_\kappa}{m})\, y_\kappa \tag{40}$$

where $\delta_{i\kappa}$ is the Kronecker delta.

Using the angular momentum integral we can eliminate completely the angle θ from the equations of motion. Thus, the relative motion of the N-body system, described by the coordinates x_1 x_3, $\ldots x_N$, y_3, $\ldots$ y_N, can be studied independently of the motion of the system XOY itself. In fact, if the relative motion is known, the angle θ can be obtained from (38) by a quadrature.

2.2 Analytic Continuation

It will be shown elsewhere (Hadjidemetriou 1975c) that if

$$m_1 \neq 0, \quad m_2 \neq 0, \quad m_3 = \dots = m_N = 0 \quad , \tag{41}$$

the system (36) - (38) becomes separable: The equations (36) and (38) represent then a two-body motion only, for the bodies P_1 and P_2. If the initial conditions are such that P_1 and P_2 describe circular motion, then the equations (37), for each value of k separately, represent a restricted circular three-body motion for the body P_k under the attraction of the primaries P_1 and P_2. Evidently, the motions of the bodies $P_3 \dots P_N$ are in this case, independent from each other.

Let us assume now that the orbits of $P_3, \dots P_N$ for the masses (41) (and circular motion of the primaries) are all periodic with periods $T_3, \dots T_N$ which are in integer dependence. This means that the N-body system (41) is periodic with respect to the rotating frame XOY. It can be proved now that if the masses of all the bodies $P_3, \dots P_N$ are slightly increased, then it is possible to obtain new initial conditions such that the new, general, N-body system is still periodic with respect to XOY.

2.3 Existence of Symmetric Periodic Orbits

We note that the system (36), (37), (38) is invariant under the transformation

$$x_1 \to x_1 \quad x_i \to x_i, \quad y_i \to -y_i, \quad t \to -t \qquad (\text{also } \theta \to -\theta).$$
$$(i=3, \dots N)$$

This means that if at $t=0$ we have

$$y_3 = \dots = y_N = 0, \quad \dot{x}_1 = \dot{x}_3 = \dots = \dot{x}_N = 0 \tag{42}$$

and at $t=T/2$ the above relations are again verified, then the motion of the N-body system is periodic with period T. Geometrically this means that if at $t=0$ the bodies $P_3, \dots P_N$ cross the x axis perpendicularly while P_1 is at rest on the x axis and after a time $t=T/2$ they cross again the x axis perpendicularly and P_1 is at rest on the x axis and after a time $t=T/2$ they cross again the x axis perpendicularly and P_1 is at rest, the motion is periodic and the orbits of the bodies $P_3, \dots P_N$ in the

XOY plane are symmetric with respect to the x axis. Thus we see that the initial conditions which determine a symmetric periodic orbit are

$$x_{10},\ x_{30},\ldots x_{N0},\ \dot{y}_{30},\ldots\dot{y}_{N0},\ \dot{\theta}_0\ ,$$

provided t=0 at the time of perpendicular crossing of the bodies $P_3 \ldots P_N$ with the x axis.

Before we proceed let us normalize the variables. This can be done by the conditions

$$m_1+\ldots+m_N = 1,\quad k^2 = 1,\quad \dot{\theta}_0 = 1 \tag{43}$$

Because of the first normalization condition (43) we need N-1 parameters to determine the masses and we take as such the ratio $q = m_1/m_2$ and the masses $m_3,\ldots m_N$.

Let us assume now that the masses of the N bodies are fixed and we take the initial conditions at t=0 as

$$x_1 = x_{10},\quad x_i = x_{i0},\quad \dot{y}_i = \dot{y}_{i0},\ (i=3,..N) \tag{44}$$

and

$$\dot{x}_1 = \dot{x}_i = y_i = 0.$$

In order for the initial conditions (44) to correspond to a periodic orbit, the relations (42) must be verified at t=T/2. Thus, we must have

$$\begin{aligned} &\dot{x}_1(x_{10},\ x_{k0},\ \dot{y}_{k0},\ q,\ m_3,\ldots m_N;\ T/2) = 0\ ,\\ &\dot{x}_j(x_{10},\ x_{k0},\ \dot{y}_{k0},\ q,\ m_3,\ldots m_N;\ T/2) = 0\ ,\\ &y_j(x_{10},\ x_{k0},\ \dot{y}_{k0},\ q,\ m_3,\ldots m_N;\ T/2) = 0\ .\\ &\qquad\qquad (j=3,..N) \end{aligned} \tag{45}$$

These are 2N-3 periodicity conditions in the 2N-3 unknown variables (44) and the period T, provided the masses are fixed. These periodicity conditions can be considered to form a system of 2N-3 equations in the 2N-3 unknown initial conditions (44). Its solution, if it exists, will provide the initial conditions of a symmetric periodic orbit with respect to the rotating frame XOY.

The solution of (45) can be expressed formally as

$$x_{10} = \Phi_1(q,\ m_3,\ldots m_N,\ T),\qquad \text{etc.} \tag{46}$$

from which we readily obtain that the periodic motion thus found belongs to a monoparametric family with parameter the period T, provided all the masses are fixed. If the masses are allowed to vary, then we come to the conclusion that the symmetric periodic orbits of the general N-body problem belong to a N-parameter family with parameters the N-1 independent masses and the period. The above result can be also generalized to nonsymmetric periodic orbits but we shall deal in the following with symmetric periodic orbits only. It should be also mentioned that the above reasoning does not constitute a rigorous proof of the existence of families of periodic orbits in the N-body problem as the existence of the solution (46) of the system (45) has not been proved. The numerical results however indicate that this solution does exist in general.

3. Numerical results

We shall illustrate the above by considering five bodies with masses $m_1 = m_2 = 0.5$, $m_3 = m_4 = m_5 = 0$ (Hadjidemetriou and Michalodimitrakis, 1975) and we shall assume that the normalization (43) is used. The initial conditions of the body P_3

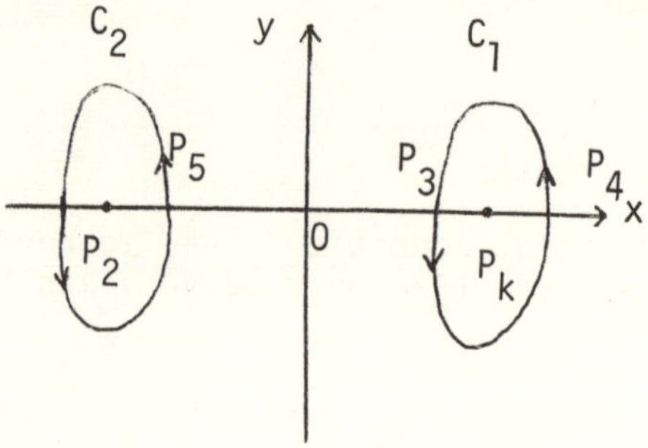

Fig. 2: The orbits of P_3, P_4, P_5 in the rotating frame (schematically)

are taken to correspond to a periodic orbit of the restricted circular three-body problem with period T and let C_1 (Fig. 2) be this orbit in the rotating frame XOY. Due to the symmetry in this case, because of the equality of the masses of m_1 and m_2, we take for P_5 symmetric initial conditions so that P_5 describes an orbit C_2 which is symmetric to C_1 with respect to the origin O. Finally the body P_4 is taken at t=0 to be at the other intersection of the orbit C_1 with the x axis and consequently it describes the same orbit C_1 but with a phase difference of half a period compared to P_3.

Evidently, the above 5-body system is periodic with period T. We increase now the masses m_3, m_4, m_5 in such a way that

$m_3=m_4=m_5>0$. The masses m_1, m_2 are then given by $m_1=m_2=\frac{1}{2}(1-3m_3)$. For a small value of $m_3=m_4=m_5=\varepsilon$ we have solved numerically the periodicity conditions, starting with the initial conditions of the 3 restricted 3-body problems mentioned above, as an approximate solution. The new solution has been an approximate solution. The new solution has been used as an approximate solution for a still larger value of the masses $m_3=m_4=m_5$, and so on. Throughout this continuation process we kept x_{30} fixed as we varied the masses and we obtained thus a mono-parametric family of periodic orbits of a 5-body system with the mass $m_3(=m_4=m_5)$ as a parameter. It was found that this continuation could be carried out for large values of the masses m_3, m_4, m_5 and we obtained thus a periodic 5-body system where the distances and the masses are all of the same order of magnitude. The numerical results are shown below. The initial conditions of the periodic orbit of the restricted circular three-body problem we stated with are, for the normalization (43) and $m_1=m_2=1/2$ $x_1 = 0.5$, $x_3 = 0.36242545$, $\dot{y}_3 = -1.77012013$ for the one crossing with the x axis and $x_1 = 0.5$, $x_3 = 0.63597175$, $\dot{y}_3 = 1.79245459$ for the next crossing in the opposite direction. Thus, the initial conditions for the 5-body problem, in the rotating frame XOY, with masses

$$m_1 = m_2 = 0.5, \quad m_3 = m_4 = m_5 = 0$$

are

$$x_1 = 0.5, \ x_3 = 0.36242545, \quad x_4 = 0.63597175, \quad x_5 = -0.36242545$$
$$\dot{y}_3 = -1.77012013, \quad \dot{y}_4 = 1.79245459, \quad \dot{y}_5 = 1.77012013 \ .$$

These initial conditions have been used as a starting point and a whole family was generated. One member of this family, corresponding to the masses

$$m_1 = m_2 = 0.3335, \quad m_3 = m_4 = m_5 = 0.111,$$

is described by the initial conditions

$$x_1=0.62279071, \ x_3=0.36242545, \ x_4=0.88387670, \ x_5=-0.33764049$$
$$\dot{y}_3=-0.95144312, \ \dot{y}_4=0.94374010, \ \dot{y}_5=0.99366432.$$

This is clearly a periodic orbit of a system of 5 bodies with distances and masses of the same order of magnitude.

We have also found that by keeping the masses fixed and varying all the initial conditions, we can obtain a family of periodic orbits of the 5-body problem for fixed masses of all bodies.

We have also used a different method, from the practical point of view, to obtain a starting point for the generation of families of periodic orbits of the N-body problem. The method is the following: Consider the body P_2 with mass $m_2=1$, to be at rest with respect to an inertial frame and the bodies P_1, P_3, .. P_N, with zero masses, to describe circular orbits around P_2, with radii R_1, R_3, ... R_N, respectively. We now use the rotating frame XOY, defined by P_2 and P_1 as in Fig. 1 and select the radii R_3,... R_N in such a way that the relative periods T_3,..T_N are in integer dependence (note that for the normalization (43) we always have $R_1=1$). This is a periodic N-body system with respect to the rotating coordinates XOY, of the planetary type. We may continue now this system by increasing simultaneously the masses of P_1, P_3,..P_N and obtain thus a periodic N-body system of the planetary type, where the masses of all planets are finite.

As a numerical example we show below the initial conditions of a planetary 4-body problem, consisting of 3 planets (bodies P_1 P_3 and P_4) and the Sun (P_2). The normalization (43) has been used and for the masses

$$m_1 = m_3 = m_4 = 0.0008, \quad m_2 = .9992$$

we have the initial conditions

$$x_1 = 0.99410289, \quad x_3 = 1.5, \quad x_4 = 7.44865711$$

$$\dot{y}_3 = -0.68245554, \quad \dot{y}_4 = -7.0815616$$

which correspond to a realistic planetary system with three planets.

The above method of generating periodic orbits of the N-body problem is very useful because one can construct models of our solar system, both for the Syn and the planets or for a planet and its satellites. And there are many such configurations in the actual case which are very close to a relative periodic orbit and can be studied in this way.

Another method to obtain a starting point for the generation of periodic orbits of the general N-body problem is to consider a periodic orbit of the r stricted circular 3-body problem and place many bodies with zero mass on this orbit, with a phase difference between them. This is clearly a periodic non symmetric system where all masses $m_3, \ldots m_N$ are equal to zero. If this orbit is continued by increasing the masses $m_3, \ldots m_N$ we shall obtain a nonsymmetric periodic orbit of the N-body problem. We have not yet applied this method, but we believe that it will yield positive results. The same method can be used by starting from planetary orbits and placing many planets, with zero masses initially, in the same orbit.

As far as stability is concerned, one can apply the same method that we have developed for the 3-body problem. It was found that the same properties of the corresponding matrix A, as in Eq. (29) are true.

REFERENCES

1. Bozis, G. and Hadjidemetriou, J. D., 1975, On the Continuation of Periodic Orbits, Celestial Mechanics, (to appear).
2. Broucke, R., 1969, NASA Technical Report 32-1360.
3. Broucke, R., and Boggs, D., 1975, Celestial Mechanics, 11, 13.
4. Broucke, R., 1975, On Relative Periodic Solutions of the Planar General Three-Body Problem, Celestial Mechanics (to appear).
5. Delibaltas, P., 1975, (in preparation).
6. Hadjidemetriou, J. D., 1975a, The Continuation of Periodic Orbits from the Restricted to the General Three-Body Problem, Celestial Mechanics, 12. (in print).
7. Hadjidemetriou, J.D., 1975b, The Stability of Periodic Orbits in the Three-Body Problem, Celestial Mechanics, (in print).
8. Hadjidemetriou, J.D., 1975c, N-Body Problem (in preparation).
9. Hadjidemetriou, J.D., 1975d, Families of Periodic Planetary-type orbits of the 3-Body Problem and their Stability, Astrophys. Space Science, (in print).
10. Hadjidemetriou, J.D., and Christides, Th., 1975, Families of Periodic Orbits in the Planar Three-Body Problem, Celestial Mechanics, (in print).
11. Hadjidemetriou, J.D., and M. Michalodimitrakis, 1975, Families of Periodic Orbits in the N-Body Problem and their Stability (in preparation).
12. Henon, M., 1974, Celestial Mechanics, 10, 375.
13. Henon, M., 1975, Celestial Mechanics, (to appear).
14. Standish, E. M., 1969, in G. E. O. Giacaglia (ed.), Periodic Orbits, Stability and Resonances, D. Reidel Publishing Co., 373.

15. Szebehely, V., 1967, "Theory of Orbits", Academic Press.
16. Szebehely, V., 1969, in G. E. O. Giacaglia (ed.), Periodic Orbits, Stability and Resonances, D. Reidel Publ. Co., 382.
17. Szebehely, V., 1971, Celestial Mechanics, 4, 116.

TRIPLE COLLISION

J. Waldvogel

Swiss Federal Institute of Technology, Zurich, Switzerland

ABSTRACT. A new theory of the triple collision and the triple close encounter in the planar problem of three bodies is developed. The basic tool is the method of matched asymptotic expansions (singular perturbations), applied to solutions neighboring triple collision solutions. First, the general ideas will be introduced by considering the binary collision in the problem of two bodies from this point of view. Application of the same ideas to triple collision allows a complete discussion of this kind of motion in a first order approximation, based on a small number of numerically calculated solutions of the three-body problem. One result is that a triple close encounter generally leads to the escape of one body with arbitrarily large asymptotic velocity. The theory also yields accurate estimates of this asymptotic velocity from conditions well ahead of the close encounter.

1. INTRODUCTION

The planar problems of two and three bodies are the simplest mathematical models of planetary and stellar systems. They idealize the celestial bodies to point masses, thus permitting arbitrarily close encounters. Only gravitational forces are taken into account, and Newton's law is assumed to be valid over the entire range of distances $0 \leq r < \infty$. Whereas the solutions of the two-body problem can be represented by closed-form parametrizations, no complete theory of the solutions of the three-body problem is known. However, extended numerical integrations have led to the conjecture that for the majority of

V. Szebehely and B. D. Tapley (eds.), Long-Time Predictions in Dynamics, 241–258. *All Rights Reserved. Copyright © 1976 by D. Reidel Publishing Company, Dordrecht-Holland.*

all mass ratios and initial conditions one of the three bodies will escape sooner or later, leaving the others in a binary escaping in the opposite direction with respect to the center of mass (Szebehely, 1967). Immediately prior to this terminal phase all three bodies approach to unusually small distances, compared to initial distances. This situation will be termed <u>triple close encounter</u>.

The purpose of this article is to develop a theory of the triple close encounter which, of course, includes the exact triple collisions, where all three bodies meet simultaneously at time $t = t_o$.

There are many classical results on triple collision, two of which will be mentioned here. Let m_1, m_2, m_3, be the three mass points and $\underline{x}_1$, $\underline{x}_2$, $\underline{x}_3$ their respective position vectors relative to the center of mass 0. Then we have $\sum_{i=1}^{3} m_i \underline{x}_i = 0$. Sundman (1907, 1912) proved that a necessary condition for the occurence of a triple collision is the vanishing of the total angular momentum,

$$\underline{c} = \sum_{i=1}^{3} (m_i \underline{x}_i * \frac{d\underline{x}_i}{dt}) = \underline{0} .$$

Then the bodies, even when free to move in space, remain in a fixed plane.

Siegel (1941) represented the entire family of solutions having a triple collision at $t=t_o$ by convergent power series in t_o-t $(t<t_o)$ or in $t-t_o (t>t_o)$, for instance

$$\underline{x}_i = \sum_{j=1}^{\infty} \underline{c}_{ij} (t_o - t)^{\nu_j} , \quad i=1,2,3,$$

where the exponents α_j (ν_j) are irrational numbers in general. There follows that a triple collision solution as a real-valued function can exist at most on half the time axis, $t<t_o$ or $t>t_o$, respectively. Siegel's series will be discussed in more detail in section 4.

In order to go beyond these results we shall consider solutions of the three-body problem that are <u>nearby</u> triple collision solutions in some sense. These solutions turn out to be free of triple collisions in the neighborhood of t_o. The remaining binary collisions may be <u>regularized</u> by well known procedures (e.g., Levi-Civita, 1906). Hence the nearby solutions exist in a neighborhood of t_o and can possibly be continued much further. In this way the triple collision solutions are embedded in a family of solutions that exist beyond the time of collision t_0.

In a first order approximation the nearby solutions satisfy the variational equation of the three-body problem. The problem of the triple close encounter is thus to match the (known) solutions of the variational equation to the motion ahead of the triple close encounter.

Before pursuing this general line of approach, the same procedure will be carried out in the simpler case of the problem of two bodies.

2. THE PROBLEM OF TWO BODIES

Let the mass points m_1, m_2 move under their mutual Newtonian attraction in the (x_1, x_2)-plane, and let $\underline{x} = (x_1, x_2)^T$ be the position vector of m_1 with respect to m_2. The motion of m_1 is governed by the differential equation

$$(1) \quad \frac{d^2\underline{x}}{dt^2} = - m \frac{\underline{x}}{r^3}, \quad r = |\underline{x}| ,$$

where

$$(2) \quad m = m_1 + m_2$$

is the total mass. The gravitational constant is assumed to be unity.

In order to introduce the family of nearby solutions we choose a comparatively simple collision solution $\underline{x}_0(t)$ with $\underline{x}_0(t_0) = 0$ as a reference solution of Eq. (1).

Assume

$$(3) \quad \underline{x}_0(t) = a\, \underline{e}\, (t-t_0)^{\nu}$$

with some appropriate positive constants a, ν and a constant unit vector $\underline{e}$, to be a solution of (1), corresponding to rectilinear motion of m_1 with respect to m_2. By using the invariance of (1) under the time shift

$$(4) \quad \tilde{t} = t - t_0$$

and under the notation

(5) $x = D(\phi)\,\underline{x}$, $D(\phi)$ $\begin{pmatrix} \cos\phi & -\sin\phi \\ \sin\phi & \cos\phi \end{pmatrix}$,

Eq. (3) may be simplified to

(6) $\underline{x}_0(t) = a\,\underline{e}\,t^{\nu}$, $\underline{e} = \begin{pmatrix} 1 \\ 0 \end{pmatrix}$,

tildes being omitted. The collision now takes place to t=0. Solutions $\underline{x}(t)$ with $\underline{x}(0) = 0$ will be called collision solutions.

Inserting (6) into the differential equation (1) proves the existence of a solution of the form (6) and uniquely yields $\nu = \frac{2}{3}$ as well as

(7) $a = (\frac{9}{2}m)^{1/3}$;

therefore

(8) $\underline{x}_0(t) = a\,\underline{e}\,t^{2/3}$.

This is the well-known parabolic fall in the problem of two bodies.

We now consider solutions

(9) $\underline{x}(t) = \underline{x}_0(t) + \underline{\xi}(t)$

that are close to $\underline{x}_0(t)$ in some interval of the time axis, the perturbation $\underline{\xi}(t)$ being small compared to $x_0(t)$ (in some norm) in this interval. Here the theory will be carried to first order perturbations only. In this approximation $\underline{\xi}(t)$ satisfies the variational equation of (1), taken at the reference solution $\underline{x}_0(t)$:

(10) $\frac{d^2\xi}{dt^2} = J\,\xi$

where J is the Jacobi matrix of the right-hand side of (1),

(11) $J_{ik} = \frac{\partial}{\partial x_i}\left(-m\,\frac{x_k}{r^3}\right)\Big|_{\underline{x} = a\underline{e}t^{2/3}} = \frac{2}{9}t^{-2}(3e_i e_k - \delta_{ik})$.

With the constant matrix

(12) $E_{ik}(\underline{e}) = 3e_i e_k - \delta_{ik}$, $i,k=1,2$

Eq. (10) becomes

(13) $t^2 \frac{d^2\underline{\xi}}{dt^2} = \frac{2}{g} E(\underline{e}) \cdot \underline{\xi}$,

a system of linear differential equations which may be solved in closed form by the power function

(14) $\underline{\xi}(t) = \underline{\gamma} t^{\nu}$,

where $\underline{\gamma}$ is a constant vector. Inserting this into (13) yields

$$E(\underline{e})\underline{\gamma} = \frac{g}{2}\,\nu(\nu-1)\underline{\gamma} \quad ;$$

hence $\underline{\gamma}$ must be an eigenvector of $E(\underline{e})$, and $4.5\,\nu(\nu-1)$ is the corresponding eigenvalue. In the special coordinate system given by (6) the matrix E is diagonal with eigenvalues $\lambda_1=2$, $\lambda_2 = -1$; to each eigenvalue λ_j there are 2 exponents ν_j, ν_j^* satisfying the quadratic equation

(15) $\nu_j^2 - \nu_j - \frac{2}{g}\lambda_j = 0.$

The eigenvectors λ_j of E are the two basis vectors of the coordinate system; hence the 4 linearly independent solutions of (13) are given by (14) with

$$\underline{\gamma}_1 = \underline{e} = \binom{1}{0}, \quad \lambda_1 = 2, \quad \nu_1 = \frac{4}{3}, \quad \nu_1^* = -\frac{1}{3}$$

$$\underline{\gamma}_2 = \underline{c} = \binom{0}{1}, \quad \lambda_2 = -1, \quad \nu_2 = \frac{2}{3}, \quad \nu_2^* = \frac{1}{3} \quad .$$

The four-parameter family of solutions nearby $\underline{x}_0(t)$, to first order, is thus

(16) $x = at^{2/3}\left\{\underline{\gamma}_1 + \underline{\gamma}_1(a_1 t^{\alpha_1} + a_1^* t^{\alpha_1^*}) + \underline{\gamma}_2\,(a_2 t^{\alpha_2} + a_2^* t^{\alpha_2^*})\right\}$

with

(17) $\alpha_j = \nu_j - \frac{2}{3}, \quad \alpha_j^* = \nu_j^* - \frac{2}{3}$,

or specifically

(18) $x = at^{2/3}\left\{\underline{e} + \underline{e}(a_1 t^{2/3} + a_1^*\, t^{-1}) + \underline{c}\,(a_2 + a_2^* t^{-1/3})\right\}$,

where a_1, a_2, a_1^*, a_2^* are integration constants small compared to 1.

Only the nearby solutions with $a_1^* = a_2^* = 0$ have a collision at $t=0$; they remain nearby $\underline{x}_0(t)$ even for $t \to 0$. The general member of the family (18), however, is no longer a collision solution; the first order perturbation becomes infinite at $t=0$. This leads to the concept of singular perturbations which will be p rsued later.

The members of the subfamily $a_1 = a_2 = 0$ are close to $\underline{x}_0(t)$ for $t \to \infty$; they are the parabolic solutions with $\underline{x} \to \infty$, $d\underline{x}/dt \to \underline{0}$ for $t \to \infty$.

In the following we shall discuss the subfamilies where only one of the parameters a_1, a_2, a_1^*, a_2^* does not vanish. Although only first order terms have been calculated, these subfamilies can easily be identified by means of the classical integrals of the problem of two bodies.

(i) $\underline{x}_1 = a\underline{e}[t^{2/3} + a_1 t^{4/3} + 0(t^{4/3})]$

The total energy of the motion, by integration of (1), is

$$h = \frac{1}{2}\dot{\underline{x}}^2 - \frac{m}{r} = \text{const} \quad (\dot{} \equiv \frac{d}{dt}) \quad .$$

Substitution of (i) yields

$$\frac{1}{2}\dot{\underline{x}}_1^2 - \frac{m}{|\underline{x}_1|} = \frac{2}{3} a^2 \cdot a_1 + 0(1).$$

Since the right-hand side is a constant we exactly have

$$h = \frac{2}{3} a^2 \cdot a_1 \ ;$$

hence the family (i) is the family of collision solutions on the line of $\underline{e}$, where a_1 is proportional to the total energy h. Obviously, the angular momentum vanishes,

$$\underline{c} = \underline{x} * \dot{\underline{x}} = 0.$$

(ii) $\underline{x}_2 = at^{2/3}(\underline{e} + a_2\underline{c} + \ldots)$

is the family of parabolic collision solutions on the line of $\underline{e} + a_2\underline{c} + \ldots$, related to $\underline{x}_0(t)$ by means of a rotation of the coordinate system, according to the invariance property (5):

$$\underline{x}_2(t) = D(\phi)\, x_0(t) \quad , \quad a_2 = \phi \ .$$

(iii) $\underline{x}_1^* = a\ \underline{e}\ [t^{2/3} + a_1^*\ t^{-1/3} + 0(t^{-1/3})]$

is the family of parabolic solutions on the line of $\underline{e}$. According to the invariance of (1) with respect to the time shift (4), (iii) may be written as

$$\underline{x}_1^*(t) = x_0(t+T), \quad a_1^* = \frac{2}{3}T.$$

For $T \neq 0$ this is no longer a collision solution with collision at $t=0$.

(iv) $x_2^* = at^{2/3}[\underline{e} + a_2^* \underline{c}t^{-1/3} + 0(t^{-1/3})]$

The angular momentum of these solutions is

$$\underline{x}_2^* * \dot{x}_2^* = -\frac{1}{3}a^2 \cdot a_2^*(\underline{e} * \underline{c}) + 0(1).$$

Since this is a constant there follows

$$\underline{c} = -\frac{1}{3}a^2 \cdot a_2^* (\underline{e} \wedge \underline{c}).$$

The family (iv) is thus the family of nondegenerate parabolic orbits $(a_2^* \neq 0)$, where a_2^* is proportional to the angular momentum of the motion. The total energy vanishes for each member of the family,

$$h = 0.$$

To govern initial position and velocity vectors $\underline{x}$, $\dot{\underline{x}}$ producing a solution nearby $\underline{x}_0(t)$, the 4 small parameters a_1, a_2, a_1^*, a_2^* may be calculated in a unique way from the system of 4 linear equations

$$\underline{x} = a\, t^{2/3}\{\underline{e} + \underline{e}(a_1 t^{2/3} + a_1^* t^{-1}) + \underline{c}(a_2 + a_2^* t^{-1/3})\}$$

(19)

$$\dot{\underline{x}} = \frac{2}{3}a\, t^{-1/3}\{\underline{e} + \underline{e}(2a_1 t^{2/3} - \frac{1}{2} a_1^* t^{-1}) + \underline{c}\,(a_2 + \frac{1}{2} a_2^* t^{-1/3})\}$$

whose determinant is

$$\frac{5}{9} a^4 \,|\underline{e} * \underline{c}|^2 = \frac{5}{9} a^4 \neq 0.$$

3. MATCHING

So far, no advantage was taken of the assumption that a_1, a_2, a_1^*, a_2^* are constants small compared to 1. In Eq. (18) the relative importance of the 5 terms is controlled not only by the value of the the variable t but also by the magnitude of these constants, which are now assumed to be roughly of the same order of magnitude.

Let $t \geq 0$ decrease from positive values to zero. For large t Eq. (18) is meaningless since the perturbation term $\underline{e}\, a_1\, t^{2/3}$ is large compared to the leading term $\underline{e}$ in the parenthesis. As t decreases, the term $\underline{e}\, a_1\, t^{2/3}$ becomes small, but further decrease of t causes $\underline{e}\, a_1^*\, t^{-1}$ to become the dominant perturbing term. Finally, for sufficiently small values of t Eq. (18) is again invalid because the perturbation is large compared to the reference solution. The regions of dominance of the various terms are determined by the magnitude of a_1, a_2, a_1^*, a_2^*.

The systematic classification of the relative importance of these terms will be done according to the ideas used in the theory of singular perturbations (matched asymptotics, see, for instance, Cole, 1968). We introduce one single small parameter $\varepsilon > 0$ of the order of a_j, a_j^* and write

(20) $a_j = \varepsilon\, \hat{a}_j, \quad a_j^* = \varepsilon\, \hat{a}_j^*, \quad j = 1,2.$

For comparing the terms in (18) the limit $\varepsilon \to 0$ of arbitrarily small constants a_j, a_j^* is considered ($\hat{a}_j$, $\hat{a}_j^*$ are fixed). By using Landau's order symbols we thus have

(21) $a_j = O(\varepsilon), \quad a_j^* = O(\varepsilon), \quad j = 1,2.$

Furthermore the order function $t = t(\varepsilon)$ with $t \to 0$ as $\varepsilon \to 0$ is introduced, thus requiring that ε and t simultaneously tend to zero. We choose

(22) $t = \varepsilon^{\kappa}, \quad t = O(\varepsilon^{\kappa}), \quad \kappa > 0$

and note that $\kappa < 0$ implies $t \to \infty$ in the limit.

Now the terms of (18) are written as order functions of ε and κ, The leading (0th order) term becomes

$$a\underline{e}t^{2/3} = O(\varepsilon^{2\kappa/3}),$$

and the first order perturbation terms are of order

$$O(\varepsilon^{1+2\kappa/3+\beta\kappa}) \ , \quad \beta = -1,\ -\frac{1}{3},\ 0,\ \frac{2}{3}.$$

The graphs of the **orders** (exponents of ε) of the various terms versus κ are shown in Fig. 1. In a given region of the (ε, t)-plane (determined by κ) a term is the larger in the limit

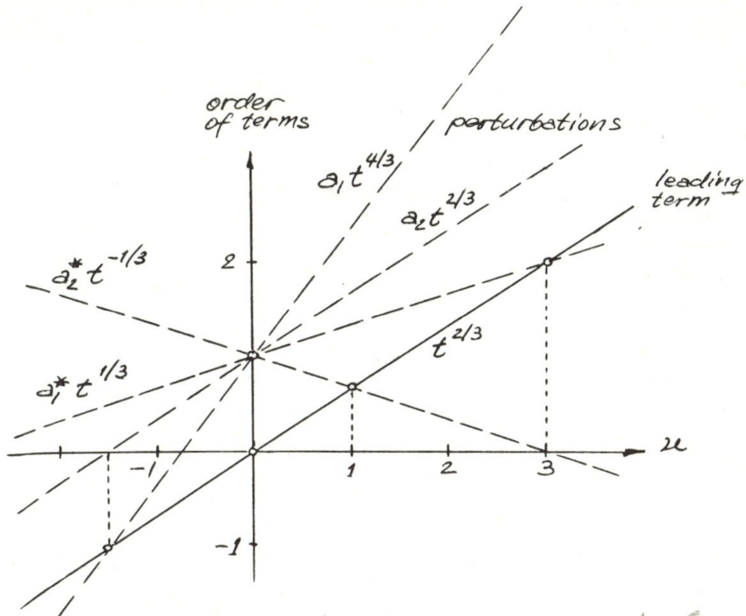

les exposants utilisés ici sont les $\gamma = \alpha + \frac{2}{3}$

Fig. 1: Comparison of terms. a_j, $a_j^* = O(\varepsilon)$, $t = O(\varepsilon^\kappa)$.

the lower its order is. As can be read from Fig. 1, the leading term is of lower order than the perturbation of $-1.5 < \kappa < 1$. This region is referred to as the matching region. For $\kappa < 0$ the term $a_1 \underline{e}\, t^{2/3}$ (energy variation) is the dominant perturbation, whereas for $\kappa > 0$ the perturbation $a_1^* \underline{e}\, t^{-1}$ (time shift) is dominant. ici : les α

It is desirable to have a matching region as large as possible. Here it may easily be extended by eliminating the term $a_1^* \underline{e}\, t^{-1}$ from Eq. (18) by means of the substitution

(23) $t = \tau - \frac{3}{2} a_1^*$

according to (iii). We obtain from (18):

(24) $\underline{x} = a\tau^{2/3}\{\underline{e}(1+a_1\tau^{2/3})+\underline{c}(a_2+a_2^*\tau^{-1/3})\} + O(a_1^{*2})$;

ce qui revient à prendre le time shift $T = 0 = a_1^*$

the second order remainder can be omitted in this first order theory. In the following the shifted time τ will be used as independent variable. In the limit $\varepsilon \to 0$, $\tau \to 0$ the term $a_2^* \underline{c}\, \tau^{-1/3}$ is now the dominant perturbation, and the matching region is $-1.5 < \kappa < 3$, according to Fig. 1.

Now the meaning of the limit $\varepsilon \to 0$ will be discussed. If τ is kept constant the nearby solution $\underline{x}(\tau)$ tends to the reference solution $\underline{x}_0(\tau)$ as $\varepsilon \to 0$. The result of the limit process is different, however, if $\underline{x}$ and τ are properly scaled in order to give the leading term and the dominant perturbation equal importance ("distinct limit").

A natural way of scaling variables in gravitational problems is the homothetic transformation ("blow-up" transformation, see Khilmi, 1961, Waldvogel, 1975)

(25) $\underline{x} = \delta^2 \tilde{\underline{x}}, \qquad \tau = \delta^3 \tilde{\tau}$,

where $\tilde{\underline{x}}$ and $\tilde{\tau}$ are the new coordinates and time, and $\delta(\varepsilon)$ is a scale factor. The differential equation (1) is invariant with respect to the transformation (25), whereas the velocity $d\underline{x}/d\tau$, the angular momentum $\underline{c}$ and the energy h are transformed as follows:

(26) $d\underline{x}/d\tau = \delta^{-1}\, d\tilde{\underline{x}}/d\tilde{\tau}$, $\underline{c} = \delta \tilde{\underline{c}}$, $h = \delta^{-2}\tilde{h}$.

By means of the transformation (25) Eq. (24) becomes

(27) $\tilde{\underline{x}} = a\,\tilde{\tau}^{2/3}\{\underline{e}(1+a_1\delta^2\tilde{\tau}^{2/3}) + \underline{c}(a_2+a_2^*\delta^{-1}\tilde{\tau}^{-1/3})\}$.

In the region $\kappa > 0$ we put $\delta = a_2^*$; then the so-called inner limit $\varepsilon \to 0$, $\delta \to 0$, $\tau \to 0$, (with $\tilde{\tau}$ fixed) yields

(28) $\tilde{\underline{x}}_{inner} = a\,\tilde{\tau}^{2/3}(\underline{e} + \underline{c}\tilde{\tau}^{-1/3})$.

In the region $\kappa < 0$, however, we put $\delta = a_1^{-1/2}$, and the outer limit $\varepsilon \to 0$, $\delta \to \infty$, $\tau \to \infty$ yields

(29) $\tilde{\underline{x}}_{outer} = a\,\tilde{\tau}^{2/3}\,\underline{e}(1 + \tilde{\tau}^{2/3})$.

In this context the nearby solution $\underline{x}$ is referred to as the intermediate solution which allows matching of the inner and the outer solution.

The inner solution (28) belongs to the subfamily (iv) of non-degenerate parabolic motions with energy $\tilde{h} = 0$, as follows also from (26). By the homothetic transformation all of its parameters have been eliminated, such that all possible inner solutions are known if only $\underline{x}_{inner}$ is known.

The independent variable $\tilde{\tau}$ describing the inner solution is given by

$$\tilde{\tau} = \delta^{-3}\tau$$

which becomes $O(\varepsilon^{\kappa-3})$ in the inner limit. Since $\kappa < 3$ in the matching region, there follows

$$\tilde{\tau} \to \infty \quad ;$$

this is the "value" of $\tilde{\tau}$ where the inner solution matches with the outer one. For $\tilde{\tau} \to \infty$ Eq. (28) asymptotically represents the inner solution ("initial conditions" for $\tilde{\tau} \to \infty$). Therefore, $\tilde{\tau}$ may potentially vary in the interval $-\infty < \tilde{\tau} < \infty$, and the parabolic motion on the entire $\tilde{\tau}$-axis describes the process of the binary close encounter.

Due to the symmetry of this motion with respect to the pericenter the past of the close encounter is the mirror image of its future. Hence, two-body motion can be confirmed beyond binary collisions.

4. THE PROBLEM OF THREE BODIES

The procedure described in the preceding sections will now be applied in an analogous way to triple collision. Only the important steps and the differences to double collision will be described, however.

Let $\underline{y}_1, \underline{y}_2$ be the relative coordinates of m_1, m_2 with respect to m_3 in an inertial center-of-mass coordinate system. The motion of m_1, m_2 is governed by the systems of differential equations

$$\ddot{\underline{y}}_1 = -(m_1+m_3)\frac{\underline{y}_1}{r_1^3} - m_2\frac{\underline{y}_2}{r_2^3} - m_2\frac{\underline{y}_1-\underline{y}_2}{r_{12}^3} \tag{30}$$

$$\ddot{\underline{y}}_2 = -(m_2+m_3)\,\frac{\underline{y}_2}{r_2^3} - m_1\,\frac{\underline{y}_1}{r_1^3} - m_1\,\frac{\underline{y}_2-\underline{y}_1}{r_{12}^3}$$

where

$$r_1 = |\underline{y}_1|\ ,\quad r_2 = |\underline{y}_2|\ ,\quad r_{12} = |\underline{y}_1-\underline{y}_2|\,.$$

for short we shall write Eq. (30) as

$$(31)\quad \ddot{y} = f(y)\quad ,\quad y = \begin{pmatrix}\underline{y}_1\\ \underline{y}_2\end{pmatrix}$$

where $f \varepsilon R^4$ is an abbreviation for the right-hand sides.

As a reference solution with a triple collision at $t = 0$ we choose the homothetic solution

$$(32)\quad y_0(t) = r(t)\,\hat{y},$$

where $r(t)$ is a scalar function and $\hat{y}$ is a constant 4-vector. By inserting this into (31) and using the property that $f(y)$ is homogeneous of degree -2, we obtain the conditions

$$(33)\quad f(\hat{y}) = -m\hat{y},$$

$$(34)\quad \ddot{r} = -\frac{m}{r^2}$$

with an arbitrary constant m. Solutions $\hat{y}$ of (33) are called central configurations, and it is well known that for three bodies there are two kinds of them, namely the collinear and the triangular central configurations. For simplicity only the latter case will be considered here.

It is easily seen that Eq. (33) is satisfied if

$$(35)\quad r_1 = r_2 = r_{12} = 1\ ,$$

and

$$(36)\quad m = m_1 + m_2 + m_3$$

Taking advantage of the rotational invariance of (30) we now choose in agreement with (35):

$$(37)\quad y = \begin{pmatrix} \underline{e}_1 \\ \underline{e}_2 \end{pmatrix}, \quad \underline{e}_1 = \begin{pmatrix} \sqrt{3}/2 \\ -1/2 \end{pmatrix}, \quad \underline{e}_2 = \begin{pmatrix} \sqrt{3}/2 \\ 1/2 \end{pmatrix}.$$

Solving Eq. 1 (34) by a power of t is analogous to section 2; the reference solution corresponding to the parabolic fall is therefore

$$(38)\quad y_0(t) = a\,\hat{y}\,t^{2/3} \quad ,$$

where the constant a is defined in (7) (with m from (36)).

In the nearby solution

$$(39)\quad y(t) = y_0(t) + \eta(t)$$

the small perturbation $\eta(t)$ satisfies in a first order approximation the variational equation

$$(40)\quad \ddot{\eta} = J\,\eta \quad ,$$

where J is the Jacobian of (30) at the reference solution (38). Since J is homogeneous of degree -3, Eq. (40) may be written as

$$(41)\quad t^2\,\ddot{\eta} = \frac{2}{g} M\,\eta$$

with the constant matrix

$$(42)\quad M = \frac{1}{m}\begin{pmatrix} (m_1+m_3)E(\underline{e}_1) + m_2E(\underline{e}_2-\underline{e}_1) & m_2E(\underline{e}_2)-m_2E(\underline{e}_2-\underline{e}_1) \\ m_1E(\underline{e}_1)-m_1E(\underline{e}_1-\underline{e}_2) & (m_2+m_3)E(\underline{e}_2)+m_1E(\underline{e}_1-\underline{e}_2) \end{pmatrix}$$

(for the matrix $E(\underline{e})$ see Eq. (12)). Eq. (41) has the particular solutions

$$(43)\quad \eta_j = \gamma_j\,t^{\nu_j} \quad , \quad j = 1, 2, 3, 4 \; ,$$

where γ_j are the eigenvectors of M, and the ν_j satisfy the quadratic equation (15), λ_j being the eigenvalues of M.

For solving the eigenvalue problem of M we observe that two of its eigenvectors are known a priori since the family (39) contains the Lagrangian solutions nearby $y_o(t)$, namely

$$\gamma_1 = \hat{y}$$

(44) $$\gamma_2 = \begin{pmatrix} \underline{c}_1 \\ \underline{c}_2 \end{pmatrix}, \quad \underline{c}_1 = \begin{pmatrix} 1/2 \\ \sqrt{3}/2 \end{pmatrix}, \quad \underline{c}_2 = \begin{pmatrix} -1/2 \\ \sqrt{3}/2 \end{pmatrix}.$$

As in section 2, γ_1 corresponds to energy variation or time shift, γ_2 corresponds to angular momentum variation or rotation. By introducing γ_1, γ_2 as new basis vectors, the matrix M is transformed into

$$\left(\begin{array}{cc|c} 2 & 0 & \\ 0 & -1 & B \\ \hline 0 & 0 & \\ 0 & 0 & A \end{array}\right),$$

using an appropriate similarity transformation.
The submatrix A has the eigenvalues

(45) $$\lambda_3 = \frac{1}{2} + \frac{3}{2}\frac{\mu}{m}, \qquad \lambda_4 = \frac{1}{2} - \frac{3}{2}\frac{\mu}{m},$$

where

(46) $$\mu = \sqrt{\tfrac{1}{2}[(m_1-m_2)^2+(m_2-m_3)^2+(m_3-m_1)^2]} \geq 0$$

With these eigenvalues and the obvious ones $(\lambda_1=2,\ \lambda_2=-1)$ the following exponents are obtained, using (15) and (17):

$$\alpha_1 = \frac{2}{3}, \qquad \alpha_1^* = -1$$

$$\alpha_2 = 0, \qquad \alpha_2^* = -\frac{1}{3}$$

(47) $$\left.\begin{matrix}\alpha_3 \\ \alpha_3^*\end{matrix}\right\} = \frac{1}{6}\left(-1 \pm \sqrt{13 + 12\frac{\mu}{m}}\right)$$

$$\left.\begin{matrix}\alpha_4 \\ \alpha_4^*\end{matrix}\right\} = \frac{1}{6}\left(-1 \pm \sqrt{13 - 12\frac{\mu}{m}}\right.$$

These are the values found by Siegel (1941) and later given by Siegel-Moser (1971). The masses enter only through the symmetric function $0 \leq \mu/m \leq 1$. In Fig. 2 the exponents α_j, α_j^* are plotted versus μ/m. There follows

$$(48) \quad 0 \leq \alpha_j \leq \frac{2}{3} \;, \quad -1 \leq \alpha_j^* \leq -\frac{1}{3} \;, \quad j = 1, 2, 3, 4.$$

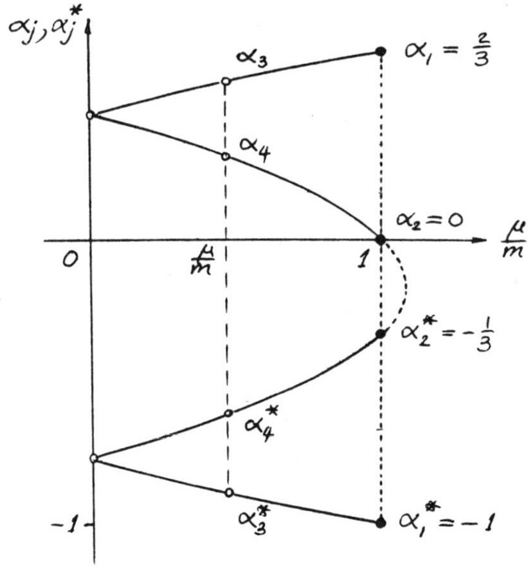

Fig. 2: Siegel's Exponents.

The solutions of the three-body problem nearby $y_0(t)$ form the 8-parameter family

$$(49)\ y(t) = at^{2/3} \quad \gamma_1 + \sum_{j=1}^{4} \gamma_j \ (a_j t^{\alpha_j} + a_j^* \ t^{\alpha_j^*})$$

in a first order approximation. A particular member is characterized by a set of 8 small constants a_j, a_j^* $(j = 1, 2, 3, 4)$ which can be uniquely determined from initial conditions producing such a solution.

The subfamily $a_j^* = 0$ $(j = 1, 2, 3, 4)$ is the family of nearby solutions with a triple collison at $t = 0$. Eq. (49) then is identical with Siegel's convergent series up to first order. On the other hand, the nearby solutions with $a_1 = a_3 = a_4 = 0$ are the parabolic solutions with zero total energy and with $\dot{y} \to 0$ as $t \to \infty$; they can be represented by convergent series similar to those found by Siegel.

Siegel's theory only considers pure triple collisions solutions, but not the nearby solutions that become singular as $t \to 0$. These solutions, however, govern the behavior in a close triple encounter. Here, they will again by investigated by means of the inner limit. In order to define the inner solution we assume $a_j = O(\varepsilon)$, $a_j^* = O(\varepsilon)$, $t = O(\varepsilon^\kappa)$ and let $\varepsilon \to 0$. Also, the perturbation term $a_1^* \gamma_1 t^{-1}$ is eliminated by a time shift; the new time will be denoted by the same symbol, t. In view of the generally irrational exponents α_j, α_j^* we let t again decrease through positive values to zero.

Due to (48) a non-empty matching region analogous to Fig. 1 can be defined; there the perturbations are small compared to the leading term. For $\kappa > 0$ the term with the smallest exponent, $a_3^* \gamma_3 t^{\alpha_3^*}$, is the dominant perturbation (the degenerate case $m_1 = m_2 = m_3$, $\mu = 0$ will be excluded).

In order to carry out the inner limit process coordinates and time are scaled according to the homothetic transformation

$$(50)\ y = \delta^2 \tilde{y}\ , \quad t = \delta^3 \tilde{t}\ ,$$

where $\delta(\varepsilon) \geq 0$ is an order function. Eq. (49) becomes

$$\tilde{y} = a\tilde{t}^{2/3} \left\{ \gamma_1 + \sum_{j=1}^{4} \gamma_j \left(a_j \delta^{3\alpha_j} \tilde{t}^{\alpha_j} + a_j^* \ \delta^{3\alpha_j^*} \tilde{t}^{\alpha_j^*} \right) \right\},\ a_1^* = 0$$

At this time we assume $a_3^* > 0$. By putting

(51) $\delta = (a_3^*)^{-1/3\alpha_3^*}$

and letting $\varepsilon \to 0$ $(\delta \to 0)$ the "distinct limit"

(52) $\tilde{y}_{inner} = a\,\tilde{t}^{2/3}(\gamma_1 + \gamma_3\,\tilde{t}^{\alpha_3^*})$

is obtained. Thereby we have again $\tilde{t} \to \infty$; hence (52) is the asymptotic representation of the inner solution for $\tilde{t} \to \infty$

The equations of motion (30) or (31) are invariant with respect to the transformation (50). The total energy $\tilde{h}$ of the transformed motion, however, tends to zero in the limit $\delta \to 0$, according to (26). Therefore the inner solution is a particular parabolic solution $(\tilde{h} = 0)$ that can be calculated by numerical integration of (30), beginning with a large value of $\tilde{t}$, such that the first order asymptotic approximation (52) is sufficiently accurate.

If the nearby solution under consideration has a close triple encounter with minimum distance of $O(\varepsilon)$, the corresponding inner solution will have a triple encounter with minimum distances of $O(1)$; therefore, the inner solution exists in $-\infty<\tilde{t}<\infty$.

According to Chazy (1922) the final evolution of a three-body system with $\tilde{h} = 0$ can be either parabolic or hyperbolic-elliptic (escape). Since the family of parabolic solutions has lower dimensionality, the final evolution of the inner solution for $t \to \infty$ is generally hyperbolic-elliptic (except for a set of measure zero).

Let $\underline{\tilde{v}}$ be the limiting velocity of the escaping body as $\tilde{t} \to -\infty$. $\underline{\tilde{v}}$ is independent of ε since the initial conditions (52) do not contain ε either. According to (26), (50), (51) the escape velocity $\underline{c}$ in the original variables is

(53) $\underline{v} = \delta^{-1}\,\underline{\tilde{v}} = (a_3^*)^{1/3\alpha_3^*} \cdot \underline{\tilde{v}}$,

which is arbitrarily large if only a_3^* is sufficiently small.

Hence, after a sufficiently close triple encounter one body generally escapes with arbitrarily large velocity in some direction; the other two particles disappear in the opposite direction as an arbitrarily close and fast spinning binary.

REFERENCES

1. Chazy, J., 1922, "Sur l'allure du mouvement dans le probléme des trois corps quand le temps croit indefiniment", Ann. Sci. Ecole Norm. Sup. (3) 39. 29-130.
2. Cole, J. D., 1968, Perturbation Methods in Applied Mathematics, Blaisdell Publishing Company.
3. Khilmi, G. F., 1961, "Qualitative Methods in the Many-Body Problem", Russian Tracts on Advanced Mathematics and Physics 3, Gordon and Breach Science Publishers.
4. Levi-Civita, T., 1906, "Sur la Resolution qualitative du probleme restreint des trois corps", Acta Math. 30, 305-327.
5. Siegel, C. L., 1941, "Der Dreierstoss", Ann. Math. (2) 42, 127-168.
6. Siegel, C. L., and Moser, J.K., 1971, Lectures on Celestial Mechanics, Springer Verlag.
7. Sundman, K. F., 1907, "Recherches sur le probléme des trois corps", Acta Soc. Sci. Fenn. (6) 34, 1-43.
8. Sundman, K. F., 1912, "Memoire sur le probléme des trois corps", Acta Math. 36, 105-179.
9. Szebehely, V., 1967, "Burrau's Problem of Three Bodies", Proc, Nat. Acad. Sci. U.S.A., 58, 60.
10. Waldvogel, J., 1075, "The Close Triple Approach", Celest. Mech. 11, 429-432.

AN INVARIANT MEASURE FOR THE PLANAR RESTRICTED THREE BODY PROBLEM

F. Nahon

Université Pierre et Marie Curie,
11 rue Pierre Curie Paris 5

ABSTRACT. In the study of hyperbolic comets, Von Zeipel has introduced an integral invariant which is related but not equivalent to Khirchin's invariant integral of Statistical Mechanics. We give a simple expression of this invariant in the planar case and study some of its properties.

I - Let $X(x,t)$ be a velocity field in R_n and consider the associated system $\frac{dx_i}{dt} = X_i(x,t)$. If X is "incompressible",

$$\text{i.e. if } \Sigma \frac{\partial X_i}{\partial x_i} = \text{div } X \equiv 0$$

the volume element

$$\delta\bar{\omega} = [\delta x_i, \delta x_2, \ldots, \delta x_n] \text{ is an invariant measure.}$$

If X is "autonomous", i.e. $\frac{\partial X}{\partial t} \equiv 0$, to every solution $x = f(t)$ we can by an arbitrary time shift τ, associate the solution

$$x' = f(t + \tau).$$

Suppose that $X(x)$ is both autonomous and incompressible and consider the surface

$$(\Sigma): \phi(x) = 0$$

Define $\psi(x) = \left(\frac{d\phi}{dt}\right)_{\phi = 0}$

V. Szebehely and B. D. Tapley (eds.), Long-Time Predictions in Dynamics, 259–266. *All Rights Reserved. Copyright © 1976 by D. Reidel Publishing Company, Dordrecht-Holland.*

In the vicinity of Σ, we introduce the family

$$\Sigma_\lambda : \phi(x) = \lambda$$

and assume that $\frac{\partial \phi}{\partial x_i} \neq 0$ for $\lambda = 0$ so that we can solve the equation $\phi(x) = \lambda$ with respect to x_1 and write

$$d\bar{\omega} = \frac{1}{\left|\frac{\partial \phi}{\partial x_1}\right|} \delta\lambda \, [\delta x_2 \ldots \delta x_n]$$

For $\delta\lambda$, i.e., for the position of the neighboring surface Σ' take the position at time $t + \delta\tau$ of the molecule which crosses Σ at time t.

$$\begin{cases} \delta\lambda = \psi \, \delta\tau \\ \delta\bar{\omega} = \delta\tau \quad \dfrac{\psi}{|\partial\phi/\partial x_1|} [\partial x_2 \ldots \partial x_n] \end{cases}$$

The quotient $\sigma = \frac{\delta\bar{\omega}}{\delta\tau}$ of the two invariants is an invariant measure for the crossing of the surfaces $\phi(x) = \lambda$ by a tube of trajectories.

In particular, if the tube initiated at some point x of Σ crosses again Σ in some x^1, Σ is a surface of section with a transformation ξ $(\Sigma \to \Sigma)$ defined in some neighborhood of x; this transformation ξ preserves the measure

$$\sigma = \psi \frac{d\Sigma}{g}$$

where $d\Sigma$ is the element of area and $g = |\overrightarrow{\text{grad}}\, \phi|$.

This is the measure related to the Khinchin invariant of statistical Mechanics, for which we have the expression:

$$\sigma = \psi \frac{d\Sigma}{g} = \frac{\psi}{\left|\frac{\partial\phi}{\partial x_1}\right|} [\delta x_2 \ldots \delta x_n] \tag{1}$$

II - Let us suppose that the differential system admits an "integral of energy"

$$F(x) = K$$

We can eliminate x_1, x_2 by solving the equations

$$\begin{cases} \phi(x) = \lambda \\ F(x) = K \end{cases}$$

if we assume that $J = \frac{D(\ ,F)}{D(x_1,x_2)}$ is different from zero for $\lambda = 0$, and the given value K of the "energy".

We have $\delta\bar{\omega} = \frac{\delta\lambda\ \delta K}{|J|}\ \delta\bar{\omega}'$ (where $\delta\bar{\omega}' = [\delta x_3\ \delta x_4 .. \delta x_n]$) and choosing again $\delta\lambda = \psi\ \delta\tau$ we obtain the new invariant

$$\sigma_K = \frac{\psi}{|J|} \qquad \delta\bar{\omega}' \tag{2}$$

Consider the particular case where the field X is Hamiltonian, with Hamiltonian $F(p,q)$.

Take: $K = F(p,q)$

$\Sigma: \quad \phi(x) = p_1 = 0$

$x_1 = p_1,\ x_2 = p_0$

We have

$$J = \frac{D(p_1,F)}{D(p_1 q_1)} = \frac{\partial F}{\partial q_1}$$

$$\psi = \frac{dp_1}{dt} = - \frac{\partial F}{\partial q_1}$$

so that $$\sigma_K = \pm\ \delta\bar{\omega}' \tag{3}$$

The sign is the sign of ψ, depending on the flow entering or leaving Σ.

Theorem 1.

For the Hamiltonian flow of energy level:

$$F(p,q) = K, \text{ and the surface of section } \Sigma\text{: } p_1 = 0$$

The transformation ξ generated by two consecutive crossing of Σ preserves the measure

$$\sigma_\alpha = [\delta p_2\ \delta q_2\ \dots\ \delta p_n\ \delta q_n]$$

Remark 1 - For this Hamiltonian flow, we know the Cartan's invariant integral

$$d\omega = \sum_1^n [dp\ dq] - [dF\ dt]$$

which

for $F = K$, and $p_1 = 0$, reduces to

$$d\omega_{\binom{F=K}{p_1=0}} = \sum_{2}^{n} [dp\, dq] .$$

We note that this $d\omega$ is identical with σ_K in the case of two degrees of freedom.

2 - The expression of σ_K is not altered if we substitute to the surface Σ: $p_1 = 0$ the conjugated surface

$$\Sigma_0: \; q_1 = \text{constant}.$$

So we can write:

Theorem 2 -

For the Hamiltonian flow of energy level $F(p,q) = K$, the transformation ξ^* generated by the consecutive crossing of the two surfaces:

$$\Sigma_0: q_1 = \text{constant}$$
$$\Sigma : p_1 = 0$$

preserves the measure

$$\sigma_K = [dp_2\, dq_2 \ldots dp_n\, dq_n]$$

III Application to the planar restricted three body problem.

Let $\xi = r \cos \alpha$, $\eta = r \sin \alpha$ the coordinates of the "comet"

in the rotating frame of reference and

$$V(r,\alpha) = \frac{1-\mu}{r_1} + \frac{\mu}{r_2} \quad \text{the function of forces.}$$

It is known that the motion is governed by the Hamiltonian

$$H = \frac{1}{2} \left(R^2 + \frac{c^2}{r^2}\right) - V(r,\alpha) - C$$

which gives:

1) the conjugate variables of r, α

$p_r = R =$ the radial velocity $\frac{dr}{dt}$

$p_\alpha = C =$ the absolute kinetic momentum:

$$C = r^2(\dot{\alpha} + 1)$$

2) the Jacobi integral defining the level of energy:

$$K = C + V(r,\alpha) - \frac{1}{2}\left(R^2 + \frac{C^2}{r^2}\right)$$

Choose for Σ the surface $R = 0$ (extrema for r) and for Σ_0 the surface $r = a$.

Theorem 1 says that the transformation ξ generated by two consecative extrema of r, for the given level α, preserves the measure $\sigma_K = \delta C\ \delta\alpha$.

Theorem 2 says that the transformation ξ generated by crossing first $r = a$ and secondly $R = 0$ preserves the same measure. Let us consider Σ: $R = 0$
or equivalently $\phi = 0$
with $\phi = \xi\dot{\xi} + \eta\dot{\eta}$

$$\text{compute } \psi = \left(\frac{d\phi}{dt}\right)_{\phi=0} = \frac{C^2}{r^2} - r\frac{\partial\phi}{\partial r}$$

Σ is divided by $\psi = 0$ in two domains
U: domain of pericenters
V: domain of apocenters

where for example V is defined by

$$\text{V(apocenters)} \quad \begin{aligned} R &= 0 &(5)\\ K &= C + V(r,\alpha) - \frac{1}{2}\frac{C^2}{r^2} &(6)\\ \psi &= \frac{C^2}{r^2} - r\frac{\partial V}{\partial r} < 0 &(7)\end{aligned}$$

V is two-dimensional, with coordinates C, α; r is the solution of (6) which satisfies the inequality (7). We have for σ_K an important property given by von Zeipel.

Theorem 3 (Von Zeipel's Lemma):

The measure $\int_V \sigma_K$ is finite.

Let $J(\mu) = \int_V \sigma_K$; we will restrict ourselves to compute $J(0)$ in the limiting case $\mu = 0$ of the Keplerian problem in the rotating frame. We have now

$$V(r) = \frac{1}{r}$$

Call $\frac{1}{2} = z$, so that:

$$\sigma_K = \delta C\ \delta\alpha$$

$$\text{V(apocenters)}\quad K = C + z - \frac{1}{2} C^2 z^2 \qquad (8)$$

$$C^2 z^2 + z < 0 \qquad (9)$$

$$(10)$$

The conditions (9) & (10) mean that the equation (9) for z must have two positive solutions and can be written equivalently

$$\left.\begin{array}{l}\text{discriminant } \Delta > 0\\ \text{product of roots} > 0\end{array}\right\} \Rightarrow C \leq K \leq C + \frac{1}{2} C^2$$

which gives in the (K,C) diagram of Figure (1) the domain of apocenter (the hatched region).

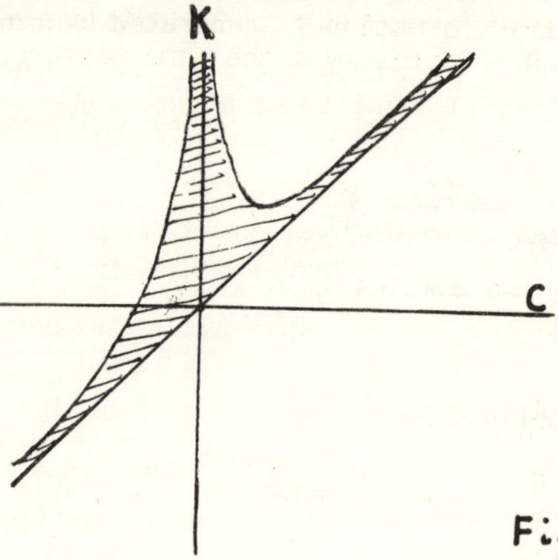

Fig. 1

Note that $K = C - h$, so the (K-C) diagram is the familiar Lindblad diagram in (K,C) variables.

$K = C$ corresponds to parabolic orbits
$K + C + \frac{1}{2C^2}$ corresponds to circular orbits.

We have

$$J = 2\pi\ \Delta C(K)$$

and find the following parametrization of $\Delta C(\alpha)$. Let $x_1 < 0 < x_2 < x_3$ the three roots of the cubic

$$2C^3 - 2\ KC^2 + 1 = 0$$

Then if $K \leq \frac{3}{2}$, $\Delta C(K) = \dfrac{1}{2x_1^{\,2}}$

if $K \geq \frac{3}{2}$ $\Delta C(K) = 2x_2$

We deduce that

ΔC has a maximum = 2 for $X = \frac{3}{2}$

$\Delta C \to 0$ for $K \to -\infty$, $\sim \dfrac{1}{2K^2}$

$\Delta C \to 0$ for $K \to +\infty$, $\sim \sqrt{\dfrac{2}{K}}$

and we have the graphic of Figure 2.

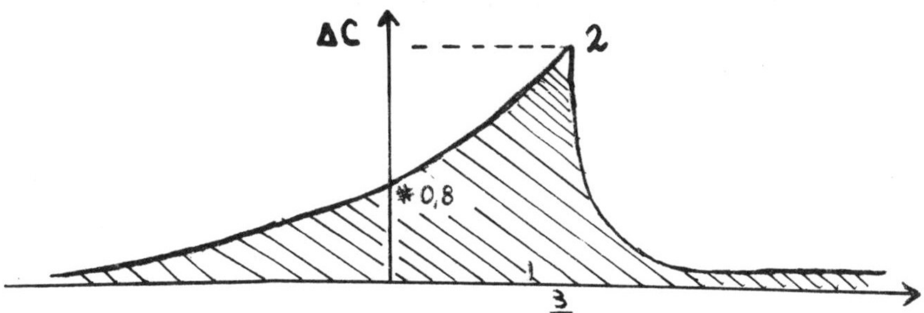

Fig. 2

IV INDICATIONS FOR FURTHER WORK.

Von Zeipel introduces the two sets:

U_0: $r = a$, $R < 0$, which for a sufficiently large define the orbits hyperbolic for $t \to -\infty$

U_1 $\{R = 0, R^\circ > 0\}$ the subset of first pericenters of the hyperbolic orbits. (In our notation, $U_1 = \xi^* U_0$) and using the generalized reversibility principle which is known to be valid for the restricted circular problem; he defines symmetrically:

U_0': $r = a$, $R > 0$

set of orbits hyperbolics for $t \to +\infty$

U_1': subset of last pericenters of hyperbolic orbits for $\delta \to +\infty$

The intersection

$X_1 = U_1 \quad U_1'$ is the set of first and last pericenters, the corresponding orbits are escape orbits both for $t \to -\infty$ and $\to +\infty$.

The remainings

$$\bar{U}_1 = U_1 - X_1, \quad \bar{U}_1' = U_1' - X_1$$

are associated by the generalized reversibility and have the same measure $J_1 < J$.

If $\mu = 0$, evidently $J_1 = 0$. It would be interesting to compute J_1, the measure of hyperbolic $(t \to -\infty)$ orbits which return at least once; and for that to study the "fronker orbits" of U_0 and U_1.

It would be useful in particular to know the locus of first pericenters for the "parabolic symmetric orbits"; that is:

the orbits for which:

$$\lim \frac{dr}{dt} = 0 \text{ for } t \to \pm \infty$$

first pericenter on the axis of primaries.

REFERENCES.

1. M.V. Zeipel: "Sur l'instabilite der mouvement des cometes"
2. Bulletin Antonomique, t. XXII, Decembre 1905
3. Khinchin: Mathematical Foundations of Statistical Mechanics, Dover Publications, 1949.

PERIODIC OSCILLATIONS OF THE FIRST GENERATION IN THE STORMER PROBLEM

C. L. Goudas, A. A. Halioulias*, V. V. Markellos and G. Macris

Department of Mechanics, University of Patras, Greece
*Technical University of Athens, Greece

ABSTRACT. A grid search for symmetric periodic oscillations of the first generation in a magnetic dipole field (the Stormer problem), inside the "thalweg", led to the discovery of six families of such motions some of which were already known. Further, a predictor-corrector steepest descent algorithm was applied for the computation of the initial conditions of a sufficient number of members of each family. This work is presently carried on in order to exhaust all families of first generation.

I. INTRODUCTION

To limit our remarks to the history of Störmer's problem to a bare minimum we remind the reader that it was formulated by Störmer (1907) in connection with auroral and cosmic rays phenomena. Störmer and many others made contributions to its qualitative investigation. Van Allen et al (1959) discovered charged particles performing motions under conditions almost identical to those postulated in Störmer's problem. These motions are "trapped" within certain "belts" and thus the Earth carries a thick permanent shield of radiation controlled primarily by its magnetic field and secondarily by particle-to-particle interactions and the solar wind. Other small effects are also present.

If we neglect all other forces that may influence the motion of a charged particle and consider only the action of the Earth's magnetic field upon it we are faced with the simplest, but reasonably close, dynamical description of the real situation. It is

V. Szebehely and B. D. Tapley (eds.), Long-Time Predictions in Dynamics, 267–276. *All Rights Reserved. Copyright* ©1976 *by D. Reidel Publishing Company, Dordrecht-Holland.*

essential to note that even under these simplifying assumptions, the equations of motion admit only to two integrals of motion (see e.g. Dragt, 1965) and thus this problem belongs to the non-soluble category. Further simplification of the problem by smoothing out the Earth's magnetic field to resemble the field of a magnetic dipole leaves unchanged the non-soluble character of the problem. In what follows we shall treat the case of the simplified field of the Earth.

A second and last essential note to be made is that without additional effort the motion of a charged particle in a magnetic dipole can receive a relativistic treatment (see e.g. Stern, 1967). To make the statement precise, we should add any that due to the energy integral the relativistic treatment is identical to the non-relativistic one.

The particular form of the energy integral in this problem establishes the existence of closed zero-velocity curves which limit the motions of charged particles of a certain range of energy. For higher values of energy the zero-velocity curves form open regions thus making escape to infinity possible.

The existence of two integrals and of one ignorable variable (see e.g. Lifshitz, 1942, p. 284) permit the study of the motion of the particle in its own meridian plane. In this plane, we shall treat the periodic families of first generation. Such are the families from which other families branch off with periods that are integral multiples of the periods of the orbits at the branching point. A family from which a branching of lower order occurs is a generation family.

II. EQUATIONS

We start with the relativistic Hamiltonian

$$H_R = \{m^2c^4 + c^2[p_z^2 + p_\rho^2 + (\frac{p_\phi}{\rho} - q\frac{M\rho}{r^3})^2]\}^{1/2}$$

where M is the magnitude of the magnetic moment, m the mass of the particle, q the charge of the particle, $\gamma = (1-\upsilon^2/c^2)^{-1/2}$, $r = (\rho^2 + z^2)^{1/2}$, and (ρ,ϕ,z) are the cylindrical coordinates employed. The momenta are defined by the relations

$$p_\rho = m\frac{d\rho}{dt}, \; p_\phi = m\rho^2\dot{\phi} + \frac{M\rho^2}{r^3}, \; p_z = m\frac{dz}{dt}.$$

Because $\partial H_R/\partial t = 0$ the energy is a constant of motion and

$$H_R = \gamma mc^2 = \text{constant}.$$

The non-relativistic Hamiltonian is

$$H = \frac{1}{2\gamma m}\left[p_z^2 + p_\rho^2 + \left(\frac{p_\phi}{\rho} - \frac{Mq\rho}{r^3}\right)^2\right].$$

We, therefore, use H to write the equations of motion which, in canonical form, are

$$m\frac{d\rho}{dt} = p_\rho \quad , \qquad \frac{dp_\rho}{dt} = -\frac{\partial H}{\partial \rho} \quad ,$$

$$m\frac{dz}{dt} = p_z \quad , \qquad \frac{dp_z}{dt} = -\frac{\partial H}{\partial z} \quad ,$$

$$m\frac{d\phi}{dt} = \frac{1}{p^z}\left(p_\phi - \frac{qM}{r^3}\right), \qquad \frac{dp_\phi}{dt} = -\frac{\partial H}{\partial \phi} \quad .$$

It is immediately found that these equations admit two integrals:

(i) Conservation of kinetic energy

$$H = \frac{1}{2}\gamma m\upsilon^2 = \text{constant}$$

(ii) Conservation of Momentum

$$p_\phi = qM\Gamma = \text{constant},$$

where Γ has the dimension of a reciprocal length. Replacing the position variables (p,z,ϕ,t), by the dimensionless ones (p',z',ϕ',t'), where,

$$z' = \Gamma z \ , \quad \rho' = \Gamma\rho \ , \quad \phi' = \phi, \ t' = \Gamma^3 qMt/\gamma m,$$

we obtain the following Canonical Equations:

$$\frac{d\rho'}{dt'} = p'_\rho \quad , \cdot \qquad \frac{dp'_\rho}{dt'} = \left(\frac{1}{\rho'} - \frac{\rho'}{r'^3}\right)\left(\frac{1}{\rho'^2} + \frac{1}{r'^3} - \frac{3\rho'^2}{r'^5}\right) \ ,$$

$$\frac{dz'}{dt'} = p'_z \quad , \qquad \frac{dp'_z}{dt'} = \frac{3\rho' z'}{r'^5}\left(\frac{\rho'}{r'^3} - \frac{1}{\rho'}\right) \quad ,$$

$$\frac{d\phi'}{dt} = \frac{1}{\rho'^2} - \frac{1}{r'^3} \quad , \qquad \frac{dp'_z}{dt'} = 0 \ .$$

The integrals of motion are:

(i) Energy

$$H' = \frac{1}{2}(p'^2_z + p'^2_\rho) + \frac{1}{2}\left[\frac{1}{\rho'} - \frac{\rho'}{r'^3}\right]^2 = \text{constant} \ ,$$

(ii) Momentum

$$p'_\phi = \text{constant.}$$

Note that ϕ' is an ignorable (cyclic) variable computed by quadratures through the expression

$$\phi' = \int \left(\frac{1}{\rho'^2} - \frac{1}{r'^3}\right) dt \quad ,$$

after the solution of the Equations of Motion is obtained. Hereafter the stresses will not be used.

A qualitative description of the motion of a charged particle in a dipole is as follows:

1) The particle gyrates about a guiding magnetic line: The Equation of the guiding line is

$$r = r_o \cos^2\lambda \ .$$

The frequency ω of gyration is

$$\omega = \frac{qB}{\gamma m}$$

where,

$$B = \frac{M}{r^3}(1 + 2\sin^2\lambda)^{1/2}$$

is the intensity of the dipole at the point (r_0,λ). The field has an axis (Oz) and a plane (Oyz) of symmetry.

2) The center of gyration moves slowly along the guiding magnetic line, first, say, toward the northern hemisphere.

3) For increasing λ the intensity of the field increases and hence the radius of gyration decreases.

4) The motion to the north vanishes becase the r.h.s. of the Eq. for dp_z/dt becomes negative.

Thus the particle is reflected back and moves to the south. The same reflection will occur at the southern hemisphere and the motion of the center of gyration will be an oscillatory one along the guiding magnetic line.

III. ZERO-VELOCITY CURVES

From the energy integral we deduce that

$$\frac{1}{2}(p_\rho^2 + p_z^2) = H - \frac{1}{2}\left(\frac{1}{\rho} - \frac{\rho}{r^3}\right)^2 ,$$

where H=C= constant and always positive. Hence the curves

$$\left(\frac{1}{\rho} - \frac{\rho}{r^3}\right)^2 = 2H = 2C \qquad (1)$$

are the zero-velocity curves.

Thus the motion of charged particles with the corresponding energies will be limited in the shaded areas of the following three Figures 1, 2 and 3. For the value C = 1/32 the zero-velocity equation (1) has a double root, when expressed in the form $z = z(\rho,C)$. For C < 1/32 the motion is restricted in the closed area (shaded) of Fig. 1 between the curves C_1 and C_2, or in the shaded area outside C_3.

The shaded area of Fig. 2 is the permissible space for motion of charged particles with energy equal to 1/32.

For C > 1/32 the area od permissible motion is the one shown in Fig. 3. In this case escape of particles is possible.

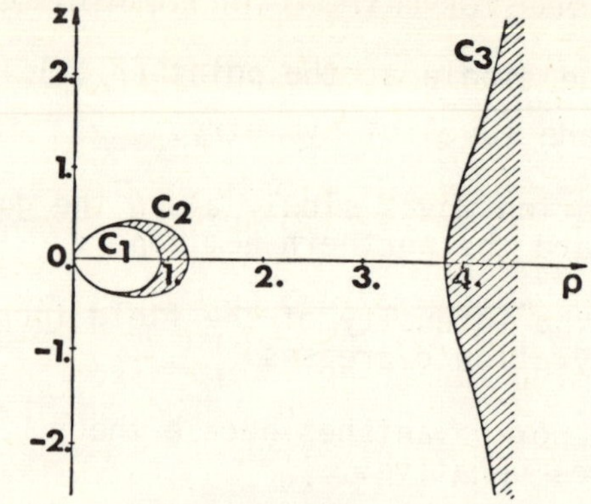

Fig. 1. Zero-velocity curves for C<1/32

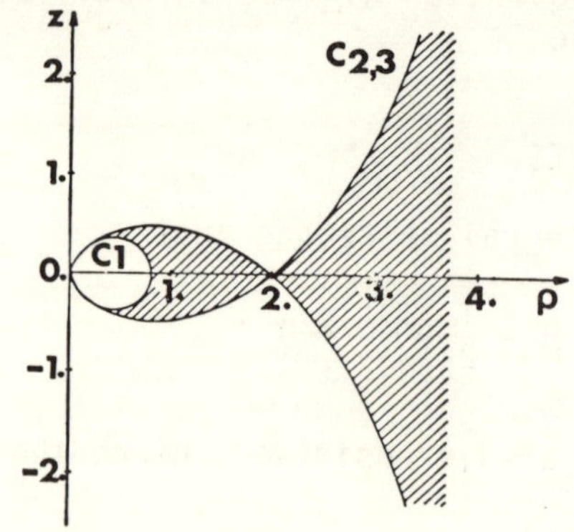

Fig. 2. The permissible space for motions with energy equal to 1/32.

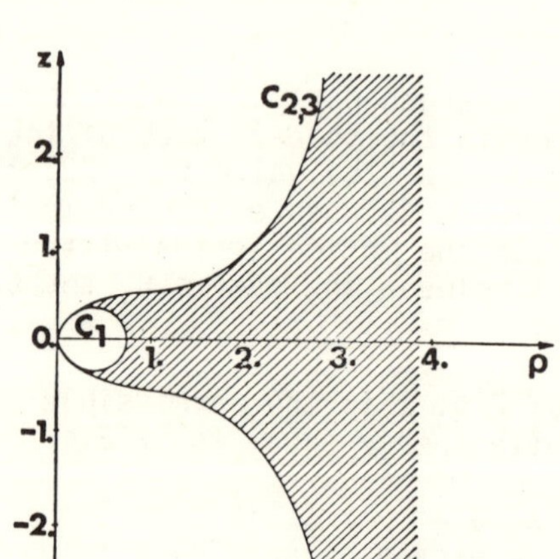

Fig. 3. For C>1/32 the area of permissible motion is open and escape of particles is possible.

IV. THE PERIODIC ORBITS OF FIRST GENERATION

The objective set for this program was the discovery and calculation of all symmetric periodic orbits of the first generation. The span of the space to be investigated was so extended that limiting it to the interior of the "thalweg" was necessary and logical. The "thalweg" is given in Fig. 4 where the permissible space for symmetric oscillations of trapped radiation is defined by the inequality

$$C \geq \frac{1}{2}\left(\frac{1}{\rho_0} - \frac{1}{\rho_0^2}\right)^2 .$$

The permissible space is shown in Fig. 4 in shading.

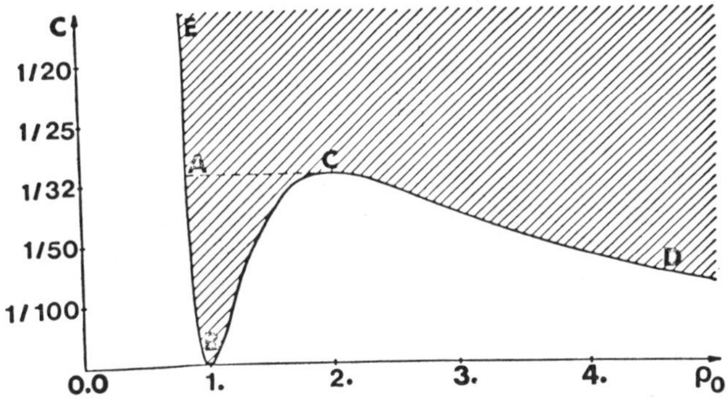

Fig. 4. The "thalweg", defined by the relation $C = (\rho_0^{-1} - \rho_0^{2})^2/2$. Trapped radiation performing oscillations symmetric with respect to the equatorial line e exist in the "well" ABC.

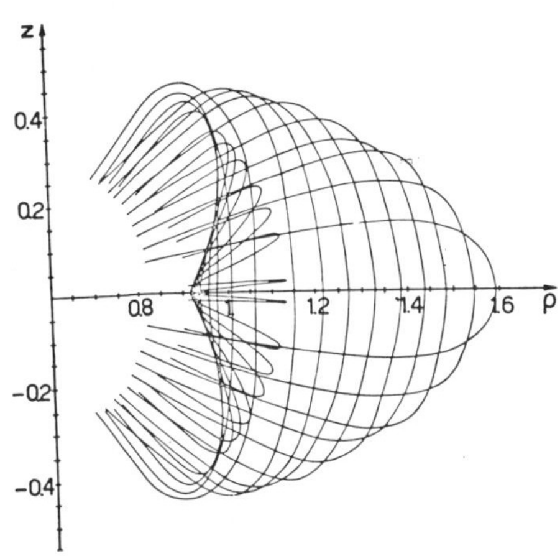

Fig. 5. A group of members of the family FII - FIII.

A grid-search program aimed at locating symmetric oscillations of charged particles was prepared, using as basic criterion for identification of the interval where such orbits exist the property of these orbits to intersect the equatorial line normally at two different times. After limiting the search inside or in any case close to the "thalweg" the work executed by the program included the following steps:

i. Give a value C_0 to the enenergy constant.

ii. Define the smallest of the three roots of the equation

$$C_o = \frac{1}{2}(\rho_o^{-1} - \rho_o^{-2})^2 \quad . \qquad (\alpha)$$

iii. Compute $p_{z,o}$ from equation

$$p_{z,o} = [2C_o - (\rho_o^{-1} - \rho_o^{-2})^2]^{1/2} \quad .$$

iv. Compute the solution corresponding to the initial state $(\rho_0, 0, 0, p_{z,o})$ until the next point of the form $(\rho, 0, p_\rho, p_z)$.

v. Increase ρ_0 by $\delta\rho_0$ and execute step iii. Repeat until the second root of Equation (a) is reached.

vi. Increase C_0 by δC and go to step i. Repeat until c becomes larger than 1.32.

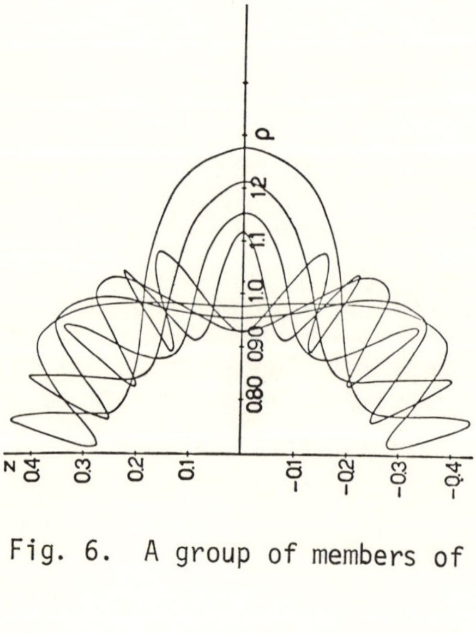

Fig. 6. A group of members of the family FVI.

While performing step iv the location of the periodic solution is established from the change of the sign of p_ρ. Once a periodic solution is located a predictor-corrector-steepest descent program (Mavroganis-Goudas, 1975) was used for the improvement of the solution and the determination of the entire family.

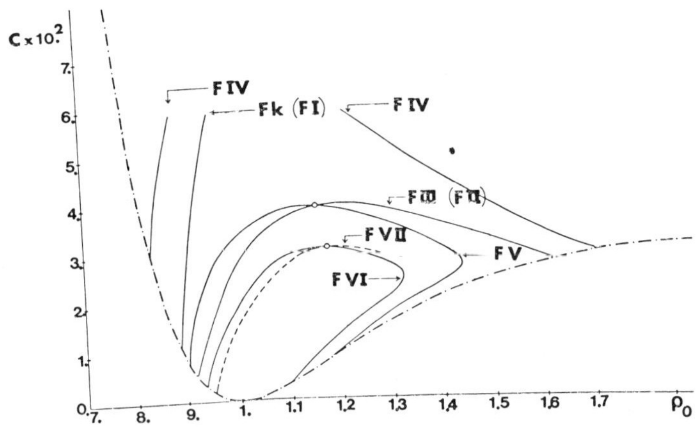

Fig. 7. The families of first generation solutions found inside and near the "well" of the "thalweg."

The above programs and the work presented in an earlier paper (Mavraganis-Goudas, 1975) led to the six families of symmetric oscillations shown in Fig. 7. The Families FII and FIII, reported earlier, are in fact one, and a few members of this Family is shown in Figure 5. On the other hand Family FIV is shown in Fig. 7 by means of two arcs. This is so because of the additional time reversibility property of the Störmer problem, on account of which, both points of normal intersection of the equatorial line are representative points of the same solution on the (C,ρ_0) - plane. We should remark here that the families FI,FII-FIII and FVII consist of solutions for which the normal intersections occur at the same point of the equatorial axis. The remaining Families, i.e. FIV,FV,FVI consist of solutions, the normal intersections of which occur at two different points. A few members of FVI are shown in Fig. 6.

The following interesting two properties are to be remarked: First, that all the families discovered up to now end on the boundary of the "thalweg". The families of solutions with two normal intersections intersect the curve defining the "thalweg" at the points resting on lines parallel to the $O\rho_0$-axis. Secondly, that each family with solutions of two normal intersections is intersected by one and only one family with solutions of one normal intersection.

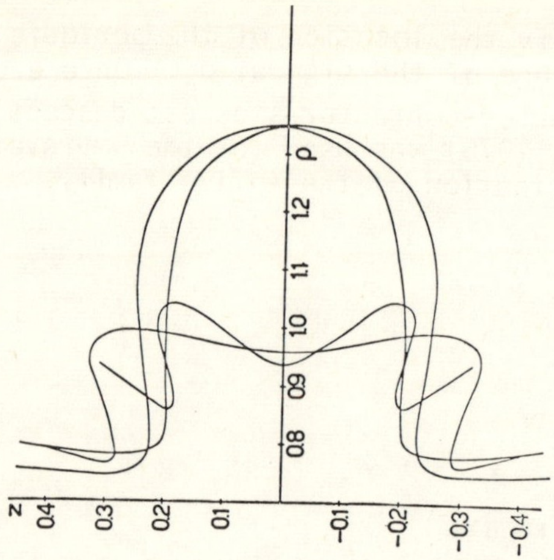

Fig. 8. A group of members of the family FVII.

Such intersections occur at extrema of the families with orbits of two normal intersections plotted in the (C,ρ_0)-plane. The intersections between families FII-III and FV as well as between families FVI and FVII are shown by small circles in Fig. 7. The intersection between FI9or Fk) and FIV rests well outside the region of interest of the present work. Family FVII was not computed until recently and the results for all solutions in numerical and graphical form will appear in a future publication. Here we show in Fig. 8 only a few members of this family.

The present work is now continued to cover all first generation solutions inside the "thalweg" and in particular inside the region enclosed by the boundary of the "thalweg" and the curved for FVI and FVII shown in Fig. 7. The rest of the interior of the "thalweg" was covered and no families other than the ones found exist there.

REFERENCES

1. Draght, A.J.: 1965, Rev. Geophys., Vol. 3, No. 2, 255
2. Lifshitz, J.: 1942, J. Math. Phys., Vol. 21, 284.
3. Mavraganis, A. and Goudas, C.L.: 1975, Astroph. and Space Sci., Vol. 32, 115.
4. Stern, D.: 1967, Pl. Sp., Vol. 15, 1525
5. Stömer, C.: 1907, Arch. Sci. Phys. et Nat Genève, Vol. 24, 350.
6. Van Allen, J.A., McIlwain, C.E. and Ludwing, G.H.: 1959, J. Geophys. Research, 64, 271.

PART IV

DYNAMICS IN ASTRONOMY

FORMAL EXPRESSIONS FOR THE MOTION OF N PLANETS IN THE PLANE, WITH THE SECULAR VARIATIONS INCLUDED, AND AN EXTENSION TO POISSON'S THEOREM.

P. J. Message
University of Liverpool

INTRODUCTION. In Poincaré's "Méthodes Nouvelles de la Mécanique Céleste", pp. 127-135, there is given a formal solution of the problem of three bodies in motion under their mutual gravitational attractions, in the case where one of the bodies (the "primary") has mass appreciably larger than the others, and where the motion of the others relative to the primary is nearly circular, and nearly coplanar. We may call this the "planetary case" of the general gravitational problem of three bodies. The solution is derived using Von Zeipel type transformations, in a number of stages, making use of infinite series in powers of the ratios of the smaller masses to that of the primary, and, in the treatment of the secular variations, powers of quantities of the order of the orbital eccentricities and orbital inclinations. In the present treatment, the simple extension to n planets is made, and the transformations employing power series are introduced making use of the Lie series method introduced by Hori (Pub. Astron. Soc., Japan, Vol. 18, pp. 287-295, 1966) which gives explicit expressions for the transformed variables in terms of the untransformed, and vice-versa. The short-period terms are removed by a single transformation at the outset, so that the elements of the matrix defining the linear transformation employed in the secular variation theory are functions of the constant transformed major axes, and so do not themselves possess short-period terms as those in Poincaré's solution do. The case of plane motion only is dealt with; no difficulties not encountered here are to be expected in extending the treatment to the case of small but non-zero orbital inclinations. It proves possible to make an extension of Poisson's theorem, that the major semi-axes are free of secular terms to the second order in the disturbing masses. We can assert that, in our formal solution, to all orders, they possess only constant and periodic terms, if, amongst

V. Szebehely and B. D. Tapley (eds.), Long-Time Predictions in Dynamics, 279–293. *All Rights Reserved. Copyright © 1976 by D. Reidel Publishing Company, Dordrecht-Holland.*

the periodic terms, are included terms with the periods of the secular variations.

1. FORMULATION OF THE PROBLEM

We consider a system composed of a primary P_0, of mass m_0, and n planets, P_1, P_2,....., P_n, of massess m_1, m_2,...., m_n, respectively, all moving under their mutual gravitational attractions only. We consider each of the bodies as a particle, and suppose that m_i is small compared to m_0, for $i \geq 1$. In some inertial frame, let the position vectors of P_0, P_1, P_2,...., P_n, be $\underset{\sim}{r}_0$, $\underset{\sim}{r}_1$, $\underset{\sim}{r}_2$,....,$\underset{\sim}{r}_n$, respectively. Then the total kinetic energy of the system is

$$T = \sum_{i=0}^{n} \frac{1}{2} m_i \dot{\underset{\sim}{r}}_i^{\,2} \quad , \tag{1.1}$$

and the potential energy is

$$V = - \sum_{i=0}^{n-1} \sum_{j=i+1}^{n} \frac{G m_i m_j}{|\underset{\sim}{r}_i - \underset{\sim}{r}_j|} \quad , \tag{1.2}$$

Where G is the constant of gravitation.

We use Jacobi's relative position vectors, $\underset{\sim}{x}_1$, $\underset{\sim}{x}_2$,....., $\underset{\sim}{x}_n$, which are defined by

$$\begin{aligned}
\underset{\sim}{x}_1 &= \underset{\sim}{r}_1 - \underset{\sim}{r}_0, & \bar{\underset{\sim}{r}}_1 &= (m_0 \underset{\sim}{r}_0 + m_1 \underset{\sim}{r}_1)/(m_0 + m_1); \\
\underset{\sim}{x}_2 &= \underset{\sim}{r}_2 - \bar{\underset{\sim}{r}}_1, & \bar{\underset{\sim}{r}}_2 &= \{(m_0 + m_1)\bar{\underset{\sim}{r}}_1 + m_2 \underset{\sim}{r}_2\}/(m_0 + m_1 + m_2); \\
&\vdots \\
\underset{\sim}{x}_i &= \underset{\sim}{r}_i - \bar{\underset{\sim}{r}}_{i-1}, & \bar{\underset{\sim}{r}}_i &= \{(m_0 + m_1 + .. + m_{i-1})\bar{\underset{\sim}{r}}_{i-1} + m_i \underset{\sim}{r}_i\}/(m_0 + \\
&\vdots & & \qquad m_1 + \ldots + m_i) \; ; \\
\underset{\sim}{x}_n &= \underset{\sim}{r}_n - \bar{\underset{\sim}{r}}_{n-1} \; ,
\end{aligned} \tag{1.3}$$

where of course $\bar{\underset{\sim}{r}}_i$ is the position vector of the mass-centre of P_0, P_1,...,P_i. Then

$$m_0 \dot{\underset{\sim}{r}}_0^{\,2} + m_1 \dot{\underset{\sim}{r}}_1^{\,2} = (m_0 + m_1)\dot{\bar{\underset{\sim}{r}}}_1^{\,2} + m_0 m_1/(m_0 + m_1)\, \dot{\underset{\sim}{x}}_1^{\,2} \quad ,$$

$$(m_0 + m_1)\dot{\bar{\underset{\sim}{r}}}_1^{\,2} + m_2 \dot{\underset{\sim}{r}}_2^{\,2} = (m_0 + m_1 + m_2)\dot{\bar{\underset{\sim}{r}}}_2^{\,2} + \frac{(m_0+m_1)m_2}{(m_0+m_1+m_2)} \dot{\underset{\sim}{x}}_2^{\,2} \quad ,$$

so that

$$T = \frac{1}{2} M\dot{\bar{r}}^2 + \varepsilon \sum_{i=1}^{n} \frac{1}{2} \beta_i \dot{x}_i^2 , \tag{1.4}$$

where $M = \sum_{i=0}^{n} m_i$ is the mass of the whole system,

$\bar{r} = (\sum_{i=0}^{n} m_i r_i)/M$ is the position vector of the mass-centre of the whole system, and

$$\varepsilon\beta_1 = m_0 m_1/(m_0 + m_1),$$

$$\varepsilon\beta_2 = (m_0 + m_1)m_2/(m_0 + m_1 + m_2),$$

$$\ldots\ldots.$$

$$\varepsilon\beta_i = (\sum_{j=0}^{i-1} m_j)m_i/(\sum_{j=0}^{i} m_j), \tag{1.5}$$

&ce.,

(the "reduced masses"). The quantity ε is still at our disposal, and we choose it to equal to the largest of the m_i/m_0 for $i \geq 1$, so that the β_i are of order unity or smaller. We find in turn that

$$r_1 - r_0 = x_1, \quad \bar{r}_1 - r_0 = m_1 x_1/(m_0 + m_1),$$

$$r_2 - r_0 = x_2 + m_1 x_1/(m_0 + m_1), \quad \bar{r}_2 - r_0 = (m_1 x_1 + m_2 x_2)/(m_0 + m_1 + m_2),$$

$$r_3 - r_0 = x_3 + (m_1 x_1 + m_2 x_2)/(m_0 + m_1 + m_2),$$

$$\bar{r}_3 - r_0 = (m_1 x_1 + m_2 x_2 + m_3 x_3)/(m_0 + m_1 + m_2 + m_3),$$

$$r_4 - r_0 = x_4 + (m_1 x_1 + m_2 x_2 + m_3 x_3)/(m_0 + m_1 + m_2 + m_3),$$

&ce., (1.6)

Thus in each case $r_i - r_0$ differs from x_i by quantities of order ε at most. Note also from (1.5) that

$$m_i = (\sum_{j=0}^{i-1} m_j)\varepsilon\beta_i/\{(\sum_{j=0}^{i-1} m_j) - \varepsilon\beta_i\} = \varepsilon\beta_i/\{1 - \varepsilon\beta_i/(\sum_{j=0}^{i-1} m_j)\},$$

so that $Gm_0 m_i = \varepsilon G\beta_i\{m_0 + \varepsilon\beta_i + O(\varepsilon^2)\}$.

Hence if we define μ_i so that

$$\varepsilon\mu_i\beta_i = Gm_0m_i, \tag{1.7}$$

then $\varepsilon\mu_i = G\{m_0 + \varepsilon\beta_i + O(\varepsilon^2)\}$, and also

$$V = -\varepsilon \sum_{i=1}^{n} \mu_i\beta_i/|\underset{\sim}{x}_i| - \varepsilon R, \tag{1.8}$$

where R is of order ε, and can be expressed in terms of the quantities $|\underset{\sim}{x}_i|$ and the $\underset{\sim}{x}_i \cdot \underset{\sim}{x}_j$.

The equation of motion for $\underset{\sim}{x}_i$ is then

$$\varepsilon\beta_i \ddot{\underset{\sim}{x}}_i = -\frac{\partial V}{\partial \underset{\sim}{x}_i} = -\varepsilon\mu_i\beta_i\underset{\sim}{x}_i/|\underset{\sim}{x}_i|^3 + \varepsilon\frac{\partial R}{\partial \underset{\sim}{x}_i},$$

(where $\frac{\partial f}{\partial \underset{\sim}{x}_i}$ is the gradient of f with respect to $\underset{\sim}{x}_i$) or, on dividing by ε,

$$\beta_i \ddot{\underset{\sim}{x}}_i = -\mu_i\beta_i\underset{\sim}{x}_i/|\underset{\sim}{x}_i|^3 + \frac{\partial R}{\partial \underset{\sim}{x}_i} \tag{1.9}$$

so that we may put the equations for the relative motion into Hamiltonian form by taking, as the momentum conjugate to $\underset{\sim}{x}_i$,

$$\underset{\sim}{p}_i = \beta_i\dot{\underset{\sim}{x}}_i ,$$

and then the equations may be derived from the Hamiltonian function

$$H(\underset{\sim}{x}_1,\underset{\sim}{x}_2,..,\underset{\sim}{x}_n;\ \underset{\sim}{p}_1,\underset{\sim}{p}_2,...,\underset{\sim}{p}_n) = \sum_{i=1}^{n} \{\underset{\sim}{p}_i^{\,2}/(2\beta_i) - \mu_i\beta_i/|\underset{\sim}{x}_i|\} - R. \tag{1.10}$$

Then, in the limit $\varepsilon \to 0$, since then $R \to 0$, the problem separates into n independent Kepler inverse square central force problems

$$\beta_i\ddot{\underset{\sim}{x}}_i = \dot{\underset{\sim}{p}}_i = -\mu_i\beta_i\underset{\sim}{x}_i/|\underset{\sim}{x}_i|^3 . \quad (i=1,2,...,n) \tag{1.11}$$

We may describe the state of motion at any time by means of the osculating orbits corresponding to these limiting Keplerian problems. Let the elements defining the osculating orbit corresponding to $\underset{\sim}{x}_i$ be a_i, the major semi-axis, e_i, the eccentricity, i_i, the orbital inclination, Ω_i, the longitude of the ascending node, $\bar{\omega}_i$, the apse longitude, and let the mean longitude be λ_i. Then

$$(\underset{\sim}{x}_i ; \underset{\sim}{p}_i) \mid\rightarrow (\lambda_i, \bar{\omega}_i, \Omega_i ; \Lambda_i, \Pi_i, N_i)$$

is a contact transformation, where

$$\Lambda_i = \beta_i\sqrt{\mu_i a_i}, \quad \pi_i = \Lambda_i\{\sqrt{(1-e_i^2)}-1\}, \text{and } N_i = \Lambda_i\sqrt{(1-e_i^2)}(\cos i_i - 1), \tag{1.12}$$

and the Hamiltonian function for the transformed problem is

$$H'(\lambda_1,\lambda_2,\ldots,\lambda_n,\bar{\omega}_1,\ldots,\Omega_1\ldots,\Lambda_1,(\pi_1,\ldots)N_1,\ldots) = -\sum_{i=1}^{n} \frac{1}{2}\beta_i^3\mu_i^2/\Lambda_i^2 - R. \tag{1.13}$$

(This arises from the solution of the Hamilton-Jacobi equation

$$1/(2\beta_i)\left(\frac{\partial S}{\partial x_i}\right)^2 - \mu_i\beta_i/|x_i| = -\beta_i^3\mu_i^2/(2\Lambda_i^2)$$

for the Kepler problem (1.11), as described, for example, in H. C. Plummer's "Introduction to Dynamical Astronomy", Cambridge University Press, 1918, 135-6; or in E. W. Brown & C. Shook, "Planetary Theory", (C.U.P., 1933, 5.9-5.13).

Properties of these Keplerian elements which we will use are:

(i) that the $|x_i|$ and $x_i \cdot x_j$, and hence R, have period 2π in each of the λ_i, $\bar{\omega}_i$, and Ω_i, and so R may be expressed as a multiple Fourier series with terms of the type

$$K\cos(j_1\lambda_i + j_2\bar{\omega}_i + j_3\Omega_i + j_4\lambda_k + j_5\bar{\omega}_k + j_6\Omega_k) \tag{1.14}$$

with $j_1+j_2+j_3+j_4+j_5+j_6 = 0$.

(ii) the "d'Alembert property", that the $|\underset{\sim}{x}_i|$ and $\underset{\sim}{x}_i \cdot \underset{\sim}{x}_j$ are expressible as power series in the $e_i \cos\bar{\omega}_i$, $e_i \sin\bar{\omega}_i$, $\sin i_i \cos\Omega_i$, and $\sin i_i \sin\Omega_i$, so that the coefficient K in (1.14) can be expressed as

$$e_i^{|j_2|}\, e_k^{|j_5|}\, \sin^{|j_4|} i_i\, \sin^{|j_6|} i_k$$

multiplied by a power series in e_i^2, e_k^2, $\sin^2 i_i$, and $\sin^2 i_k$, the coefficients in that series being functions of a_i and a_k.

(See, for example, E. W. Brown & C. Shook, loc. cit., 3.4, 4.15, 6.4, or D. Brower & C. M. Clemence, "Methods of Celestial Mechanics", Academic Press, 1961, p. 79).

2. THE USE OF LIE SERIES TRANSFORMATIONS

Now $(q_1, q_2, \ldots, q_n; p_1, p_2, \ldots, p_n) | (Q_1, Q_2, \ldots, Q_n; P_1, P_2, \ldots, P_n)$ is a contact transformation, where

$$q_i = Q_i + \frac{\partial S}{\partial P_i} + \frac{1}{2}\{\frac{\partial S}{\partial P_i}, S\} + \frac{1}{3!}\{\{\frac{\partial S}{\partial P_i}, S\}, S\} + \ldots..$$

and

$$p_i = P_i - \frac{\partial S}{\partial Q_i} - \frac{1}{2}\{\frac{\partial S}{\partial Q_i}, S\} - \frac{1}{3!}\{\{\frac{\partial S}{\partial Q_i}, S\}, S\} - \ldots..$$

$$(i = 1, 2, \ldots, n) \quad (2.1)$$

where S is a function of the Q_i and the P_i with all derivatives, and $\{f, g\}$ is the Poisson bracket expression, given by

$$\{f, g\} = \sum_{i=1}^{n} \left(\frac{\partial f}{\partial Q_i} \cdot \frac{\partial g}{\partial P_i} - \frac{\partial f}{\partial P_i} \cdot \frac{\partial g}{\partial Q_i}\right) ,$$

within the range of validity of the series (see Hori, loc. cit.).

Further, for any function $f(Q, P)$ of all (or some) of the Q_i and the P_i, (where Q denotes the n quantities $(Q_1, Q_2, \ldots, Q_n)$, with corresponding meanings for P, q, p),

$$f(q;p) = f(Q;P) + \{f, S\} + \frac{1}{2}\{\{f, S\}, S\} + \ldots \quad (2.2)$$

and conversely

$$f(Q;P) = f(q;p) - \{f, S\} - \frac{1}{2}\{\{f, S\}, S\} - \ldots \quad (2.3)$$

where, in the right-hand side of the latter equation, each of the functions is evaluated with the q_i and p_i substituted for the Q_i and P_i. We suppose that S, and therefore the transformation, is independent of the time explicitly, so that if $H(q;p)$ and $H'(Q;P)$ are the respective Hamiltonian functions, then

$$H'(Q;P) \equiv H(q;p) = H(Q;P) + \{H, S\} + \frac{1}{2}\{\{H, S\}, S\} + \ldots. \quad (2.4)$$

where, on the right-hand side, the Q_i and P_i have been substituted for the q_i and p_i.

The objective sought in the use of this type of transformation is to choose S so as to render H' independent of the Q_i, that is, to make each Q_i and ignorable co-ordinate, the equations of motion then becoming $\dot{P}_i = 0$, $\dot{Q}_i = \frac{\partial H'}{\partial P_i}(P)$, so that in the solution each P_i is a constant of integration, and so each $\dot{Q}_i$ is a constant, say ν_i, so that $Q_i = \nu_i t + \delta_i$ where each δ_i is a further constant of integration. Then the equations (2.1) give explicit expressions for the original variables q_i and p_i in terms of the constants P_i and the linear functions Q_i of the time. We are concerned with the type of situation in which we may express H in the form

$$H = H_0 + H_1 + H_2 \dots .$$

where H_0 is independent of the small parameter ε, H_1 is of first order in ε, H_2 of second order, and so on. We require that H_0 be independent of the q_i. We seek to construct an expansion of the form

$$S = S_1 + S_2 + S_3 + \dots ,$$

where S_1, S_2, S_3, ... are respectively of first, second, third, etc., order in ε, and likewise

$$H' = H'_0 + H'_1 + H'_2 + \dots .$$

Equating terms of zero order in equation (2.4) then shows that $H'_0(P) = H_0(P)$, so that H'_0 must be the same function of the P_i that H_0 is of the p_i. Equating terms of first order gives

$$H'_1 = H_1 - \sum_{i=1}^{n} \frac{\partial H_0}{\partial P_i} \frac{\partial S_1}{\partial Q_i} .$$

Corresponding to a term of the type $K(p) \cos(\sum_i j_i q_i + \delta)$ in H_1, we include in S_1 the term $\dfrac{K(P) \sin (\sum_i j_i Q_i + \delta)}{\sum_i j_i \dfrac{\partial H_0}{\partial P_i}}$,

so that the net contribution to the right-hand side from this term is zero, and the term can therefore be omitted from H'_1, which can therefore be chosen free of terms periodic in the Q_i, provided that none of the denominators vanish. If H_1 is a periodic function of the q_i, then H'_1 will be the constant term in the Fourier series for H_1 in terms of the q_i. In like manner each of S_2, S_3,.etc., can be chosen so that H'_2, H'_3, etc., are independent of the Q_i.

In the problem of n planets before us, we have $H_0 = -\sum_{i=1}^{n} \dfrac{\beta_i^3 \mu_i^2}{2\Lambda_i^2}$ and in fact $\dfrac{\partial H_0}{\partial \Lambda_1} = \beta_i^3 \mu_i^2 / \Lambda_i^3 = \mu_i^{1/2} a_i^{-3/2} = n_i$, the osculating mean motion. The method just outlined may therefore be employed to remove from H' those terms of R which are periodic in one or more of the λ_1 (the "short-period" terms). However, since $\dfrac{\partial H_0}{\partial \Pi_1}$ and

$\frac{\partial H_0}{\partial N_i}$ are zero, we cannot in the same way remove those terms which involve the ϖ_i and the Ω_i but which do not involve the λ_i. For such terms, called "secular terms", a different treatment is necessary.

3. THE ELIMINATION OF THE SHORT-PERIOD TERMS

We will deal with the case of motion of the planets, relative to the primary, in a fixed plane, for simplicity, since the essential difficulties are still met with in this case, and the methods employed may readily be extended to encompass the case of non-zero orbital inclinations. Then R will be independent of the Ω_i, (which will in fact be undefined), and the λ_i and the ϖ_i will be the only effective co-ordinates in the problem. The N_i will also be identically zero. Corresponding to each Jacobi vector $\underset{\sim}{x}_i$, therefore, the effective canonical variables are $(\lambda_i, \varpi_i; \Lambda_i, \Pi_i)$. We begin with a transformation

$$(\lambda, \varpi; \Lambda, \Pi) \mapsto (\lambda', \varpi'; \Lambda', \Pi') \quad ,$$

(where λ' denotes $(\lambda'_1, \lambda'_2, \ldots, \lambda'_n)$, with corresponding meanings for ϖ', Λ', and Π'), defined by

$$\lambda_i = \lambda'_i + \frac{\partial S}{\partial \Lambda'_i} + \frac{1}{2}\{\frac{\partial S}{\partial \Lambda'_i}, S\} + \frac{1}{3!}\{\{\frac{\partial S}{\partial \Lambda'_i}, S\}, S\} + \ldots..$$

$$\varpi_i = \varpi'_i + \frac{\partial S}{\partial \Pi'_i} + \frac{1}{2}\{\frac{\partial S}{\partial \Pi'_i}, S\} + \ldots.$$

$$\Lambda_i = \Lambda'_i - \frac{\partial S}{\partial \lambda'_i} - \frac{1}{2}\{\frac{\partial S}{\partial \lambda'_i}, S\} - \ldots.$$

$$\Pi_i = \Pi'_i - \frac{\partial S}{\partial \varpi'_i} - \frac{1}{2}\{\frac{\partial S}{\partial \varpi'_i}, S\} - \ldots.$$

$$(i = 1, 2, \ldots.,n) \tag{3.1}$$

where S is to be chosen. If $H'(\lambda', \varpi'; \Lambda', \Pi')$ is the Hamiltonian function giving the equations of motion for the transformed variables, then by (2.4).

$$H' \equiv H + \{H, S\} + \frac{1}{2}\{\{H, S\}, S\} + \ldots.. \tag{3.2}$$

Now we may put $H = H_0 - R_1 - R_2 - R_3 - \ldots.$ (3.3)

where R_p is of degree p in ε, and seek expansions of the form

$$H' = H'_0 - R'_1 - R'_2 - R'_3 - \ldots.$$

and $$S = S_1 + S_2 + S_3 + \ldots\ldots. \tag{3.4}$$

As before, equating terms of order zero gives

$$H'_0 \equiv H_0 \tag{3.5}$$

and equating terms of first order gives

$$- R'_1 \equiv - R_1 + \{H_0, S_1\} = - R_1 - \sum_{i=1}^{n} \frac{\partial H_0}{\partial \Lambda_i} \frac{\partial S_1}{\partial \lambda_i}$$

$$= - R_1 - \sum_{i=1}^{n} n_i \frac{\partial S_1}{\partial \lambda_i} \tag{3.6}$$

Corresponding to each term in R_1 of the form

$$K \cos(j_1\lambda_i + j_2 \bar{\omega}_i + j_3\lambda_k + j_4\bar{\omega}_k)$$

we include in S_1 a term

$$- \{K \sin(j_1\lambda_i + j_2\bar{\omega}_i + j_3\lambda_k + j_4\bar{\omega}_k)\}/(j_1 n_i + j_3 n_k),$$

so that the total contribution to the right-hand side of (3.6) from this term is zero, and it is omitted from R'_1, which will therefore consist of just those terms of R_1 which are independent of the λ_i. In turn, each S_p may be chosen so that each R'_p contains only terms which are independent of the λ'_i. We are then left with a Hamiltonian function H' which is independent of the λ'_i, and which corresponds to the "secular problem".

4. TREATMENT OF THE SECULAR PROBLEM.

We note that the λ'_i are now ignorable, so that their conjugate momenta Λ'_i are constants of the motion. In view of the properties of R already listed, we see that its secular terms are of the form

$$K_{ijk} \cos k(\bar{\omega}_i - \bar{\omega}_j), \tag{4.1}$$

where the coefficient K_{ijk} has the form $(e_i e_j)^k$ multiplied by a series in positive powers of e_i^2 and e_j^2, the coefficients being functions of Λ_i and Λ_j. Thus R'_1 consists of a series of such terms, and the higher order functions R'_2, R'_3,. &ce., consist of terms of this type and products of terms of this type. Now

$$\Pi_i = \Lambda_i\{-\frac{1}{2}e_i^2 + \text{terms in } e_i^4, e_i^6, \text{etc.,}\},$$

so the term (4.1) can be expressed in terms of $\sqrt{(\Pi_i\Pi_j)}\cos(\bar{\omega}_i - \bar{\omega}_j)$, Π_i, Π_j, Λ_i, and Λ_j. Now the Laplace and Lagrange theory of the secular variations consists of the solution of that problem in which only those secular terms of R of degree 2 in the e_i are retained. We begin, as do Laplace and Lagrange, by making a transformation to rectangular-type variables. Let us in fact introduce

$$\xi_i = \sqrt{(-2\Pi'_i)}\cos\bar{\omega}'_i,$$

$$\text{and} \quad \eta_i = \sqrt{(-2\Pi'_i)}\sin\bar{\omega}'_i, \text{ for } i = 1, 2,\ldots,n, \tag{4.2}$$

so that $(\bar{\omega}'_i; \Pi'_i) \mapsto (\xi_i; \eta_i)$ is a contact transformation. We note that, to second order in e_i, $\xi_i = \sqrt{(\Lambda'_i)}\, e'_i \cos\bar{\omega}'_i$, and $\eta_i = \sqrt{(\Lambda'_i)}\, e'_i \sin\bar{\omega}'_i$.

Then R'_1, R'_2, etc., can be expressed in terms of the quantities $\xi_i^2 + \eta_i^2$, $\xi_i\xi_j + \eta_i\eta_j$, and products thereof, with coefficients which are functions of the Λ'_i. The terms of lowest order are of the form

$$\sum_{(i,j)} K_{ij}(\xi_i\xi_j + \eta_i\eta_j), \tag{4.3}$$

and these are the terms retained in the Laplace and Lagrange theory, the equations for which are therefore,

$$\dot{\xi}_i = -\sum_j K_{ij}\,\eta_j, \quad \dot{\eta}_i = \sum_j K_{ij}\,\xi_j, \quad (i = 1, 2, \ldots, n). \tag{4.4}$$

We now make a linear transformation, to diagonalize the quadratic form (4.3), by defining

$$\sigma_k = \sum_{i=1}^{n} m_{ki}\,\xi_i, \text{ and } \tau_k = \sum_{i=1}^{n} m_{ki}\,\eta_i, \tag{4.5}$$

for $k = 1, 2,\ldots, n$, where the m_{ki} are the elements of a square matrix to be chosen. We wish to ensure that

$$(\lambda', \xi; \Lambda', \eta) \mapsto (\lambda'', \sigma; \Lambda'', \tau)$$

(where ξ denotes $(\xi_1, \xi_2,\ldots, \xi_n)$, with similar meanings for η, σ, τ) is a contact transformation. This requires that the differential form

$$\sum_i \eta_i\, d\xi_i + \sum_i \Lambda'_i\, d\lambda'_i - \sum_k \tau_k\, d\sigma_k - \sum_i \Lambda''_i\, d\lambda''_i$$

be a total differential. It is equal to

$$\sum_i \{\eta_i d\xi_i + \Lambda'_i d\lambda'_i - \Lambda''_i d\lambda''_i - \sum_k m_{ki}\eta_i(\sum_j m_{kj} d\xi_j + \sum\sum_{j\ell} \frac{\partial m_{kj}}{\partial \Lambda_\ell} \xi_j d\Lambda_\ell)\},$$

and so we must have that

(i) the matrix (m_{ki}) is orthogonal, that is,

$$\sum_k m_{ki}\, m_{kj} = \begin{cases} 0 \text{ if } j \neq i, \\ 1 \text{ if } j = i, \end{cases}$$

(ii) $\Lambda'_i = \Lambda''_i$,

and

(iii) $$\lambda'_i = \lambda''_i + \sum\sum\sum_{jkr} m_{kr}\,\eta_r \frac{\partial m_{kj}}{\partial \Lambda_i}\,\xi_j = \lambda''_i + \sum\sum\sum_{jkp} \tau_k \frac{\partial m_{kj}}{\partial \Lambda_i} m_{pj}\sigma_p. \tag{4.6}$$

The matrix may also be chosen so that

$$\sum\sum_{ij} m_{ki}\, K_{ij}\, m_{rj} = \begin{cases} 0 & \text{if } r \neq k \\ \frac{1}{2}A_k & \text{if } r = k, \end{cases} \tag{4.7}$$

provided that the eigenvalues, $\frac{1}{2}A_k$, of the matrix (K_{ij}) are all different. Now R'_1, R'_2, etc., are expressible in terms of the $\sigma_k{}^2 + \tau_k{}^2$, and the $\sigma_k\sigma_r + \tau_k\tau_r$, the terms of lowest order being

$$\frac{1}{2}\sum_k A_k(\sigma_k{}^2 + \tau_k{}^2),$$

so that the equations for the Laplace and Lagrange theory are

$$\dot{\sigma}_k = - A_k\tau_k \text{ and } \dot{\tau}_k = A_k\sigma_k, \quad (k = 1, 2,\ldots n). \tag{4.8}$$

We now carry out a further contact transformation, $(\sigma; \tau) \mapsto (\omega; \Omega)$ defined by

$$\sigma_k = \sqrt{(-2\Omega_k)} \cos \omega_k,$$

$$\text{and} \quad \tau_k = \sqrt{(-2\Omega_k)} \sin \omega_k, \text{ for } k = 1, 2,\ldots, n. \tag{4.9}$$

Now R'_1, R'_2, etc., are expressible in terms of the quantities $\sqrt{(\Omega_k\Omega_r)} \cos(\omega_k - \omega_r)$, and the Ω_k. The terms of lowest degree are

$$- \sum_k A_k \Omega_k,$$

and the equations for the Laplace and Lagrange theory are

$$\dot{\omega}_k = A_k, \text{ and } \dot{\Omega}_k = 0, \quad (k = 1, 2, \ldots, n). \tag{4.10}$$

so that the Ω_k are constants of integration in the Laplace and Lagrange theory, in which theory the ω_k are linear functions of the time, their rates of change A_k being in fact all positive. (The periods of the component oscillations in their theory are the quantities $2\pi/A_k$.)

But in the full problem R'_1, R'_2, etc., have terms of higher than first degree in the Ω_k, each term having the form

$$K_j(\Omega_1, \Omega_2, \ldots, \Omega_n) \quad \cos(\sum_{i=1}^{n} j_i \; \omega_i),$$

where here $j = (j_1, j_2, \ldots, j_n)$, and in each case $\sum_{i=1}^{n} j_i = 0$, and K_j has a factor $\sqrt{(\Omega_i)^{j_i}}$ for each i.

The Ω_i are quantities of the order of the squares of the eccentricities of the orbits, which we suppose to be small, and we proceed by constructing series in powers of ρ, chosen to be the largest of the Ω_k. We then put $\Omega_k = \rho\tilde{\Omega}_k$, for $k = 1, 2, \ldots, n$, and, after making this substitution,

$$R'_1 + R'_2 + \ldots = R' = R'^{(1)} + R'^{(2)} + R'^{(3)} + \ldots . \tag{4.11}$$

where $R'^{(p)}$ is of degree p in ρ, and in fact $R'^{(1)} = - \rho \sum_k A_k \tilde{\Omega}_k$. The equations of motion for the ω_k and $\tilde{\Omega}_k$ are in fact

$$\dot{\Omega}_k = \frac{\partial R'}{\partial \omega_k}, \text{ and } \dot{\omega}_k = - \frac{\partial R'}{\partial \Omega_k}, \text{ (for } k = 1, 2, \ldots, n), \tag{4.12}$$

so if we write $R'(\omega, \Omega) = \rho \; \tilde{R}(\omega, \tilde{\Omega})$ (ω denoting $(\omega_1, \omega_2, \ldots, \omega_n)$, and correspondingly for Ω and $\tilde{\Omega}$)

$$\dot{\tilde{\Omega}}_k = \frac{\partial \tilde{R}}{\partial \omega_k}, \text{ and } \dot{\omega}_k = - \frac{\partial \tilde{R}}{\partial \tilde{\Omega}_k}, \text{ (for } k = 1, 2, \ldots, n). \tag{4.13}$$

We may write $\tilde{R} = \tilde{R}^{(0)} + \tilde{R}^{(1)} + \tilde{R}^{(2)} + \ldots$,
where $\tilde{R}^{(p)} = \rho R'^{(p+1)}$ is of degree p in ρ. We seek now a transformation $(\omega, \tilde{\Omega}) \mapsto (\omega^*, \tilde{\Omega}^*)$ (ω^* denoting $(\omega_1, , \ldots, \omega^*_n)$ and $\tilde{\Omega}^*$ correspondingly)

$$\omega_k = \omega^*_k + \frac{\partial \tilde{T}}{\partial \tilde{\Omega}^*_k} + \frac{1}{2}\{\frac{\partial \tilde{T}}{\partial \tilde{\Omega}^*_k}, \tilde{T}\} + \ldots .$$

$$\tilde{\Omega}_k = \tilde{\Omega}^*_k - \frac{\partial \tilde{T}}{\partial \omega^*_k} - \frac{1}{2}\{\frac{\partial \tilde{T}}{\partial \omega^*_k}, \tilde{T}\} - \ldots. \qquad (k = 1,2,\ldots n) \qquad (4.15)$$

where $\tilde{T}(\omega^*, \tilde{\Omega}^*)$ is to be chosen. In fact we express

$$\tilde{T} = \tilde{T}^{(1)} + \tilde{T}^{(2)} + \ldots \qquad (4.16)$$

where $\tilde{T}^{(p)}$ is of degree p in ρ. If the Hamiltonian function for the $(\omega^*_k, \tilde{\Omega}^*_k)$ is $\tilde{R}^*(\omega^*_k, \tilde{\Omega}^*_k)$, then we have

$$\tilde{R}^* \equiv \tilde{R} + \{\tilde{R}, \tilde{T}\} + \frac{1}{2}\{\{\tilde{R}, \tilde{T}\}, \tilde{T}\} + \ldots. \qquad (4.17)$$

We expect to obtain an expansion of the form

$$\tilde{R}^* = \tilde{R}^{*(0)} + \tilde{R}^{*(1)} + \tilde{R}^{*(2)} + \ldots. \qquad (4.18)$$

and then equating terms of degree zero in ρ in equation (4.17) gives $\tilde{R}^{*(0)} = \tilde{R}^{(0)}$. Equating terms of first degree gives

$$\tilde{R}^{*(1)} \equiv \tilde{R}^{(1)} + \{\tilde{R}^{(0)}, \tilde{T}^{(1)}\} = \tilde{R}^{(1)} + \sum_k A_k \frac{\partial \tilde{T}^{(1)}}{\partial \omega^*_k} . \qquad (4.19)$$

As in §2, we may ensure that $\tilde{R}^{*(1)}$ is independent of the ω^*_k by including $\tilde{T}^{(1)}$, for each term K $\cos(\sum_{i=1}^{n} j_i \omega_i)$ of $\tilde{R}^{(1)}$, the term $\{K \sin(\sum_{i=1}^{n} j_i \omega^*_i)\}/(\sum_{i=1}^{n} j_i A_i)$. This may be extended to terms of each successive degree in ρ, so that $\tilde{R}^*$ becomes a function of the $\tilde{\Omega}^*_k$ only.

We then introduce $\Omega^*_k = \rho\tilde{\Omega}^*_k$, for k = 1, 2,..., n, and consider the transformation from the original variables $(\omega_k; \Omega_k)$ to the $(\omega^*_k; \Omega^*_k)$. Put $T(\omega^*; \Omega^*) \equiv \rho\tilde{T}(\omega^*; \tilde{\Omega}^*)$. Then if, for any function $f(\omega^*; \Omega^*) \equiv \tilde{f}(\omega^*; \tilde{\Omega}^*)$, we have

$$\{\tilde{f}, \tilde{T}\} = \sum_k (\frac{\partial \tilde{f}}{\partial \omega^*_k}\frac{\partial \tilde{T}}{\partial \tilde{\Omega}^*_k} - \frac{\partial \tilde{f}}{\partial \tilde{\Omega}^*_k}\frac{\partial \tilde{T}}{\partial \omega^*_k})$$

$$= \sum_k (\frac{\partial f}{\partial \omega^*_k}\frac{\partial T}{\partial \Omega^*_k} - \frac{\partial f}{\rho\, \partial \Omega^*_k}\rho\frac{\partial T}{\partial \omega}\frac{1}{\rho}) = \{f, T\},$$

$$\text{so that } \Omega_k = \rho\tilde{\Omega}_k = \rho(\tilde{\Omega}^*_k + \frac{\partial \tilde{T}}{\partial \omega^*_k} - \frac{1}{2}\{\frac{\partial \tilde{T}}{\partial \omega^*_k}, T\} - \ldots.)$$

$$= \Omega^*_k - \frac{\partial T}{\partial \omega^*_k} - \frac{1}{2}\{\frac{\partial T}{\partial \omega^*_k}, T\} - \ldots.. \quad (k,1,2,\ldots,n) \qquad (4.20)$$

$$\text{and also } \omega_k = \omega^*_k + \frac{\partial\tilde{T}}{\partial\tilde{\Omega}^*_k} + \frac{1}{2}\{\frac{\partial\tilde{T}}{\partial\tilde{\Omega}^*_k}, \tilde{T}\} + \ldots.$$

$$= \omega^*_k + \frac{\partial T}{\partial\Omega^*_k} + \frac{1}{2}\{\frac{\partial T}{\partial\Omega^*_k}, T\} + \ldots. \quad (k=1,2,\ldots,n), \quad (4.21)$$

so that $(\omega; \Omega) \mapsto (\omega^*; \Omega^*)$ is a contact transformation. If we put $R^*(\Omega^*) \equiv \rho\tilde{R}^*(\tilde{\Omega}^*)$, then by (4.17)

$$R^* \equiv \rho\{\tilde{R} + \{\tilde{R}, \tilde{T}\} + \frac{1}{2}\{\{\tilde{R}, \tilde{T}\}, \tilde{T}\} + \ldots.$$

$$= R' + \{R', T\} + \frac{1}{2}\{\{R', T\}, T\} + \ldots. \quad (4.22)$$

so that R^* is the Hamiltonian function for the ω^*_k and the Ω^*_k. Now T will also depend on the Λ_i, through the coefficients K, and the A_k, so our transformation is in fact

$$(\lambda'', \omega; \Lambda'', \Omega) \mapsto (\lambda^*, \omega^*; \Lambda^*, \Omega^*),$$

where we have

$$\lambda''_i = \lambda^*_i + \frac{\partial T}{\partial\Lambda^*_i} + \frac{1}{2}\{\frac{\partial T}{\partial\Lambda^*_i}, T\} + \ldots., (i=1,2,..,n) \quad (4.23)$$

and, since T is independent of the λ^*_i,

$$\Lambda''_i = \Lambda^*_i \qquad (i = 1, 2, \ldots, n), \quad (4.24)$$

The Hamiltonian function for the λ^*_i, ω^*_k; Λ^*_i, Ω^*_k is

$$H_0(\Lambda^*) - R^*(\Omega^*),$$

so that all of the co-ordinates, λ^*_i, ω^*_k are now ignorable, and are therefore linear functions of the time in the motion, and their conjugate momenta Λ^*_i, Ω^*_k, are constants of the motion. The equations of transformation, (3.1), (4.2), (4.9), (4.20), and (4.21), now provide the explicit solution of the original problem, in formal series, in terms of these linear function of time, and constants of integration. The rates of change of the λ^*_i are constants differing from the mean motions by quantities of order ε and ρ, and the rates of change of the ω^*_i differ from the A_i by quantities of order ρ.

5. EXTENSION OF THE POISSON RESULT ON THE MAJOR SEMI-AXES.

It has been known for some time (see F. Tisserand, Mem. Acad, Sci. Toulouse, ser, 7, vol. 7, 1876) that, to second order in the disturbing masses, the major semi-axes of the planetary orbits have no true secular terms, that is to say, in terms of the present treatment, if, in equations (3.1) there were to be substituted expressions for the λ'_i, $\bar{\omega}'_i$, Λ'_i, Π'_i, arrived at by solving the secular problem by power series in the time, then the resulting expressions for the major semi-axes a_i would contain no terms proportional to the time, to second order in ε. The work of Haretu (Ann. Obs. Paris, Memoires, vol. 18, 1855), Eginitis (loc. cit., vol. 19, 1889), and Meffroy, (Bull. Astron., vol. 19, pp.1-224, 1955), shows that such terms do appear in the third order. We have seen, however, that the solution of the secular problem can be expressed formally entirely in terms of mutiply- periodic expressions, the periods being those arising in the Laplace and Lagrange theory of the secular variations, that is, in the case of the solar system, periods of between 58,000 and 2,100,000 years.

We have also seen that the λ'_i are ignorable co-ordinates, so that the Λ'_i are constants of the motion. From (3.1) we have the expansion

$$\Lambda'_i = \Lambda'_i - \frac{\partial S}{\partial \lambda'_i} - \frac{1}{2}\{\frac{\partial S}{\partial \lambda'_i}, S\} - \ldots. \quad (i=1, 2, ..,n) \qquad (5.1)$$

where S is a series of terms of the form

$$\{K \sin(j_1\lambda'_i + j_2\lambda'_k + j_3\bar{\omega}'_i + j_4\bar{\omega}'_k)\}/(j_1n_i + j_2n_k),$$

that is, S is a series of short-period terms. However, the products of derivatives of such terms arising in the third and later terms on the right-hand side of equation (5.1) will result in some terms of the form $L \cos(\sum_k k\bar{\omega}'_k)$. In view of the d'Alembert property described before, such a term must be expressible in terms of sums and products of the quantities Π'_i and $\sqrt{(\Pi'_i\Pi'_j)}\cos(\bar{\omega}'_i - \bar{\omega}'_j)$, and therefore in terms of the $\xi_i^2 + \eta_i^2$ and $\xi_i\xi_j + \eta_i\eta_j$, and therefore of the $\sigma_k^2 + \tau_k^2$ and $\sigma_k\sigma_r+\tau_k\tau_r$, and therefore in terms of the Ω_k and the $\sqrt{(\Omega_k\Omega_r)}$ $\cos(\omega_k - \omega_r)$, and therefore finally in terms of the quantities Ω^*_k and $\sqrt{(\Omega^*_k\Omega^*_r)}$ $\cos(\omega^*_k - \omega^*_r)$. We recall now that the Ω^*_k are constants of the motion, and that the ω^*_k are linear functions of the time, so that we have here periodic functions of the time, the periods arising from linear combination of the frequencies in the secular variation theory. Thus the Λ_i are expressible as constants, together with a series of periodic terms, including some with these long periods. This is true to all orders in ε. Now since $a_i = \Lambda_i^2/(\mu_i\beta_i^2)$, the same is true of the major semi-axes a_i.

THE PRINCIPLE OF LEAST INTERACTION ACTION

M. W. Ovenden

Department of Geophysics and Astronomy, and Institute of Astronomy and Space Science, University of British Columbia, Vancouver, Canada.

ABSTRACT. The resonant structures, the regular spacing of major semi-axes and the distribution of orbital eccentricities found in the solar system are to be expected for any planetary (or satellite) system capable of evolution under point-mass gravitational forces. Whether the actual solar system, in its present state, is capable of such evolution is still an open question. The time-scales calculated in previous papers are spurious, since the so-called 'secular' terms of Meffroy are not, in fact, true secular terms.

1. INTRODUCTION

At the NATO Advanced Study Institute in Dynamical Astronomy held in Cortina d'Ampezzo in 1972, I reported some speculations that I had made concerning the possible significance of the observed distribution of the orbits in the planetary and satellite systems (Ovenden 1973). This work led, as a corollary, to the suggestion that there had existed, between Mars and Jupiter, a planet (Aztex) of mass $\sim 90\ M_{\oplus}$, and that this planet had 'disappeared' in the relatively recent past, $\sim 10^7$ years ago.

As was to have been anticipated, this conclusion has led to some scientific controversy. It has also become entangled with a controversy of a quite different character, namely the status of the theories of Immanuel Velikovsky (see e.g. Nieto; Bass 1974). In this connection I would make only two comments:

(i) I recall for you the remark of Betrand Russell (1928), to be applied as you will, viz: "The opinions that are held with passion

V. Szebehely and B. D. Tapley (eds.), Long-Time Predictions in Dynamics, 295–305. *All Rights Reserved.*
Copyright ©1976 *by D. Reidel Publishing Company, Dordrecht-Holland.*

are always those for which no good ground exists; indeed the passion is a measure of the holder's lack of rational conviction."

(ii) There is controversy as to whether the theories of Velikovsky conflict with the principles of celestial mechanics, or whether they do not. However, Velikovsky explicitly made no claim to have deduced his results from celestial mechanics. I do for mine. Whatever may have been the approximations, the simplifications and indeed the errors which I have made, such plausibility as may adhere to my speculations does so because they were derived from mechanical principles.

I have deliberately left the subject to "simmer" for a while, so that the issues involved may be more clearly seen. It is my purpose to survey, in this paper, the present status (as I see it) of my general dynamical argument. A subsequent paper to the Institute will deal with the plausibility of the missing planet Aztex.

The subject of this Institute is "Long-time Prediction in Dynamics". We have discussed analytical methods of solution and numerical methods of integration - and their limitations. We have not discussed hitherto the possibility of using analogue computers. For the N-body gravitational problem, I would like to propose the Solar System as an analogue computer.

It has the advantage of having been "in operation" for several thousand million years - a period of time far longer than the known period of validity of other methods of analysis. It has the grave disadvantage that we do not know the "initial conditions". Nevertheless, if there are any general long-term features of an N-body planetary gravitational system, the Solar System should exhibit them.

There are two important questions:

(i) Is the Solar System "rigorous" (i.e. to what extent can the present properties of the system be attributed to the long-time action of point-mass gravitational forces)?

(ii) Is the Solar System "interesting" (i.e. are there any regularities in the structure of the system that are not to be expected by "chance")?

Regarding (i), we may envisage the following possible departures from rigour:

(a) The Solar System is not isolated. Have the nearby stars, the general galactic gravitational field or passing interstellar clouds had any effect on the structure of the system?

(b) The planets and satellites are not rigid spheres or point masses. What are the long-term effects of forces such as tidal forces?

(c) Cosmogonic processes; to what extent is the present structure of the system due to processes which operated during the formation of the system, but which are no longer operating, and therefore are not accessible to direct observational analysis?

There is a prevalent tendency to seek explanation of the

manifest regularities of the solar system in such cosmogonic processes. There is no apriori reason why one of the many scenarios that have been proposed for the origin of the solar system should not be correct. But we have no direct access to the past. Such explanations must therefore proceed by deduction from hypotheses to observations. Of necessity, all such explanations have so strong an ad hoc character that it is not clear just how far they do in fact constitute genuine explanations. In these circumstances, it seems to me to be only prudent to proceed from the least ad hoc hypothesis - namely that the present state of the solar system is primarily a result of the long-time operation of the simple laws of gravitation. Only when or if this hypothesis is found wanting should one proceed to the next stage - the effect of tidal forces, etc.

The following regularities in the planetary system are recognised:

(a) Near-coplanarity
(b) Nearly circular orbits
(c) Common sense of revolution
(d) Frequent near-commensurabilities of mean motion
(e) Quasi-logarithmic distribution of major semi-axes
(f) High eccentricities for the nearest and farthest planets (and for Mars, adjacent to which a "gap" exists in the distance distribution).

Generally speaking, apart from the sense of revolution of the retrograde satellites, the same regularities apply to the satellite systems of Jupiter, Saturn and Uranus. They do not apply to the asteroid belt, suggesting a significantly different dynamical evolution for the latter.

I do not discuss in this paper the extent to which regularities (a), (b) and (c) might be due to mutual perturbations subsequent to the formation of the planets. I contend that (d), (e) and (f) may be subsumed under the single statement that the bodies in the solar system are found as far away from each other as possible, and that this condition is precisely to be expected under the effects of mutual point-mass perturbations.

The main part of this paper is concerned with the problem of the time-scale required for such dynamical evolution to take place (including the possibility that the required time is infinite).

2. PRINCIPLE OF LEAST INTERACTION - SUMMARY

My interest in this subject began many years ago, when A. E. Roy and I showed that the number of near-commensurable pairs of mean motions in the planetary and satellite systems was in significant excess over the number that reasonably could be expected by chance (Roy and Ovenden 1954). In a subsequent paper (Roy and Ovenden 1955) it was argued that the reason for such a preference for near-commensurability was that, if in its dynamical evolution a system of satellites (or planets) came close to a configuration corresponding to a short-term periodic solution, then the system would linger near this configuration, simply because the mutual perturbations of the satellites would be nearly short-term periodic. Of course, the preferred configuration of the actual system could not (without the action of dissipative forces) be exactly that of a short-term periodic solution, since the solution would be periodic in the past as well as in the future. While tidal forces may, in practise, be important in locking-on to a periodic solution, it is of interest to examine the problem first without invoking tidal dissipation.

The condition for such short-term periodicity, for two satellites with mean motions n_1, n_2 is

$$A_1 n_1 + A_2 n_2 + A_3 \dot{\tilde{\omega}} = 0$$

where the A_i's are integers such that $\sum_i A_i = 0$, and where $\dot{\tilde{\omega}}$ is the angular velocity of either the line of nodes or the apse line, according to the character of the resonance associated with the periodic orbit. These arguments have been amplified by Dermott (1973).

Subsequently, numerical investigations by Hills (1970) and Ovenden (1975) confirmed that a preference for near-commensurable configurations was exhibited by simple point-mass satellite systems, acting under their mutual gravitation alone.

I have given the background to the extension of these ideas to systems of more than two satellites (Ovenden 1975). Briefly, the argument is that, in a similar way, a system of satellites will linger close to a configuration where the satellites are as far from each other as possible, compared with the other configurations making up the evolutionary sequence of the system. The criterion of "as far away as possible" was taken to be "that the time-mean of the action associated with the mutual interaction of the satellites is a minimum".

Becuase it is still not practicable to integrate accurately N-body simulations for times comparable with the age of the solar system, the isolation of the least interaction action configuration for a given system could be only approximate. Nevertheless, the existing satellite systems seem to be in good agreement with the calculations (Ovenden, Feagin and Graf 1974; Ovenden 1973).

Applied to the planetary system, the principle of least interaction action that there had been a massive planet between Mars and Jupiter which disappeared a time ago which is short compared with the time required for the system to reach a new "minimum" configuration.

It should be noted (Ovenden 1975) that the principle of least interaction action is consistent with the two-satellite arguments, since a resonant situation in this case does indeed correspond to minimum interaction action.[1]

The principle also makes sense of the fact that the orbital eccentricities of Mercury and Pluto are large. In a system of satellites or planets, the mutual interaction is least (in general) if a planet whose orbit lies between two others has a small eccentricity, while extreme satellites may have large eccentricities, taking up the "slack" in the angular momentum integral of the system. That Mars also has a large orbital eccentricity is to be explained by the gap between Mars and Jupiter. If this gap had once contained a sizeable planet, then the time-scale for evolution of a large eccentricity in Mars would be short compared with the dynamical time-scale for variation of major semi-axes.

An interesting precursor of the Principle of Least Interaction Action (of which I was unaware at the time of writing my earlier papers) was given by E. W. Brown. In one of his rare excursions into generalizations, in his Presidential Address to the A.A.S. in 1931 (Brown 1932), he said.

"The general effect of the mutual gravitational forces appears to be, not a recurrent motion, but a development into types of configurations which are least affected by small internal or external forces. If this is the case, it follows also that we can predict the future course of the system for a given interval more accurately than we can learn of the past during the same length of interval."[2]

The principle is related to the so-called "exclusion principle of stellar dynamics" discussed by Miller (1972).

It should be noted that the minimum configuration for a set of planets depends upon the masses of the planets as well as on their initial positions and velocities. For, example, on the approximation of circular orbits used in the analysis of Ovenden, Feagin and Graf (1974), a set of equal masses starting with a linear distribution of major axes has this linear distribution as the minimum one. It may be considered a limitation of the proposed explanation of "Bode-type laws" that the critical distribution of major semi-axes depends upon the distribution of planetary masses. In this connection it is worth noting that, with three satellites of different masses, where the system is allowed to evolve with interchange of order of the satellites, the deepest interaction action minimum is given by the sequence in which the most massive satellite lies between the other two. Thus the general progression of

---

[1]More strictly, it corresponds to stationary interaction action, since (unstable) periodic solutions also exist corresponding to maximum interaction action, in the two-satellite case.

[2]I am indebted to Professor R. W. Bass for drawing my attention to this remark by E. W. Brown. Professor Bass had independently formulated (in an unpublished paper) his own Principle of Least Mean Absolute Potential Energy which has a strong resemblence to the Principle of Least Interaction Action.

planetary masses in the solar system may also be explicable on the present hypothesis.

3. THE PROBLEM OF TIME-SCALES

The Achilles Heel of the whole hypothesis may be the problem of time-scale. The general integrals of motion enable one to gain some general idea of the patterns through which a system might evolve. No such general argument (as far as I am aware) has given reliable estimates of the rate of evolution through the sequence of configurations. For, while the arguments summarized above might be logically correct, they might be rendered nugatory if the time-scale of the actual systems were much greater than the age of the solar system (including the limiting case of infinite time-scale, when no such evolution actually takes place).

Now there is no necessity, as far as these arguments are concerned, for the solar system to have strictly secular terms in any algebraic analysis. All that is necessary is that there should be sufficiently large variations of the major semi-axes of the satellite and planetary orbits on a sufficiently short time-scale. Whether these variations are, in a mathematical sense, secular or not is immaterial, and may in fact depend solely upon the mathematical techniques adopted in the analysis.

However, in an attempt to cut this Gordian knot, and on the strength of the fact that formally-secular terms appeared in the expansion of the disturbing function in the third order of relative mass of the disturbing planet, a supposed order-of magnitude estimate of evolutionary time-scales was made (see Ovenden, Feagin and Graf 1974). In estimating the rate of passage of a system through its evolutionary sequence, it was assumed that the energy available to power secular terms was proportional to $(\bar{R} - \bar{R}_0)$, where $\bar{R}$ was the time-mean interaction action (on the circular approximation) at the configuration of the epoch, and $\bar{R}_0$, that for the minimum interaction configuration. In making this assumption, it was equivalently supposed that the evolution of the system had a high "effective eccentricity", so that the asymptotic approach to minimum configuration was a reasonable approximation. That the actual approach cannot (except for a class of zero measure) be strictly asymptotic, follows from the theorems of E. Hopf (1930). The assumption was implicitly made that a significant fraction of the energy represented by the m^3 terms went into changes of the major semi-axes. The formula so deduced was in agreement with that deduced by Hills (1970) from a quite different argument. More importantly, it gave evolutionary times for both the Jovian and Uranian satellite systems $\sim 10^9$ years, in good agreement with the accepted age of the solar system.

Nevertheless I have concluded that the derivation of these time-scales was spurious (which does not, of course, necessarily imply that the time-scales are wrong). The derivation assumed the significance of secular terms in the third power of the relative mass of the disturbing planet. The existence of these terms was first proved by Heratu (1885); they were evaluated by Eginitis (1889) and, more completely, by Meffroy (1955). Meffroy gives the following expression for the 'secular' terms, for the perturbation of the unprimed planet by the primed planet:

$$\delta^3 a = \frac{3^3 m'^3 tae'^3}{2^5 n^3 a'^9} \left\{ \frac{1}{(2n-n')^2} + \frac{3^2}{(n-2n')^2} - \frac{3.5}{2^2(n-n')^2} \right\} \cdot$$
$$\cdot \sin 2(\tilde{\omega}-\tilde{\omega}') .$$

Except close to a situation where $n'=n$, $n'=2n$ or $n'=(1/2)n$, this gives a characteristic evolution-time for the major planets $\sim 10^{12}$ years.

However, problems arise in the adoption of the time-scales given by the Meffroy terms. (i) The whole question of the validity of the series development of the disturbing function over such a great length of time is involved, and is unsolved. Thus Brown (1932) says, "The validity of the proofs of many of the earlier theorems which have been quoted, and are still quoted, has disappeared with the increase of the estimate of the time-scale to some twenty times its former value. As long as the interval required for the attainment of the present configuration of the solar system was thought to be less than 10^8 years, there was some hope that an approximate idea of initial conditions could be obtained from the present configuration. With intervals of at least 2×10^9 years, most of these deductions appear to have lost all meaning." (ii) It is not obvious that the so-called 'secular' term given by Meffroy actually leads to 'secular' behaviour, since the size of $\delta^3 a$ depends upon the value of $(\tilde{\omega} - \tilde{\omega}')$, and the longitudes of perihelion have cyclic terms with characteristic periods much less than the characteristic time-scale of the coefficient of the term itself. Furthermore, as given by Meffroy, the term does not depend upon the longitudes of the interacting planets, but only on the orientation of their orbits. Yet the behaviour of the system depends very much on whether the longitudes of the planets are such as to give frequent mirror configurations, or not. While the Meffroy term goes to zero if $\tilde{\omega}=\tilde{\omega}' + \pi$, resonance and periodic behavior requires the correct longitude relationship as well. (iii) The time-scales of evolution of some numerically-integrated systems are shorter than the time-scales calculated from either Meffroy's term or the formula of Ovenden, Feagin and Graf. For example, the system shown in Figure 3 of Ovenden, Feagin and Graf (1974), middle row, extreme left, shows evolution of the major semi-axis of satellite No. 3 on a time-scale $\sim$ 500 periods. This is at least one order of magnitude shorter than that given by the formula of that paper, and several orders of magnitude shorter than the characteristic time-scale of the Meffroy terms (depending upon the value taken for the mean eccentricities).

It was proved by Poisson that, up to terms of the second power in the relative mass of the disturbing planet, there were no secular terms in the variation of the major semi-axes. Recently, Message (1975) has used the method of asymptotic series to extend Poisson's result to all powers of the disturbing mass. This is an important generalization. However, if the major semi-axes of the two planets are sufficiently similar, the method of analysis breaks down, as numerical integrations clearly show. The range of validity of the asymptotic series used remains undetermined.

It might seem that the only reliable method of estimating the time-scale of evolution under gravitational forces alone is to accurately integrate numerically. Since the original work on the Principle of Least Interaction Action, Cohen, Hubbard and Oesterwinter (1973a) have published their monumental integration of the motion of the outer planets for one million years. In a later paper, Cohen, Hubbard and Oesterwinter (1973b) published planetary elements for an interval in 10^7 years. However, these were obtained, not from numerical integration, but from the 'secular theory' of Brouwer and van Woerkom. Since this theory does not develop the disturbing function far enough to include 'secular terms', it is not clear how reliable the elements calculated from the theory would be for periods $\sim 10^7$ years.

Reference to Figures 8 and 9 of Cohen, Hubbard and Oesterwinter (1973a) show that no secular trends in the major semi-axes are clearly discernable. Of course, in the presence of significant oscillations, the exact significance of 'mean major semi-axis' is itself not clear. Furthermore, since the hypothesis being tested is that a planetary system spends <u>most of its time close</u> to the configuration of least interaction action, and since the age of the solar system $\sim 10^9$ years, the characteristic time-scale for evolution <u>at the present time</u> might be expected to be very long, say $\sim 10^{12}$ years. Integrations over 10^6 years might be expected, under those circumstances, to show any detectable secular trend.

However, the indication of my previous analysis was that the planetary system is, at the present time, not exactly at its minimum configuration. It was this fact that led to the postulation of Aztex. It is clear from the published plots by Cohen, Hubbard and Oesterwinter that variations of the mean major semi-axes in their integrations are significantly smaller than those calculated by me (Ovenden 1973). We must thus conclude either (i) the techniques of isolating configurations of minimum interaction action are in error in this case, and that the arguments for the former existence of Aztex are spurious, or (ii) that the actual time-scale of variation in the neighbourhood of minimum configuration is much longer than that previously calculated, meaning that Aztex disappeared longer than $\sim 10^7$ years ago.

4. ARNOLD'S THEOREM

It is clear that, at the present time, numerical integrations by themselves have not solved the question of whether or not, over the time-scale of the age of the solar system, significant changes can take place in the major semi-axes of the satellite and planetary systems as a result of simple point-mass perturbations. Some additional guidance is required from theory.

As discussed above, systems with relatively large perturbations may exhibit rapid evolution, as shown by numerical integrations. In the case of Birn (1973), drastic evolution in the Jupiter-Saturn-Uranus-Neptune system was obtained if the perturbations were magnified by starting the system far from minimum interaction. Significant changes in the major semi-axes (including a 'switch' of Uranus and Neptune) occurred in times as short as $\sim 10^3$ years. Indeed, in this time the system has already reached a quasi-steady state far from the original configuration. Nacozy (1975) has taken the system of Jupiter and Saturn, and varied the masses of the planets (keeping their ratio constant). He showed that (over relatively short integration times) the major semi-axes of the planets were bounded about the zero-mass values, until the masses reached about 29 times their actual values, at which point the character of the motions changed, and rapid evolution took place.

This is the crux of the matter. A system of strictly-circular, strictly-coplaner orbits has the property that every conjunction and every opposition is a 'mirror configuration' (Roy and Ovenden 1955). Therefore the mutual planetary perturbations are strictly short-term periodic. However, the mutual perturbation of finite planets destroys the strict circularity. As the perturbation is increased, is there a qualitative change in behaviour for some critical value of the perturbation, or does the time-scale for significant evolution of an arbitrary system simply get smaller monotonically?

Arnold (1962) made an important advance in perturbation theory. His theorem is sometimes stated to be that a solar system consisting of a Sun and a set of planets of sufficiently-small relative masses, moving in orbits sufficiently close to being circular and coplaner, will remain in that condition for all time. However, this is not a precise statement of Arnold's theorem.

Consider a system disturbed from integrable motion by a small perturbation, the Hamiltonian being given by

$$H(p,q) = H_0(p) + \varepsilon H_1(p,q) + \ldots .$$

where ε is small and H_1 has period 2π in q. In 1954, Kolmogorov proved (see Arnold 1963a) "that if

$$\det \left| \frac{\partial^2 H_0}{\partial p_i \partial p_j} \right| \neq 0$$

then under a small analytic perturbation, the majority of the invariant tori do not collapse but are only slightly deformed. These tori form a nowhere dense, closed set whose complement has measure small in ε. Apart from this, little is known about the behaviour of the trajectories. "For [number of degrees of freedom] $n>2$, the invariant n-dimensional tori do not separate the (2n-1)-dimensional energy levels H= constant, and trajectories from the gaps can apriori go to infinity."

Of his application of this theorem to planetary motions, Arnold (1963b) writes that the question, "'do small perturbations of planets on each other give rise after a sufficiently long time to collisions or departures to infinity?', remains unsolved even today. It can be supposed that topological instability is a typical case: certain trajectories beginning in the gaps between invariant tori can go a long way, since all the gaps merge into a connected set extending to infinity. The topological instability of planetary motions would, in particular, follow from the validity of this hypothesis. The results of Chapter III do not exclude the possibility that an arbitrarily small change in the initial conditions can completely change the character of the motion for an infinite time. In Chapter III we proved only that such a change in the initial conditions must be of a highly special form: the majority of changes in the initial conditions leave the motion conditionally periodic." A discussion of topological instability is given by Arnold and Avez (1968).

If the solar system had started with initial conditions such that extensive evolution were possible, then it would tend to be found close to its present resonant structure, although, among situations with neighbouring orbital and mass parameters, it would be a member of the 'exceptional' set. This is a matter that requires more topological, analytical and numerical investigation.

5. SUMMARY

The resonant structures, the regular spacing of the major semi-axes, and the distribution of orbital eccentricities found in the solar system are to be expected for any planetary (or satellite) system capable of evolution under point-mass gravitational forces.

Whether the actual solar system, in its present state, is capable of such evolution, is still an open question.

The time-scales calculated in previous papers are spurious, since the so-called 'secular' terms of Meffroy are not, in fact, true secular terms.

ACKNOWLEDGEMENTS

I am indebted to Dr. John Byl and Mr. Jay Shelton for many

useful discussions on the subject-matter of this paper. This work was supported in part by a grant from the National Research Council of Canada.

REFERENCES

Arnold, V. I., 1962: Dokl. Akad. Nauk. 145, 487. (Sov.Math. Dokl. No. 3, 1008).
Arnold, V. I., 1963a: Usp. Math. Nauk. 18, 9. (Russ. Math. Surv. 18, No. 5).
Arnold, V. I., 1963b: Usp. Math. Nauk. 18, 85. (Russ. Math. Surv. 18, No. 6).
Arnold, V. I. and A. Avez, 1968: Ergodic Problems of Classical Mechanics, Benjamin, New York.
Birn, J., 1973: Astron. & Astrophys. 24, 283.
Brouwer, D. and A. J. J. van Woerkom, 1950: Ast. Pap. Am. Eph. XIII, pt. II.
Brown, E. W., 1932: Pub. Ast. Soc. Pac. 44, 21.
Cohen, C. J., E. C. Hubbard, and C. Oesterwinter, 1973a: Ast. Pap. Am. Eph. XXII, pt. I.
Cohen, C. J., E. C. Hubbard, and C. Oesterwinter, 1973b: Celes. Mech. 7, 438.
Dermott, S. F., 1968: Mon. Not. R. Astr. Soc. 141, 349.
Dermott, S. F., 1973: Nature Physical Sci. 244, 18.
Eginitis, M. D., 1889: Ann. Obs. de Paris, Mémoires 19, I.
Heretu, S., 1885: Ann. Obs. de Paris, Mémoires 18, I.
Hills, J. G., 1970: Nature 225, 840.
Hopf, E., 1930: Math. Ann. 103, 710.
Meffroy, J., 1955: Bull. Astr. xix, 1.
Message, P. J., 1975: (Communication to NATO Institute)
Miller, R. H., 1972: Astrophys. J. 172, 685.
Nacozy, P., 1975: (Communication to NATO Institute)
Nieto, N. M. and R. W. Bass, 1974: Pensée 4, No. 3.
Ovenden, M. W., 1973: Recent Advances in Dynamical Astronomy (eds. B. Tapley and V. Szebehely) (Reidel) 319.
Ovenden, M. W., 1975: Vistas in Astronomy 18, 473.
Ovenden, M. W., T. Feagin, and O. Graf, 1974: Celes. Mech. 8, 445.
Roy, A. E. and M. W. Ovenden, 1954: Mon. Not. R. Astr. Soc. 114, 232.
Roy, A. E. and M. W. Ovenden, 1955: Mon. Not. R. Astr. Soc. 115, 296.
Russell, B., 1928: Sceptical Essays.

ROTATIONAL MOTIONS OF THE TERRESTRIAL PLANETS

R. O. Vicente

Department of Applied Mathematics, Faculty of Sciences, Lisbon, Portugal.

We are interested in the study of the motion of the planets Mercury, Venus, Earth, and Mars around their centers of mass. These rotational motions depend on the influence of the disturbing bodies and on the internal structure of these planets. Our knowledge about the bodies that disturb the motions of these planets is good, but the knowledge about the internal structure varies depending on the planet considered. Mercury and Venus are the ones for which we have less knowledge about their internal composition. We are also interested in the long time prediction of these motions. As you can already see, this subject implies the knowledge not only of the equations of dynamics, but also of the equations which relate to the internal structure of the planets. In this case, we cannot consider the bodies as point masses, but we have to deal with their finite dimensions.

We can arrange these planets in different groups depending on the property we are interested in examining. The existence of an atmosphere is important for long-time predictions of these motions, and so we have a group formed by Venus, Earth, and Mars, in the order of decreasing importance of their atmospheres, and another group without atmospheres (Mercury and the Moon).

We can consider the planets with satellites (Earth, Mars) and without satellites (Mercury, Venus) that might affect their motions

V. Szebehely and B. D. Tapley (eds.), Long-Time Predictions in Dynamics, 307–320. *All Rights Reserved.*
Copyright ©1976 *by D. Reidel Publishing Company, Dordrecht-Holland.*

around the center of mass.

Another important aspect of their structures is the possible existence of a core, especially if it can be considered as liquid in the sense that the secondary seismic waves do not propagate through such a core. We know that the Earth has a liquid core, but we do not yet have such detailed knowledge about the internal structure of the other planets under consideration.

There are certain common properties to all these planets, such as the elasticity of their internal structures and the figure of revolution of their surfaces, corresponding to an ellipsoid. All of them also rotate around an axis, denoted the polar axis in the case of the Earth; but their periods of rotation are quite different.

Another important aspect for the long time prediction of the motions aroung the center of mass is the consideration of the rheology of the planet. One of the basic problems in the study of the Earth is to understand its deformations, but the deformations can occur in two different ways: continuously or discontinuously. A continuous deformation corresponds to a condition where neighboring points in a material always remain neighboring points, while in discontinuous deformations that does not happen.

Some branches of the theory of continuous deformation, like the theory of elasticity and the hydrodynamics of viscous fluids, have been intensively developed; but the materials that constitute the terrestrial planets are not likely to fit any of these theories. The materials that constitute the interior of the Earth, for instance, show a very complicated behavior possibly of the types described by the theories of rheology.

It is possible to establish a connection between some kinematical quantities (like the strain $\underline{e}_{ik}$) and dynamical quantities (like the stress $\underline{p}_{\alpha\beta}$). Such relations are sometimes called the rheological conditions. There are many types of rheological equations relating the stress $\underline{p}_{\alpha\beta}$, the strains $\underline{e}_{ik}$, the temperature T, and such physical constants as may belong to the medium. The behavior of this rheological function for long time intervals is very important for the motions aroung the center of mass of the terrestrial planets.

The values of the masses and the radii of these planets are the following:

Table 1. System of Planetary Masses (includes satellites)

	Planet Reciprocal Mass		Equatorial Radius (km)
Mercury	6,023,600.0		2434
Venus	408,523.5		6052
		Earth	6378.140
Earth and Moon	328,900.5		
		Moon	1736
Mars	3,098,710.0		3402

The radii refer to the planet's crust. A value for Venus including the height of the cloud layer is 6110 km. The ratio of the mass of the Moon to that of the Earth is $\mu = 0.0123000$.

These are the values recommended, at the moment, by the I.A.U. Working Group on Ephermerides. These values will be submitted to the I.A.U. General Assembly next year.

Table 1 shows that, besides the special case of the Moon, there are two planets (Mercury and Mars) smaller than the other two (Venus and Earth) but three of them (Venus, Earth, Mars) have atmospheres.

The Moon, being our natural satellite, has an appreciable mass for a satellite, and we can see that by comparing its mass (supposed equal to unit m =1) to the mass of Mercury (4.50) and Mars (8.7). This is important for the study of the motion of the Earth because we have to consider as a disturbing body not only the Sun, but also the Moon.

Let us consider a rectangular system of coordinates X_i fixed in space, and another system of coordinates x_i, rotating with angular velocity ω_i relative to the system X_i, and coinciding with it for the moment. Then the equations of motion of the body moving around the center of mass can be written in the following way

$$\frac{dH_i}{dt} + \varepsilon_{ijk}\omega_j H_k = L_i \quad (i=1,2,3)$$

where L_i are the components of the moments of the effective forces about the moving axes; H_i, the components of the angular momentum, and ε_{ijk} the alternating tensor. We are considering in this expression the summation convention, that is, any repeated suffix is given all possible values and the results are then added.

We can separate the angular momentum in the following manner:

$$H_i = C_{ij}(t)\ \omega_j + h_i(t) \tag{1}$$

where $C_{ij}(t)$ is the tensor of inertia and $h_i(t)$ is a relative angular momentum due to the motion of particles $\underline{u}_i$ relative to the coordinate system $\underline{x}_i$.

These equations are not restricted to a rigid body and can be employed when the system is a collection of particles moving among themselves. They therefore apply to a body which is changing its shape, for instance, due to elasticity or the transference of heat.

Sometimes we know the mode in which the body is changing its shape; and if we can find the motion in space of any set of rectangular axes connected in a known manner with the changing body, then the motion of every part of the body is known. These types of axes are called, by some authors, axes of the body.

There are several ways of choosing the system of axes that have the advantage of simplifying the equations of motion As an example we can examine the special case of the Earth considered as a rigid body in a first approximation. We take a set of rectangular axes OX_i fixed in space, so that OX_3 is directed towards the north pole of the ecliptic, and another set of axes Ox_i, referring the motion to the principal axes of inertia, and, therefore, moving with it.

The relative position of the two sets of rectangular axes can be defined by the three Eulerian angles θ, ψ, ϕ, and Euler's kinematical equations are

$$\begin{aligned}
\omega_1 &= -\dot{\theta}\cos\phi - \dot{\psi}\sin\theta\sin\phi \\
\omega_2 &= \dot{\theta}\sin\phi - \dot{\psi}\sin\theta\cos\phi \\
\omega_3 &= \dot{\psi}\cos\theta + \phi \quad ,
\end{aligned} \tag{2}$$

where $\underline{\omega}_i$ are the angular velocities in relation to the system of axes x_i.

The dynamical equations of motion can be written in the following way:

$$\begin{aligned}
A\dot{\omega}_1 + (C-B)\omega_2\omega_3 &= L \\
B\dot{\omega}_2 + (A-C)\omega_3\omega_1 &= M \\
C\dot{\omega}_3 + (B-A)\omega_1\omega_2 &= N
\end{aligned} \tag{3}$$

where (A,B,C) are the principal moments of inertia of (L,M,N) are the moments of the external forces about the axis x_i. These equations are sometimes called the Eulerian equations of motion.

These equations can be considered as the projections, on the set of axes Ox_i, of the vector equation

$$\frac{d\vec{H}}{dt} = \vec{G} \tag{4}$$

which shows that the rate of change of the vector angular momentum $\vec{H}$, about the center of mass, is equal to the vector moment $\vec{G}$ of the resultant couple.

The motion of the Earth, relative to its center of mass, is represented at every instant, by the instantaneous rotation $\vec{\omega}$ around an axis through the center of mass. This is a very elegant way to describe the motion of the Earth, within the conditions stated, because the moments of inertia are the only parameters that have a dynamical meaning, and so it is not necessary to consider the geometrical form of the Earth.

The external forces are due to the Sun and Moon, considered as rigid bodies, acting on the solid body of the Earth. As is well known, the theories of the motion of the Sun and Moon present some inaccuracies which limit the final precision obtained for the theory of the motion of the Earth around its center of mass.

This system of reference has the advantage that the products of inertia are zero, and we can consider that the principal axes of inertia of the mantle coincide with those of the core. We can assume, therefore, the core-mantle boundary to be at rest in this system of reference, thus facilitating consideration of the boundary conditions and integration of the equations of motion.

Even in this simplified theory we can see that there are several axes that can be adopted for defining different reference systems: (1) the axis Ox_3, called the axis of figure, about which the moment of inertia is a maximum; (2) the fixed axis OX_3 directed towards the north pole of the ecliptic; (3) the instantaneous axis of rotation, or simply, the axis of rotation, along which is the instantaneous rotation $\vec{\omega}$ that represents the motion of the Earth at every instant and intersects the surface of the Earth at two points called the geographical poles. (This is the name which has been given to this axis in celestial mechanics for a long time, but currently it is referred to as the spin axis,as we also are getting used to the new name of astrodynamics for our branch of science); (4) the axis along the vector angular momentum $\vec{H}$, through the center of mass, called the invariable axis in the dynamics of rotating rigid bodies. As the axis of rotation, the axis of angular momentum, and the axis of figure all pass through the center of mass of the Earth, it is easy to show that they lie in the same plane.

The problems associated with the adoption of a convenient system of reference for the Earth have an important role for studies in geodesy (including satellite geodesy), positional astronomy, and geophysics because new techniques are increasing the precision of observations for the determination of distances on the Earth. We can compare the precision of the classical methods of astronomy for the determination of the coordinates of a point with accuracy within about 100 cm. to the precision of the new astronomical techniques with accuracy in the order of 10 cm. and, some, in the order of a few centimeters. This increase in precision of two orders of magnitude obliges us to think very carefully about the systems of reference to be adopted in studying the Earth's dynamics and the possibilities of changing from one system to another.

Another important aspect of the problem of adopting adequate systems of reference refers to the long-time prediction of the movements of the Earth around its center of mass for example, to determine what happened a hundred years ago, a million years ago or even further in the past.

We have to remember that we can consider the Earth as a rigid body, as an elastic body or as still more complicated models that are nearer to the knowledge we have currently about the internal structure of our planet.

There is agreement that we have to adopt systems of rectangular coordinates for the study of the motion of the Earth and, following the tradition of celestial mechanics, the origin of any system of coordinates will be the center of mass of the system, considered as a fixed point. I should like to point out that a few geodesists would prefer a system of reference considering some points on the Earth's crust as fixed and the center of mass as not fixed, however that idea was not accepted at the IAU Collogium held last year in Poland which included in its final recommendations, the adoption of the center of mass as the origin for any system of reference. The importance of the problem of choosing an adequate system of reference resides in the fact that the great majority of our observations are done at the surface of the Earth, and I think it is very important that we get an agreement on the best possible system that can be used by any scientist interested in these problems. Only in this way can we easily compare the observations done at several places and by different techniques.

Let us now consider a more complicated case, nearer to the structure of our planet, composed of a lithosphere, a hydrosphere and an atmosphere. Remembering the system of equations (1), we can consider the following systems of reference:

(a) The mean axes of the Earth defined in such a way that $h_i = 0$. This system of reference was employed by Tisserand and corresponds to the case where we make the hypothesis that all the relative motions do not exist, for instance, the general circulation of the atmosphere or of the oceanic currents. This system of coordinates has the advantage of simplifying the equations of motion but has the inconvenience of ignoring any relative motions that produce changes of these axes in relation to the observing stations.

(b) The principal axes of inertia, or axes of figure, corresponding to the case where the products of inertia $C_{ij} = 0$ $(i \neq j)$. This system of coordinates, as we have seen before, has the advantage of simplifying the expressions of the equations of motion, but it has the inconvenience of changing its position in relation to the observatories because of the relative motions.

(c) A system of axes, sometimes called geographic, linked to some observatories in a well defined manner. Certainly these observatories should be located at points that will not have any known geophysical motions. The disadvantage of this system lies in precisely determining fixed points at the Earth's surface when we now know a great deal about local movements of the stations, such as, polar motion, earth tides, and seismological and geological displacements.

The present day knowledge about the motions of the Earth's crust makes it difficult to envision a system of coordinates that does not suffer from any disadvantage, especially if we consider precisions of the order of a few centimeters.

We have been speaking only about the definition of the motion of the system of axes, but we have not chosen the position of the axes other than to define the origin as the center of mass of the Earth and the Ox_3 axis as coinciding with one of the axes we mentioned before. For the Ox_1 axis we can consider, for instance, the Greenwich meridian, and for the Ox_2 axis, the meridian 90° east of Greenwich. But this positioning of the axes has not been adopted by all scientists interested in these problems.

This difficulty does not yet appear for the other terrestrial planets because of the lack of knowledge about their structures.

Let us now go back to the system of equations (3), valid within the simplified hypotheses adopted, to define the main motions of the Earth around its center of mass. The actual motion of the Earth can be considered as a steady state of rotation perturbed by the external forces. We can consider the motion in this way because the effects of the external forces (due to the Sun and the Moon) are comparatively small.

Assuming that the external forces do not exist, we have $\vec{G}=0$, and equation (4) shows immediately that

$$\frac{d\vec{H}}{dt} = 0 \qquad \vec{H} = \text{const.} \quad ,$$

that is, the angular momentum is constant, and the vector $\vec{H}$ is fixed in space. This is an important result for the free motion of the Earth, called this because we neglected the existence of external forces. The result obtained for the Earth can be applied to any of the other terrestrial planets because the dynamical quantities appearing in this equation are valid for them, that is, the vector $\vec{H}$ is fixed in space.

We now introduce a simplifying hypothesis into these equations, corresponding to the case when the principal moments of inertia A and B are equal. The third of equations (3) shows immediately that

$$\omega_3 = \text{const.} = n$$

and means that assuming these simplifications, the rotation of the Earth around this axis is constant. This corresponds to the daily rotation of the Earth, and we can judge the precision of this simplified theory from the fact that only in the last decades has it been possible to detect irregularities in the daily rotation of the Earth. This hypothesis can also be applied to Mercury, Venus, and Mars, showing that their rotations around the Ox_3 axis can be considered as constant.

The irregularities in the daily rotation of the Earth are caused by the existence of motions in the atmosphere, in the hydrosphere, and inside the Earth, but we do not yet know with reliable accuracy the proportion contributed by each of these motions. As Mercury has no atmosphere or hydrosphere, its rotational period will not be affected, and the same applies to Mars because the atmosphere is not very appreciable. The case of Venus is different, and we can forecast that the rotational period will be much affected by the existence of such a dense atmosphere. If it is valid to extrapolate the results obtained for the Earth, some of these disturbances will be periodic depending on the periods of the global circulation of the Venusian atmosphere.

There is an interesting fact about Venus, referring to the perturbation of the Earth, leading to a synodic resonance period of 243.16 days. However,observations show a slightly different sidereal rotation period of 243.05± 0.05 days. The possible explanation for this disagreement resides in the fact that the theory does not take into account the existence of a dense atmosphere and the probable influence of the internal structure of the planet. Looking at the values in this way, we can consider that the perturbation of the Earth affects the motion of Venus, but the composition of the planet alters the theoretical value. If this hypothesis is valid, it will be the first time that we have evidence about the influence of the structure of Venus on some dynamical quantities.

It is interesting to notice that the Earth and Mars have fast periods of rotation while Mercury and Venus have longer periods of rotation. We can see that in the more rapidly rotating planets, we have one of the larger and one of the smaller planets; and the same happens with the more slowly rotating planets. We can note that the more slowly rotating planets have no known satellites and that Venus has the largest atmosphere of all the terrestrial planets.

These different characteristics of the terrestrial planets do not seem to explain why some planets have fast rotations and others have slower rotations, and, therefore, it is difficult to have the same explanation for the behavior of the terrestrial planets during long intervals of time. We know there is a secular retardation of the Earth's rotation caused by the existence of a hydrosphere and a natural satellite; but, in spite of that, the Earth's rotation is faster than that of Venus, where there exists a very dense atmosphere (that might make an important contribution for the slower period of rotation) but no known satellites.

On the other hand, Mercury and Mars, being the smallest planets, have quite different periods of rotation, and Mars rotates faster, having two known natural satellites of very small masses. If it is difficult to explain the rotation of Mars, it is still more difficult to understand the fast rotation of the Earth, a bigger planet with a liquid core, a hydrosphere and a sizeable natural satellite. We might attribute the slower rotation of Mercury to the proximity of the Sun, as occurs in the Moon-Earth system. Or does part of the explanation rest in the different rheological behavior or these planets for long intervals of time? We can also notice that the more rapidly rotating planets (Earth and Mars) are farther away from the Sun and, therefore, their rheological behavior will not be so disturbed by the Sun's mass.

In spite of the difficulties of the problem, I think that it is possible to forecast the future slowing down of the rotational motions of the planets that have a dense atmosphere, a hydrosphere or a liquid core. This applies, at least, to Venus and the Earth, but it is difficult to give any reliable values. It has been stated that the increase of the duration of the day is about 2 seconds per 10^5 years for the Earth. During the early history of the Earth there must have been an interesting period when the friction of the hydrosphere started acting in an appreciable way leading to an increase in the length of the day.

The two remaining equations of (3) can be written in the following way

$$\dot{\omega}_1 + \frac{C-A}{A} n\omega_2 = 0$$

$$\dot{\omega}_2 - \frac{C-A}{A} n\omega_1 = 0 \quad ,$$

and the solutions are

$$\omega_1 = \sigma n \cos(pt + \beta)$$

$$\omega_2 = \sigma n \sin(pt + \beta)$$

where σn, β are constants of integration, and $p = n\,\frac{C-A}{A}$. The period τ associated with this motion is

$$\tau = \frac{2\pi}{p} = \frac{2\pi}{n} \cdot \frac{A}{C-A}$$

As we can see, the period of this free motion depends on the values of n and of the moments of inertia $\underline{A}$, $\underline{C}$. We know that for all the terrestrial planets $A \neq C$ and, therefore, we can say that all of them will have a free motion.

For the case of the Earth, considering the sideral day as the unit of time, then $n = 2\pi$ and $\tau = \frac{A}{C-A}$. So far, it has been considered more accurate to determine this ratio from the period of the precession of the equinoxes and not from the values of the moments of inertia. The value obtained is 305 days, that is, about 10 months and this motion has been called the Free Eulerian Nutation or, simply, free nutation. One consequence of this motion is that the astronomical latitude of any point should exhibit a variation with a period of about 10 months.

The search for this period of the variation of latitude was initiated at the end of the last century but the calculations obtained from the astronomical observations revealed a value of about 14 months. This value is called the Chandler period. The disagreement between the theoretical value, computed from this simplified theory, and the observed value is due to the fact that we considered a simplified model for the Earth. If we consider a model which is nearer to the known structure of our planet, it is possible to explain this disagreement. We must take into consideration the elasticity and the existence of a liquid core in the Earth.

This elementary theory gives us the explanation for the existence of what is called "the polar motion", that is, the wandering of the geographic poles over the surface of the Earth. It is possible to compute the actual departure of the axis rotation from the axis of figure, and the order of magnitude of the angle between these two axes is about 0".3 or 10 meters at the surface of the Earth.

It is easy to verify for any of the terrestrial planets that, during this motion, the axis of figure, the axis of rotation and the axis of angular momentum lie in the same plane. Since the vector $\vec{H}$ is fixed in space, we can conclude that the plane, where these three axes exist, rotates in space around $\vec{H}$.

The subject of polar motion is important at the present time because we are interested in determining the distances and position of points at the Earth's surface with greater accuracy. Unfortunately, the actual behavior of the motion of the pole is very complicated because of the fact that the Earth is composed of an atmosphere, hydrosphere and lithosphere, and it is very difficult to forecast the influence of all these relative motions on the motion of the pole.

As we know that the terrestrial planets are not rigid bodies, we can say that the elasticity of these planets will alter the values computed for the free periods. The problem will be to know how much the elasticity will alter these values. In the case of the Earth it has been computed that the difference between a rigid and an elastic model will correspond to a lengthening of about 5 per cent in the value of the free period, that is, instead of 305 days we will have 319 days. If we make the hypothesis that the elasticity of the terrestrial planets is similar, we can say that the lengthening of the period will be of that order of magnitude.

In the case of the Earth, what does contribute to the lengthening of the period is the existence of a liquid core giving, therefore, values nearer to the 14 months period. The observations made so far have not revealed any liquid core for Mercury, Venus and Mars, and if future observations do not detect any liquid core of appreciable dimensions, we can forecast that there is not any greater lengthening for the free periods of these planets. The possible existence of a solid core of higher density does not affect our computations because any elastic model only increases slightly the values of the free periods.

Consider the existence of the external forces, due to the Sun and any other disturbing body (for instance, the Moon in the case of the Earth). We have now to study the system of equations (3), corresponding to equation (4) where $\vec{G} \neq 0$.

This motion is called the forced motion, and it has been studied by expressing the equations of motion in terms of the Euler angles (θ, obliquity, and ψ, longitude). As a first approximation, we can consider the motion represented by Poisson's equations

$$\frac{d\theta}{dt} - \frac{1}{C\ n \sin\theta}\frac{\partial U}{\partial\psi}$$

$$\frac{d\theta}{dt} \quad \frac{1}{C\ n \sin\theta}\frac{\partial U}{\partial\theta} \tag{5}$$

where U is the gravitational potential due to the external forces. The expression for U depends on the external forces that we consider for the different terrestrial planets, and the development of U can be expressed in several ways. For instance, it may be written in terms of the rectangular coordinates or the orbital elements of the disturbing bodies.

The integration of equations (5) be represented by

$$\theta = \theta_p + \Theta$$

$$\psi = \psi_p + \Psi$$

following the traditional separation of the terms into two types;

1) terms that increase continuously with time

   $at + bt^2 + ct^3 + \ldots$ (a,b,c..constants)
   corresponding to the precessions;

2) terms that are periodic functions of the time

   $a_1 \sin(\alpha_1 t+\beta_1) + a_2 \sin(\alpha_2 t+\beta_2) + \ldots$
   ($a_1, a_2 \ldots \alpha_1, \alpha_2 \ldots \beta_1, \beta_2 \ldots$ suitable constants) corresponding to the nutations.

In the case of the Earth, the disturbing bodies are the Sun and Moon, and, therefore, we have luni-solar precessions and nutations. The most recent theory has been developed by Kinoshita. For Mercury and Mars we have to consider only the Sun and solar precessions and nutations . In the case of Venus, if the Earth will also be considered as a disturbing body, we will have precessions and nutations caused by the Sun and Earth.

The computations of the solar annual precession for Mercury, Venus or Mars depends, for instance, on the knowledge of the oblateness, the eccentricity of the orbit and the sidereal year of the planet.

The rotational motions that we have described are an important factor for the structure of the terrestrial planets leading to a number of symmetrical features. We can see that in the layered structure and spheroidal form of theEEarth. The conclusion referring to the spheroidal form applies to the other terrestrial planets, but we do not have enough observations of the seismology of the other planets to draw conclusions about the layered structure of their interiors.

Mercury appears, like the Moon and Mars, to have an asymmetric surface half of it heavily cratered and the other half relatively smooth. Considering the tendency to have symmetrical features con itioned by the rotation of the planets, it seems difficult to explain these asymmetries unless we postulate that they were originated at a very early stage and with material already sufficiently solid to prohibit alteration by the rotational motion. This hypothesis would rule out the possibility of the planets being in a fluid state at those epochs. The rheology of the materials constituting the planets at that stage of their evolution might also contribute to these asymmetries.

The particular case of the Earth has been computed in greater detail because the precessions and nutations affect the position of our reference systems and, therefore, are important for any branch of astronomy that needs to compare observations made at different epochs.

The rigid body values computed by this theory do not agree with the observed values, because of the behavior of the Earth. Therefore, we have to consider more complicated models as we have said in the case of the free motion. It is interesting to notice that the precessions do not have any appreciable alteration when we consider the existence of the liquid core of the Earth.

We have to employ the equations of motion of an elastic body that can be expressed in rectangular coordinates by a system of three differential equations of the type

$$\rho \frac{\partial^2 u}{\partial t^2} = \rho \frac{\partial V}{\partial u} + \frac{\partial Xx}{\partial x} + \frac{\partial Xy}{\partial y} + \frac{\partial Xz}{\partial z} ,$$

where (u,v,w) are the components of the displacement of the point, ρ the density, V the potential of the forces acting on the Earth, and $Xx, Xy \ldots$ the components of stress.

We have also to consider Poisson's equations

$$\nabla^2 \kappa = 4\pi f \left(d_r \frac{d\rho_0}{dr} + \rho_0 \theta\right)$$

where κ is the potential of the external bodies (Sun and Moon) and of the additional variations of density, f the gravitation constant, d_r the radial displacement, θ the cubical dilatation and ρ_0 the density in the initial state.

The solution of this system of differential equations permits us to determine not only better values for the precession and nutation but also values of the bodily tide and free oscillations of the Earth. Of course, the values computed depend on the model adopted for the internal structure of the Earth.

If we remember that the Earth can be considered as composed of a lithosphere, a hydrosphere, an atmosphere, and a magnetosphere, we should examine the effects of the interactions of these different parts on the motions around the center of mass. In order to do that, we have to employ the equations not only of dynamics but also of hydrodynamics, thermodynamics, and magnetohydrodynamics. Only in this way we can have a better picture of the motions we have described by means of a first approximation theory. We would than have a general theory of the motion of the Earth around its center of mass applying to the present conditions of our planet.

Let us now examine the problem of determiing the proper rheological conditions applying to the Earth. The difficulties we encounter arise from two facts: (1) it is not easy to envision the state of the materials inside the Earth from the uppermost few kilometers, (2) the time elements involved are, for the greater part, far longer than the scientific observations we have been able to make so far.

This time variable in the rheological equations is very important, and often it has not been sufficiently emphasized. For instance, the behavior of part of the Earth as "rigid" or "fluid" has meaning only when it explicity states what are the time intervals in which the stresses in question are applied.

It is convenient to make a distinction between various typical duration intervals for the stresses, and we can classify them as short, intermediate or long. We shall consider stresses as short if they have a duration of the order of three seconds with an upper limit of about four hours, as intermediate if their duration is of the order of three years with an upper limit of 15,000 years, and as long, if their duration is of the order of 10^8 years. These figures are chosen in such a way that their ratio is about 3×10^7, and, as you can judge, it is a very large one.

This fact leads us to the conclusion that the rheological behavior deduced for one type of duration cannot be extrapolated to another, and hence they must be studied separately.

LONG-TERM ORBIT DETERMINATION AND PREDICTION FOR GEODYNAMIC INVESTIGATIONS

Bruce E. Douglas

National Oceanic and Atmospheric Administration, National Ocean Survey, Rockville, Maryland

ABSTRACT. Comparison of artificial Earth satellite orbits determined with different data types and/or force model parameters indicates that the error in such orbits is essentially short periodic with respect to the satellite motion. Therefore the mean elements of satellite orbits can be expected to be very precisely known. A precision of a few centimeters in mean semi-major axis and a few hundredths of an arc second in mean inclination has been achieved for geodetic satellite orbits using the combined analytical-numerical technique discussed in this paper. Another semi-analytical technique for analyzing these mean elements to determine a wide variety of geodynamic parameters is also presented.

1. DETERMINATION OF MEAN ORBITAL ELEMENTS

The use of mean orbital elements to follow the evolution of an orbit and determine various parameters is classical. For the case of artificial Earth satellites, geopotential, tidal, atmospheric and other parameters have been determined by many authors from sequences of mean elements.

Mean elements are usually defined as the osculating elements less the short-periodic perturbations. The latter are typically removed using analytic formulas. The completeness and accuracy of these formulas are serious questions if the best possible results are to be obtained. Some consideration will suggest that the mean

V. Szebehely and B.D. Tapley (eds.), Long-Time Predictions in Dynamics, 321–326. *All Rights Reserved. Copyright © 1976 by D. Reidel Publishing Company, Dordrecht-Holland.*

elements of a well-observed artificial Earth satellite orbit are inherently extremely well determined, so well in fact that computation of mean elements by pure analytic techniques is a formidable task if this inherent accuracy is to be maintained.

Consideration of Kepler's Third Law for the case of a typical near-Earth (a ≈ 7400 km) satellite orbit shows that an error of 10 cm in a semi-major axis at epoch causes an error along-track to propagate at the rate of about 13 m/day. Geodetic and navigation satellites are seldom in error by more than this when their orbits are determined from a few days or less of data from a global network of stations, so a sequence of mean (or osculating) semi-major axes should have an inherent precision of the order of a few centimeters. Going further, it is known (Marsh and Douglas, 1971) that the error in such orbits is mostly caused by errors in the force model and tends to be short periodic, with the result that the mean orbit may be significantly better determined than the instantaneous one.

It is clear from these rematks that any analytic scheme for converting osculating to mean elements must be very complete and accurate. Newton (1965, 1968) attempted to overcome the need for an elaborate analytic theory by employing a purely numerical method for computing mean elements. His method was to fit a numerically integrated satellite ephemeris by least squares to a secularly precessing Kepler ellipse. This procedure amounts to averaging, and is limited by the fact that there is no single interval (of reasonable length) over which all perturbations average exactly. Newton obtained a precision in mean inclination of 0.″5 (about 15 m), better than previously obtained by others, but not equal to the inherent accuracy of the navigation satellite orbits he used.

Douglas et al (1973) modified Newton's procedure and obtained for the GEOS-1 and 2 satellites mean elements that approached the inherent accuracy of the osculating orbit elements. Their procedure differed from Newton's in that the secularly precessing Kepler ellipse was fitted by least squares to an ephemeris of Kepler elements (one minute interval for one day) from which the major first-order high-frequency geopotential perturbations had been subtracted. This ephemeris of first-order mean elements had a variation of about 10-20 meters due largely to unaccounted-for second-order effects of oblateness. Thus only the small second-order effects were averaged rather than the 10-km effects of first-order perturbations as done by Newton.

The efficacy of this new procedure is shown in Fig 1, taken from the 1973 paper of Douglas et al. The figure shows the evolution of the mean semi-major axis of GEOS-1 during late 1965 and most of 1966. The precision of the mean a's was about 8 cm, as determined from a least-squares fit to a sequence of them. A precision of about 0".1 was obtained in the mean inclination of the GEOS satellites. This precision was good enough to show conclusively (Douglas et al.,1974) that tidal parameters obtained from satellite orbit perturbations are a function of orbital inclination due to ocean tide effects (Lambeck et al.,1974).

With the discovery that ocean tide effects can be detected from orbit perturbations, new requirements are placed on the precision of mean elements to accurately determine ocean tidal model parameters. For example, the perturbation of the semi-diurnal lunar (M_2) tide is less than 0".05 on the inclination of typical geodetic satellite orbits. (Fortunately, many other components of interest have larger effects.) Experiments with a U.S. Navy navigation satellite orbit by J. G. Marsh (private communication) strongly suggest that further improvement is possible. He has obtained a precision of 2-3 cm in mean semi-major axis and about 0".03 in mean inclination by the combined analytic-numerical scheme described above. This success would appear to result from the fact that the satellite he used orbits the Earth an integer (27) number of times in two days with the result that those perturbations driven by the satellite motion and those driven by the rotation of the Earth average exactly over two days. Of course this is an unusual circumstance. In general there is no conveniently small integer number of days over which the orbital period and the Earth's rotation period are commensurate. To avoid depending on the vagaries of simple averaging, C. C. Goad of the National Ocean Survey is experimenting with digital filtering of the ephemeris of first-order mean elements. This method is advantageous in that long periodic variations are not averaged. The filter cutoff is set to remove high frequency oscillations alone.

2. ANALYSIS OF MEAN ELEMENT VARIATIONS

Having obtained a sequence of mean elements, it remains to analyze their variation to recover geodetic and other parameters. In many cases, the perturbations of interest have very long periods, so that first-order analyses do not suffice. My colleagues and I have chosen to integrate numerically all long periodic and secular perturbations. Thus second-order long periodic and secular effects arising from the interaction of first-order long-periodic and secular variations are obtained automatically. Long-periodic and secular effects arising from interactions of first-order short-periodic effects are important for close Earth satellites only in the case of the perturbations due to the second zonal harmonic (J_2) of the geopotential. These effects are obtained analytically following the development of Brouwer (1959).

Isolation of short-periodic perturbations and computation of long-periodic and secular effects are facilitated if the formulations of the satellite problem given by Kaula (1966) are used. The geopotential is expressed in the form

$$V = \sum_{\ell=2}^{\infty} \sum_{m=0}^{\ell} \sum_{p=0}^{\ell} \sum_{q=-\infty}^{+\infty} V_{\ell mpq}$$

$$V_{\ell mpq} = \frac{\mu a_e^{\ell}}{a^{\ell+1}} J_{\ell m} F_{\ell mp}(i)\, G_{\ell pq}(e)\, S_{\ell mpq} ,$$

$$S_{\ell mpq} = \begin{bmatrix} \cos \\ \sin \end{bmatrix}_{\ell - m \text{ odd}}^{\ell - m \text{ even}} \{(\ell - 2p)\omega + (\ell - 2p + q)M + m(\Omega - \theta - \lambda_{\ell m})\}$$

where a , e , i , Ω , ω , M are the usual Kepler elements; a_e is the Earth's radius, μ is the product of the gravitational constant G and the mass of the Earth, $J_{\ell m}$ and $\lambda_{\ell m}$ are geopotential coefficients of degree and order (ℓ, m), θ is the sidereal time at Greenwich, and $F_{\ell mp}(i)$ is a trigonometric polynomial and $G_{\ell pq}(e)$ is a power series of order e^q. Note that short-periodic terms are those for which $(\ell - 2p + q) \neq 0$, i.e., there is dependence on mean anomaly. Deep resonances occur when

$$(\ell - 2p)\dot{\omega} + (\ell - 2p + q)\dot{M} + m(\dot{\Omega} - \dot{\theta}) \simeq 0$$

or, since $\dot{\omega}$, $\dot{\Omega}$ are small compared to $\dot{M}$ and $\dot{\theta}$, when

$$(\ell - 2p + q)\dot{M} \simeq m\dot{\theta} .$$

So called m-daily effects occur when $(\ell - 2p + q) = 0$ and $m \neq 0$ simultaneously. An analogous formulation exists for third-body perturbations (Kaula, 1964). Thus we integrate numerically equations of the form

$$\left(\frac{da}{dt}\right)_{\substack{\text{long}\\\text{periodic}}} = \frac{2}{na}\left(\frac{\partial V}{\partial m}\right)_{\substack{\text{long}\\\text{periodic}}} ,$$

etc.

The long periodic effects of atmospheric drag are obtained by numerical averaging with

$$\left(\frac{da}{dt}\right)_{\substack{\text{long}\\\text{periodic}}} = \frac{1}{2\pi}\int_0^{2\pi}\left(\frac{da}{dt}\right)_{\text{drag}} dM$$

done by Gaussian quadrature. Other perturbations modeled are radiation pressure, solid Earth tides, and precession and nutation.

Because the equations of motion contain no high frequency components local truncation error is minimized and very large integration steps (up to 10 revolutions) are possible. Douglas et al.(1974) successfully analyzed a 450-day arc of GEOS-2 mean inclinations using only about two minutes ot IBM 360/95 computer time to integrate the orbit over this time period.

3. COMMENTS ON DATA ANALYSIS

In these geodetic and geodynamic analyses we usually confine our attention to only one or two of the mean orbital elements. For example, it is well known that force model error causes much

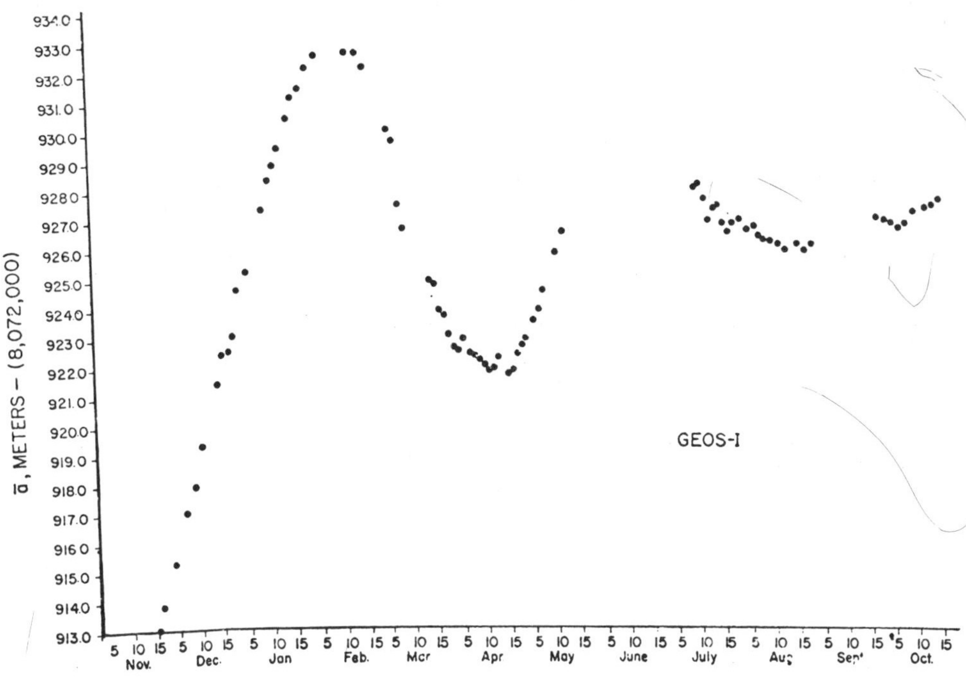

Fig. 1. Mean semi-major axes of GEOS-1.

larger error along-track than in other directions, so that variations of mean anomaly are seldom studied. In contrast, the mean inclination of an orbit is very stable and is little affected by atmospheric drag, for example, with the result that any reasonable drag force model contributed negligible error to the variation of inclination King-Hele (1974) and Wagner and Klosko (1975) have used this phenomenon for some very important investigations of satellite orbit resonances with the geopotential.

Another case where consideration of an individual element is useful is the mean semi-major axis. It is well-known that there are no long-periodic variations of the semi-major axis due to gravitational forces (other than resonances). Thus the very precise mean semi-major axes are an important source of information concerning drag and radiation pressure effects. Note in Fig 1 that radiation pressure initially caused tye semi-major axis of GEOS-1 to increase. Douglas et al. (1974) evaluated the adequacy of their radiation pressure model for the GEOS satellites by comparing observed and predicted values of the mean semi-major axis of GEOS-1. Careful examination of Fig 1 will also show that some of the mean semi-major axes are inconsistent. Subsequent investigation revealed that the inconsistent elements were due to data preprocessing error. This error would have gone unnoticed with mean elements of lesser accuracy.

REFERENCES

Brouwer, D., Astron. J. 63, 433, 1959.

Douglas, B. C., Marsh, J. G., Mullins, N. E., Cel. Mech. 7, 195-204, 1973.

Douglas, B. C., Klosko, S. M., Marsh, J. G., Williamson, R. G., Cel. Mech. 10, 165-178, 1974.

Kaula, W. M., Rev. Geophys. 2, 661, 1964.

Kaula, W. M., Theory of Satellite Geodesy, Blaisdell, Waltham, Mass., 1966.

King-Hele, D. G., Proc. IAG/COSPAR International Symposium on the Use of Artificial Satellites for Geodesy and Geodynamics, ed. G. Veis, National Technical Univ., Athens, 875-886, 1974.

Lambeck, K., Cazenave, A., Balmino, G., Rev. Geophys. Space Phys. 12, 421-434, 1974.

Marsh, J. G., Douglas, G. C., Cel. Mech. 4, 309-325, 1971.

Newton, R. R., J. Geophys. Res. 70, 5983-5989, 1965.

Newton, R. R., Geophys. J. Roy. Astron. Soc. 14, 505-539, 1968.

Wagner, C. A., Klosko, S. M., NASA/Goddard Space Flight Center Rpt. X-921-75-187, Greenbelt, Md., 1975.

PART V

ABSTRACTS OF SEMINAR-CONTRIBUTIONS

DYNAMICAL SYSTEMS INVOLVING UNCERTAINTY*

G. Adomian

The University of Georgia, Department of Mathematics,
Athens, Georgia.

ABSTRACT. Dynamical systems involving uncertainties or fluctuations and properly modelled by nonlinear stochastic differential equations are solved by a method of successive approximations, involving no closure approximations or artificial assumptions on the nature of the processes, to determine statistics to second order.

* This research was supported by the Sloan Foundation.

V. Szebehely and B.D. Tapley (eds.), Long-Time Predictions in Dynamics, 327–358. *All Rights Reserved.*
Copyright ©1976 *by D. Reidel Publishing Company, Dordrecht-Holland.*

PERIODIC ORBITS AROUND THE EQUILIBRIUM POINTS OF A CENTRAL GRAVITATIONAL POTENTIAL PERTURBED BY THE J_{22} TERM

P. A. Baxa

Istituto di Geodesia e Geofisica, University of Trieste, Trieste, Italy.

ABSTRACT. Assuming $V = V_o + J_{22} V_{22}$ as the potential of a planet, in this paper we study the position and the characteristics of the equilibrium points as functions of the J_{22} parameter. The variational equations are solved in the neighborhood of these points to determine the possible periodic orbits.

TIME REGULARIZATION AND STABILIZATION OF AN ADAMS-MOULTON-COWELL ALGORITHM

N. Borderies

Groupe de Recherche de Géodésie Spatiale, Centre National d'Etudes Spatiales, Toulouse

ABSTRACT. In geodesy we have multistep integrators such as Cowell's of fixed order 8. Unfortunately, these methods suffer from a large error along the trace. Therefore, it is not possible to integrate the motion of an artificial satellite over thousands of revolutions with sufficient accuracy.

On the other hand, the Adams-Moulton-Cowell multistep integrator with an order varying from 3 to 16, as described by Cohen and Oestwinter in [1], may possess the desired accuracy.

In order to evaluate the accuracy of the latter method we tested it in the case of the unperturbed two-body motion.

It is well known that the truncation errors are most strongly propagated if the eccentricity is large ($e<1$) and if the starting point of the integration is near the perigee. The reason for this is the irregular distribution of the computed points of the orbit.

The accuracy is improved by applying Sundmann's transformation of the independent variable. Instead of the time, we take $\sigma = \int \frac{a_0}{r}\, dt$ as the new independent variable; note σ is proportional to the eccentric anomaly. The equations of motion are then rewritten with respect to σ. Now the time is a function of σ which can be expressed in the form $t(\sigma) = \sigma + 0_0 + \mu(\sigma)$, 0_0

being a constant and $\mu(\sigma)$ is periodic in σ. In order to know t at each step of the integration procedure we add a differential equation for $\mu(\sigma)$.

As we might hope, numerical tests show an improvement of the accuracy, which now is independent of the eccentricity of the orbit and of the position of the starting point of the numerical integration.

With the above transformation of the independent variable, each component of the solution has the following form in the case of unperturbed motion: $x(\sigma) = R + S \cos n\sigma + T \sin n\sigma$ (1), where R, S, T are constants and n is the mean motion. We want to establish an algorithm which integrates functions of the type (1) exactly. In 1971, Moynot succeeded in doing this for Cowell's method, (Ref. [2]), and we invented an analagous modification for the method of Adams-Moulton-Cowell. Thus the equations of pure Keplerian motion, with respect to σ, are integrated without truncation error.

Note however that D. G. Bettis and E. Stiefel produced already in [3], a similar method.

Remember, for example, the formula of the Adams-Moulton-Cowell algorithm, predicting the position:

$$x_p = 2x_{p-1} - x_{p-2} + h^2 \sum_{j=0}^{m} \alpha'_j \ddot{x}_{p-j-1} .$$

The modified formula will be

$$x_p = 2x_{p-1} - x_{p-2} + h^2 \sum_{j=0}^{m} \alpha'_j \ddot{x}_{p-j-1} + h^2 A_p \ddot{x}_{p-1} + hB_p \dot{x}_{p-1} . \quad (2)$$

Due to the linearity of (2), it is sufficient to determine the constants A_p and B_p such that (2) holds for the functions 1, $\cos n\sigma$, $\sin n\sigma$. It turns out that these coefficients do not depend on p but only on $\frac{h}{T}$ where h is the step and T is the period.

For an exhaustive description of the modified algorithm the reader is referred to a forthcoming paper (Ref. [4]). In this paper we also describe numerical experiments which show the considerable improvement of accuracy and the reduction of the computing time.

ON THE THEORY OF AN ARTIFICIAL SATELLITE

H. Claes

Facultes Universitaires de Namur, Belgium.

ABSTRACT. A little survey of the current research in the Department of Mathematics of the Facultes Universitaires de Namur. They carry on the determination of the trajectory of an artificial satellite by Doppler observations, common work with the Royal Observatory of Belgium (O.R.B.), and on the numerical and, especially, analytical integration of the equations of motion. The Lie techniques applied to the Poisson series and the averaging method are improved for our analytical theory.

THE EVOLUTION OF PLANETARY ORBITS IN THE CAPTURE THEORY OF THE ORIGIN OF THE SOLAR SYSTEM

J. R. Dormand

Department of Mathematics and Statistics, Teeside, Polytechnic, Middlesbrough, Cleveland, England.

ABSTRACT. A byproduct of the capture process leading to the formation of the planetary system would be a resisting medium. This medium would serve to round-off the initially elliptic planetary orbits in 10^4 to 10^8 years, the time being dependent on the planetary mass.

GEODYNAMIC INVESTIGATIONS USING VERY LONG TERM OBSERVATION OF NEAR-EARTH SATELLITES

B. C. Douglas and C. C. Goad

National Oceanic and Atmospheric Administration National Ocean Survey, Rockville, Maryland.

ABSTRACT. Orbital arcs of 6000 or more revolutions of near-Earth satellites have been used by several investigators to recover estimates of polar motion and solid and fluid Earth tidal parameters. Techniques involving both osculating and mean Kepler elements have been used, the former yielding results (in terms of satellite position in the out-of-plane direction) good to about 1 meter and the latter about 3 meters. The use of osculating elements obtained from direct numerical integration of the equations of motion is extremely attractive because of the ease of modeling the force field, especially in consideration of the few-centimeter accuracy required for current and future geodynamic research projects. The need for improved efficiency and accuracy of numerical integration techniques is therefore obvious. Techniques used to minimize the effect of such errors will be discussed.

A NEW TRANSFORMATION FOR THE PERTURBED PROBLEM OF TWO BODIES

T. Feagin

University of Tennessee Space Institute, Tullahoma, Tennessee

ABSTRACT. A transformation, $(x, \dot{x}, t) \rightarrow (\theta, \theta', s)$, of the dependent and independent variables is introduced. This transformation allows the perturbed motion of two bodies in a plane to be described by a second-order differential equation of the form:

$$\theta'' = f(\theta, \theta', s)$$

where f vanishes identically for the unperturbed problem. Although the new equations are singular for rectilinear motion, they are well-behaved otherwise. The numerical advantages of the transformed system are discussed.

KOZAI'S THEORY OF AN ARTIFICIAL SATELLITE

J. Henrard

Facultes Universitaires de Namur, Namur, Belgium

ABSTRACT. The analytical theory of an artificial satellite published in 1962 by Kozai (Astronomical Journal 67, 446-461) is still one of the most complete and the most referred to theory. We know of only one attempt (by Dr. Sconzo) to check the very large amount of analytical computations entering in this theory. The computations were carried out partly by computer and partly by hand. Although some misprints and some mistakes in the work of Kozai have been corrected by Sconzo they have never been published and it is not known if the totality of the computations has been checked. Using its subroutines for analytical computations on a computer (Poisson Series Processor) the author is attempting to recompute completely automatically Kozai's theory. At this date the work is not yet completed but some of Sconzo's corrections have already been confirmed and some new corrections can be proposed. It is hoped that for this summer a complete revision of Kozai's work can be available.

CASSINI'S LAWS

H. Kinoshita

Smithsonian Institute, Astrophysical Observatory
Cambridge, Massachusetts.

ABSTRACT. The rotational motion of the moon is in a state of double resonance: the mean motion of the orbital motion is equal to that of the rotational motion and the precessional motion of the orbital plane is equal to that of the moon's equatorial plane. After eliminating short periodic terms, the rotational motion of the moon becomes a system of two degrees of freedom with two critical arguments. From the condition that such a system has a stationary solution, the relationship between the obliquity of the moon's equator to the ecliptic and the moon's gravitational coefficients is derived. The effects on the obliquity due to the higher degree harmonics $(n \geq 3)$, the sun, and the motion of the ecliptic will be discussed.

THEORY OF THE ROTATION OF THE RIGID EARTH

H. Kinoshita

Smithsonian Institute, Astrophysical Observatory
Cambridge, Massachusetts.

ABSTRACT. An analytical theory is developed for the three axes of the triaxial rigid earth (the angular-momentum axis, the figure axis, and the rotational axis). A theory on rigid earth rotation will be important to one on a nonrigid earth, since the latter gives only the ratios between the amplitudes of rigid nutation and those of nonrigid nutation. The present theory is characterized by the use of Andoyer variables, Hori's averaging method, and a moving reference plane. A comparison with Woolard's theory shows that 1) the maximum difference in nutation of the angular-momentum axis, calculated from the same constants as Woolard adopted, reaches 0".0017; 2) the maximum discrepancy in Oppolzer terms is 0".0012: and 3) the present theory has no secular term either in the first power of time in the obliquity referred to the fixed ecliptic or in the angle between the figure axis and the rotational axis; both these secular terms appeared in Woolard's theory and have an effect on the establishment of a reference system. Nutation amplitudes as large as 0".0001 are calculated for all three axes. Use is made of Brown's theory of the moon as improved by Eckert and of numerical values recommended at the working meeting of the International Astronomical Union held in Washington in September 1974. Any future revision of the lunar theory will not alter the values of the amplitudes of the nutational terms derived here.

PERIODIC OSCILLATIONS OF THE SECOND GENERATION IN THE STÖRMER PROBLEM

V. V. Markellos, S. Klimopoulos, and C. L. Goudas

Department of Mechanics, University of Patras, Greece.

ABSTRACT. Four new families of multiple-periodic oscillations are given in the Störmer problem. They are branches of the basic family k of simple-periodic oscillations, the members of which consist of the same arc traversed twice. Another branch of k consisting of asymmetric periodic oscillations has been detected.

THREE-DIMENSIONAL BRANCHINGS OF PLANE PERIODIC SOLUTIONS

V. V. Markellos and P. G. Kazantzis

Department of Mechanics, University of Patras, Greece.

ABSTRACT. In the case of the restricted three-body problem with small mass parameter a family of plane symmetric periodic orbits of the direct type around the large primary is found to have branches of three-dimensional periodic orbits. One such branch has been established consisting of stable orbits for small deviations from the plane.

ON STABILIZATION USING THE ENERGY INTEGRAL

Paul E. Nacozy

Department of Aerospace Engineering and Engineering Mechanics, University of Texas at Austin, Austin, Texas.

ABSTRACT. A clarification and discussion of the use of the energy integral to stabilize gravitational systems is presented. In a previous paper (Nacozy, Astro. and Space Sci., 1971, Vol. 14, No. 1, p. 40. See also: Baumgarte, Celestial Mechanics, 1972, Vol. 5, p. 490.), a method is described that employs the energy integral of the system as a constraint. At the completion of each (or several) integration steps, the solution is adjusted so that the error in the energy integral vanishes. And, even though the local or per step error in the energy integral is often several orders of magnitude smaller than the actual local error in the solution, satisfaction of the energy integral generally reduces the total or global error in the solution of Liapunov unstable systems by several orders of magnitude.

It is intesting to examine closely the adjustment to the solution. It is shown that the largest adjustment is to the frequency of the component of the solution tangent to the trajectory. The adjustment or correction to the energy is along the normal to the energy manifold and hence normal to the energy surface. And this adjustment produces a correction to the rate of change of a component of the solution tangent to the trajectory and tangent to the energy manifold. For a two-body gravitational system, removing errors in the energy negates errors in the mean motion of the system and achieves Liapunov stabilization. By removing errors in the frequency of the tangent or in-track component, the in-track, unbounded error growth is considerably reduced.

A discussion is given relating results of applying the method to satellite motion, including zonal and tesseral harmonics. Here, the energy integral is replaced by an integral that exists in a coordinate system that rotates with the rotation of the Earth. For an Earth satellite with a semi-major axis of 136,000 km, eccentricity of 0.8, and inclination of 40°, the method improves the accuracy of the solution by two orders of magnitude after three revolutions.

The full paper has been submitted to the journal Celestial Mechanics.

RESONANCE EFFECTS IN SATELLITE ORBITS

J. Osorio

Observatorio Astonomico, Universidade de Porto, Portugal.

ABSTRACT. If a satellite were moving in an axially symmetric gravitational field, free from other disturbing forces, two integrals of the equations of motion would be known: the energy per unit mass of the satellite and the third component of the angular momentum per unit mass. For a terrestrial satellite one cannot write such integrals since the field depends on the longitude. If a set of canonical variables, such as Delaunay variables, is chosen, the time can easily be removed from the Hamiltonian and an integral of the new set of equations of motion is obtained. The use of that integral when studying the earth's gravitational field is investigated for orbits whose mean motion is almost commensurable with the Earth's angular velocity. Numerical results were derived by using orbital data of a synchronous satellite in a nearly circular orbit and with a small value for the ratio of mean cross-sectional area/mass.

ON THE RESONANCE 1 - 1 FOR HAMILTONIAN SYSTEMS NEAR AN EQUILIBRIUM POINT

J. Roels

Institute of Mathematics, University of Louvain, Belgium.

ABSTRACT. The planar restricted 3-body problem, linearized in the neighborhood of Lagrangian equilibra L_4 and L_5, has in general two distinct eigenvalues and their opposites. When they are pure imaginary and not multiples of each other, they generate two families of periodic solutions called long and short periodic families. This is essentially a consequence of the famous theorem of Liapunov (Siegel, 1956). We showed (Roels, 1971b) how to solve the problem when the eigenvalues are multiples of each other in building series with negative exponents instead of the integer expansions of Siegel (Roels and Lauterman, 1970). When the eigen-values are equal, which is the case for the mass ratio of Routh, the problem was solved by Deprit and Henrard (1968) using formal series in ordinary unnormalized variables. That leads to very com-plicated series because of the use of variables that are not well adapted to the problem. The convergence of the series was proven by Meyer and Schmidt (1971). In this paper we solve the problem by using normalized variables. This brings us to build expansions with fractional exponents. So in summary, normalized variables generate integer series in the non-resonant cases, series with negative exponents in the case of resonance $k \geq 3$, and series with fractional exponents when the resonance is 1.

PERIODIC ORBITS NEAR AN EQUILIBRIUM POINT: THE LIAPUNOV CENTER THEOREM

D. S. Schmidt

Division of Mathematical and Physical Sciences and Engineering, University of Maryland, College Park, Maryland.

ABSTRACT. Based on Siegel's example in his book Celestial Mechanics, we present a Hamiltonian system where all solutions are periodic in the linearized system but the nonlinear system has no periodic solutions at all. This shows that certain conditions have to be met before the existence of periodic orbits can be proven in resonance cases. Then we indicate how Newton's method can be used to find periodic orbits, with a predetermined period, numerically.

ANALYTICAL AND NUMERICAL STUDY OF A PERTURBED PENDULUM

C. Simó

Cátedra de Astronomia, Universidad de Barcelona,
Spain.

ABSTRACT. We study a perturbed pendulum described by the equation $\ddot{x} + \sin x = \varepsilon \sin t$. For $\varepsilon = 0$ let $x = c \sin \ell \sum_{k=1}^{\infty} f_k(c) \sin (2k+1)\ell$ be its solution, where $\ell = pt + \beta$, $p = p(c)$, with c and β constants. Defining $\gamma = t - \ell$, we obtain the equations for the perturbed problem (averaged w.r.t.time):

$$\dot{c} = \frac{\varepsilon}{2} \sin \gamma + O(c^4, \varepsilon c^2)$$
$$\dot{\gamma} = \frac{c^2}{16} + \frac{\varepsilon}{2c} \sin \gamma + O(c^4, \varepsilon c^2) \quad . \qquad (1)$$

If we consider the truncated equation of (1), we find: $c^4/64 + \varepsilon/2 \cos \gamma = \kappa$(constant). The equilibrium position is given by $\gamma = \pi$, $c = 2\,\varepsilon^{1/3}$. For $c_o \simeq 0$, $\gamma_o = 0$, we reach a value of $c \simeq (32\varepsilon)^{1/3}$ when $\gamma = \pi$. For ε small enough the motion is quasi-periodic and the long period $T(\kappa, \varepsilon)$ is given in terms of elliptic integrals. For $\kappa \simeq 0$, $T(\kappa,\varepsilon) = O(\varepsilon^{-2/3})$. Comparing the truncated model and the numerical integration of the exact problem, minor errors are found in the period and maximum amplitude for $c < .5$, $\varepsilon < .2$. A numerical method well suited for the integration of an anharmonic oscillator is introduced. Truncation error for an harmonic oscillator is given by $T = |\Phi(\zeta) - \sum_{i=1}^{N} \alpha_i \zeta^i|$ where $\zeta = 1 - e^{-2i\sigma}$, $\sigma = \frac{\omega h}{2}$

(ω = frequency, h = step). For a fixed step $T = T(\omega_i\ \alpha_i)$. Let α_i^o be optimal parameters in the following sense: $\| T(\omega_i \alpha_i^o)\|_\infty \leq \| T(\omega\ \alpha_i)\|_\infty$ for all values of α_i, where $\| f(\omega)\|_\infty = \sup |f(\omega)|$. This produces a reduction of the truncation error if we know ω_a and ω_b. For instance, $\omega \varepsilon [\omega_a, \omega_b]$ in the Störmer case, if h = .1, $0<c<\frac{\pi}{2} \rightarrow \omega_a = .85$, $\omega_b = 1$ and the bound obtained is 1/1000 of the classical one. The cases of ε or c not small are studied. For initial values $c_o = .1$, $\gamma_o = 0$, and $\varepsilon = .34$, transitions from libration to circulation are found. This happens also if ε is very small and the initial conditions are near the separatrix of the unperturbed problem. A quasi-random transition from the circulation to the libration regions is suspected in this last case.

ALGEBRAICAL ASPECTS OF PERTURBATION THEORIES

Franz W. Spirig

Seminar fur Angewandte Matematick
Swiss Federal Institute of Technology, Zurich,
Switzerland

These are quantum mechanical perturbation theories in which the perturbation equations have the same formal structure as those established by Hori [1]. In quantum mechanics we have to deal with linear operators A, B, ..., that is, with an associative algebra. Hori's analytical perturbation theory, however, is concerned with differential vector fields a(x), b(x), ... and a multiplication

$$a \cdot b = \frac{\partial a}{\partial x} b, \text{ where } \frac{\partial a}{\partial x} \text{ is the Jacobian,}$$

that is, with a non-associative algebra. The goal of this note is to explain why Hori's theory resembles the perturbation procedure performed in quantum mechanics. Let us consider the abstract algebraical structure of these algebras. The set $\{a,b,\ldots,d,e,\ldots\}$ is a linear space, that means we can carry out the operation $\lambda a + \mu b = c$ $(\lambda, \epsilon \in \mathbb{R})$. Furthermore, we have a bilinear multiplication: ab=e. Therefore the set $\{a,b,\ldots\}$ forms a non-associative algebra. The associative law does not hold here, but we have the weaker relation

$$(ab)c - (ac)b = a(bc) - a(cb). \qquad (1)$$

Finally there exists a left 1 with the property

$$1a = a \qquad (2)$$

By virtue of (1) a Lie algebra can be constructed by introducing a Lie product in the following manner:

$$a \otimes b = ab - ba$$

Then the Lie conditions hold true: bilinearity, skewsymmetry and the Jacobian identity.

On the other hand the set {A,B,...} of the linear operators acting on {a,b,...} is a vector space, too. With the multiplication $A^*B = BA$ (i.e., the composition B and A) we get an associative algebra. The corresponding Lie algebra is established in the usual way by the definition

$$A \otimes B = A^*B - B^*A$$

Let r_a denote the right translation of a, that is $r_a x = ra$ for $x \varepsilon \{a,b,...\}$. Then the set $\{r_a, r_b,...\}$ is a linear subspace of {A,B,...}. The following equation is obvious:

$$\lambda r_a + \mu r_b = r_{\lambda a+\mu b} \qquad (3)$$

By virtue of (2) and (3) the vector spaces {a,b,...} and $\{r_a, r_b,...\}$ are isomorphic. The set of the right translations does not form an algebra, but a Lie subalgebra of the Lie algebra of the linear operators. Actually we have, by equation (1), the following important relation

$$r_a \times r_b = r_{a \otimes b} \quad . \qquad (4)$$

Therefore the mapping $a \rightarrow r_a$ establishes an isomorphism between the Lie algebra of the right translations and that of the vector fields. The connection between the algebra {a,b,...} and set $\{r_a, r_b ...\}$ is given by

$$(\ ((ab)c)d...)\ f = r_a{}^*r_b{}^*r_c{}^*r_d{}^*...{}^*r_f\ 1 \quad . \qquad (5)$$

By translating operator formulae with the aid of the above mentioned Lie algebra isomorphism and (5) we obtain the following examples of applications.

(1) Given a differential equation $\dot{x} = f(x,t)$. The Lie transformation $\dot{x} = [x+u+\frac{1}{2!} u^\circ u+...]_{x = \bar{x}}$, where $u=u(x,t)$, leads to $\dot{\bar{x}} = \bar{f}(x,t)$.

$$\text{Then } u_t = f \otimes u + \sum_{n=0}^{\infty} \frac{B_n}{n!} (f-\overline{f}) \underbrace{\otimes u \otimes u \ldots \otimes u}_{n \text{ factors}}, \tag{6}$$

where B_n denotes the well-known Bernoulli numbers and u_t stands for the partial derivative with respect to t. If we insert for f, $\overline{f}$ and u power series in ε in (6), we obtain the perturbation equations, formulated for the autonomous case by Kirchgraber [2]. In particular, if $\overline{f} = 0$ an iterative integration yields the so-called continuous Hausdorff-Campbell formula, first derived for operators by Magnus [3].

(2) The Hausdorff-Campbell formula [4]:

$$[x + u + \frac{1}{2!} u \circ u + \ldots]_{x = [x + v + \frac{1}{2!} v \circ v + \ldots]} =$$

$$x + w + \frac{1}{2!} w \circ w + \ldots$$

$$\text{where } w = u + v + \frac{1}{2} u \times v + \frac{1}{12} u \times v \times v + \frac{1}{12} v \times u \times u + \ldots \tag{7}$$

References:

1. G. Hori, Publ. Astro. Soc., Japan 23, 567 (1971)

2. U. Kirchgraber, Some Theorems in Hori's Perturbation Theory, paper presented at the 4th Conference on Ordinary Differential Equations in Oberwolfach, 1974.

   U. Kirchgraber/E. Stiefel, Methoden der aualysisdun Horwgssechuung, Teubner, in Dorbereiheng.

3. W. Magnus, "On the Expontential Solution of Differential Equations for a Linear Operator", Comm. Pure Applied Math., Vol. 7, ,649, (1954).

4. F. Hausdorff, Die symbolische Exponentialformel in der G Gruppentheorie, Ber Verh. Sächs. Akad. Wiss. Leipzig, Math.-Phys. Kl. Vol. 58, 19 (1906).

STABILIZATION AND TIME ELEMENTS FOR KEPLER'S PROBLEM

A. Stokes

Department of Mathematics, Georgetown University,
Washington, D.C.

ABSTRACT. In the two body problem, let x,y denote the momentum and position vectors respectively, and let the energy h be introduced as an additional coordinate, with its conjugate t (time). Here $x,y \in R^3$(or in the general case $R^n, n>2$). Let a new time variable s_1 be given by $dt = (|y|/\sqrt{-2h})ds_1$, ($h<0$ for the elliptic orbits of interest here). By a slight extension of a canonical transformation of Moser, one can transform the variables x, y, h, t into variables ξ, η, ρ, τ, where now $\xi, \eta, \in R^4, \rho, \tau$ scalars.

The image vectors ξ, η satisfy $|\xi|=1$, $(\xi,\eta) = 0$. The variables t and τ are related by $t = -\tau + (x,y)/(-2h)$. The new Hamiltonian (after being extended from the manifold $|\xi|=1$, $(\xi,\eta)=0$ to R^4XR^4) is given by $G = |\xi||\eta| - K^2/(2\rho)$. The resulting system of ten equations $d\xi/ds_1 = G_\eta$, $d\eta/ds_1 = -G_\xi$, $d\rho/ds_1=0$, $d\tau/ds_1 = -K^2/(2\rho)^{3/2}$ has as integrals G and (ξ,η). If $(\xi,\eta) = 0$, then solutions $\xi(s_1),\eta(s_1)$ satisfy $|\xi|=1$, $|\eta| = k^2/(2\rho)$. It results that ξ, η satisfy the equations $d^2\xi/ds_1{}^2+ \xi = 0$, $d^2\eta/ds_1^2 + \eta = 0$.

Stabilization is achieved by setting $d\tau/ds_1= -k^2/(2\hat{\rho})^{3/2}$ $-\hat{\rho}$ is the (constant) initial energy. The equivalent stabilized Hamiltonian is $G = |\xi||\eta| + k^2\rho/(2\hat{\rho})^{3/2}$. Upon returning to x, y, h, t coordinates, a Lyapounov stable system is obtained, very similar to the equations obtained by Baumgarte and Stiefel.

Returning to the original equation, let $f(u)$ be a continuous real valued function of a scalar u, with $f(u) > 0$ for $u > 0$. Consider a new time variable s_f given by $ds_1/ds_f=f(2\rho|y|/K^2)$.

By using the new coordinates introduced by Moser, one finds that $ds_1/ds_f = f(1-R\cos(s_1+\Phi))$, where R,Φ are constants determined by the initial conditions, and $0<R<1$ if the orbit is not a collision orbit. Defining $F(u)$ by $F(u) = \int_0^u dw/f(1-R\cos w)$, this equation can be integrated to give $F(s_1+\Phi)=s_f+F(\Phi)$. But integrating the τ equation above, one sees that there are constants C_0, C_1, depending upon the energy, so that $s_1=C_0t+C_1(x,y)$. Accordingly, we define a time element τ_f by $\tau_f = F(C_0t+C_1(x,y)+\Phi)$, $\tau_f(0)=F(\Phi)$. Clearly, $d\tau_f/ds_f = 1$ in the unperturbed problem, and so for the perturbed problem, $d\tau_f/ds_f=1+O(\text{perturbations})$, that is, τ_f is a time element. The precise equation for τ_f is easily obtained, and t is then given in terms of $F^{-1}(\tau_f)$ and (x,y).

To apply this, one needs to determine F^{-1}. For the special case $f(u)=u^{n-1}$, (here we write $s_f=s_n$, $\tau_f=\tau_n$, $F=F_n$), corresponding to $dt/ds_n=c|y|^n$, $F_n(u) = \int_0^u dw/(1-R\cos w)^{n-1}$ can be integrated explicitly for n a multiply of $1/2$. For $n=1,3/2,2,5/2$, F_n^{-1} can also be obtained explicitly. For $n=3,7/2,4,\ldots$etc, $F_n(u)$ has the form $F_n(u)=h^{-1}(u)+p(u)$, where h^{-1}, h and p are known. Then if we set $u_0=h(z)$, $u_{k+1}(z)=h(z-p(u_k(z)))$, $u_k(z)\to F_n^{-1}(z)$ as $k\to\infty$, so F_n^{-1} can be obtained through iteration. Other methods can of course be used, e.g., Newton's method.

REFERENCES

J. Moser, Comm. Pure and Appl. Math, 23, (1970), 609-636.
J. Baumgarte and E. Stiefel, Celestial Mech. 10, (1974). 71-85.

A NOTE ON THE HÉNON-HEILES PROBLEM

Ferdinand Verhulst

Department of Mathematics, University of Utrecht,
Utrecht, The Netherlands.

ABSTRACT. To study the existence of isolating integrals of motion besides the energy integral, Hénon and Heiles (1904) considered a two degrees of freedom Hamiltonian system. For low energy orbits the system can be written as

$$\ddot{x} + x = -2\varepsilon\, xy$$
$$\ddot{y} + y = -\varepsilon(x^2 - y^2)$$

where ε is a small parameter.

We put

$$x(t) = A(t)\cos(t + \rho(t))$$
$$\dot{x}(t) = -A(t)\sin(t + \rho(t))$$
$$y(t) = B(t)\cos(t + \psi(t))$$
$$\dot{y}(t) = -B(t)\sin(t + \psi(t))$$

to apply van der Burgh (1975) theory on second order resonance problems. We recover the energy integral (~ indicates approximate quantities)

$$\tilde{A}^2 + \tilde{B}^2 = 2E_0$$

with the estimate

$$I_3(A, B, \rho, \psi) - I_3(\tilde{A}, \tilde{B}, \tilde{\rho}, \tilde{\psi}) = O(\varepsilon)$$

on the time-scale $1/\varepsilon^2$

If one constructs with the aid of these integrals the Poincaré mapping (surface of section) as given by Hénon and Heiles, one obtains topologically the same picture as found numerically for $E_o = 0.08333$; one also recovers the periodic orbits found for other values of the energy.

In studying the behavior of the solutions with time some other interesting features arise. It turns out that dense sets of orbits exist with arbitrary long period. The area covered by these orbits in, for instance, a surface of section, increases with the energy. One of the consequences of this result is that, numerically integrating, one should always find an "ergodic sea" (of small size) which does not really exist but is generated by the set of orbits with periods longer than the integration time.

Another consequence of the theory is that in studying real galaotic orbits one should keep third order terms in the potential for second order resonances one should truncate after the fourth order terms.

An extensive account of this work will appear in the Proceedings of the International Conference on Nonlinear Oscillations, Berlin, 1975.

REFERENCES

A. M. P. van der Burgh (1975), these proceedings.

M. Hénon and C. Heiles (1969); A. T. 69. 73-79.

NEWTONIAN COSMOLOGY WITH VARYING G

J. P. Vinti

M. I. T. Measurement Systems Laboratory, Cambridge Massachusetts.

ABSTRACT. If all isolated galaxies or clusters of galaxies are treated as equivalent points, the cosmological principle states that all observers at such points observe the same cosmic density $\rho(t)$ and the same Hubble function $H(t)$, on the assumption of homogeneity and isotropy. If $\underset{\sim}{S}(t)$ is the vector separation of two equivalent points, there results $\dot{\underset{\sim}{S}}=H(t)\underset{\sim}{S}$ for the non-random motion. Let $|\underset{\sim}{S}(t)|\equiv S(t)=R(t)S_0$. Here t is the age of the universe and and the subscript 0 denotes a value at a recent epoch time τ. $R(t)$ is then the scale factor; the universe is called closed or open accordingly as R does or does not reach a maximum value. Poisson's equation and the fluid equation of continuity then lead to the fundamental equation $R^2\ddot{R}=-(4\pi/3)\,G(t)\rho_0$. Also $H(t)=\dot{R}/R$, where $R_0=1$.

If G obeys $G=A(k+t)^{-1}$ and $x=k+t$, then $R^2\ddot{R}=-Bx^{-1}$, where $B=(4\pi/3)\,G_0\rho_0x_0$. The substitution $R=xu$ leads to $x^2\dot{u}=\pm(2Bu^{-1}+c_1)^{1/2}$, where $c_1=(x_0H_0-1)^2-2Bx_0$. Given $R_0=1$ and $\dot{R}_0=H_0$, the equation has a unique solution, despite the double sign. Theorems are developed to find the sign in any given case.

If $c_1=0$ and $p=1-x_0H_0$, R reaches a maximum value $R_m=p(3-=2)^{-1/3}$ at $x_m=x_0p(3p-2)^{-1}$ if $0<x_0H_0<1/3$ and then vanishes at $x_1=3x_m$.

Any extremum of R must be a maximum. Thus for $x>x_0$, R must either grow indefinitely or go through one maximum and later reach zero. Reaching $R=0$ at $x_1>x_0$ is thus necessary and sufficient that R attain a maximum at x_m; then $x_0<x_m<x_1$. If $c_1<0$, let x_c be the value of x for which $2Bu^1+c_1=0$. We obtain Table 1.

Table 1

x_0H_0	c_1	Sign of $\dot{u}$	$H(x)$	Maximum R_m of R
≤ 1	any c_1	always -	$<x^{-1}$ if $x>x_0$	Possible for $x>x_0$
>1	≥ 0	always +	$>x^{-1}$ if $x>x_0$	Impossible
>1	<0	+ if $x_0\leq x<x_c$	$>x^{-1}$ if $x_0<x<x_c$	Impossible if $x_0<x<x_c$
		- if $x>x_c$	$>x^{-1}$ if $x>x_c$	Possible if $x>x_c$

Conditions are developed for the existence of an R_m in the two possible cases above and equations for R_m. Methods are then devised for finding corresponding values of x_0H_0, $2Bx_0$, R_m, x_m, and x_1, without explicitly solving the resulting transcendental equations.

The boundary curve of $2Bx_0$ versus x_0H_0, or of ρ_0 versus H_0, indicates that the values of ρ_0 necessary to close the universe exceed the mean density ρ_L of luminous matter by a factor of the order 100. The corresponding deceleration constant $q_0=-\ddot{R}_0H_0^{-2}$ ranges from 2 at $H_0=33$ km-sec^{-1} per MPC to 1.2 at $H_0=100$. This is on the assumption that $x_0=10^{10}$yr. If we assume a large k in $x_0=k+\tau$, still estimating $\tau \approx 10^{10}$yr, we might be able to bring the closure ρ_0 down to (say) twice ρ_L. But then the corresponding $H_0\approx 10$ km-sec^{-1} per MPC. Since recent estimates of H_0 range from 45 to 120, it thus appears that the proposed model favors an open universe.

THE EFFECTS OF INTEGRALS ON THE TOTALITY OF SOLUTIONS OF DYNAMICAL SYSTEMS

K. Zare

Department of Aerospace Engineering and Engineering Mechanics, University of Texas at Austin, Austin, Texas

ABSTRACT. Regions of possible motions are established for dynamical systems possessing time-independent Hamiltonians or for systems which are reducible to that form by means of integrals of the motion using only extended point transformations. The method is applied to the problem of three bodies in a plane and surfaces of zero velocity are found. The governing parameters are the energy, angular momentum and the masses of the participating bodies. The analytical and geometrical properties of these surfaces provide qualitative results for given constants of the motion.